THE
BEST
OF
MAKE

First Edition

The Editors of MAKE

O'REILLY®
BEIJING • CAMBRIDGE • FARNHAM • KÖLN • SEBASTOPOL • TAIPEI • TOKYO

The Best of MAKE
by the Editors of MAKE Magazine

Copyright © 2007 O'Reilly Media, Inc. All rights reserved.
Printed in Canada.

Published by Make:Books, an imprint of Maker Media,
a division of O'Reilly Media, Inc.
1005 Gravenstein Highway North, Sebastopol, CA 95472.

O'Reilly books may be purchased for educational, business,
or sales promotional use. For more information, contact our
corporate/institutional sales department: 800-998-9938
or corporate@oreilly.com.

Print History: October 2007: First Edition

Publisher: Dale Dougherty
Associate Publisher and Executive Editor: Dan Woods
Editor-in-Chief: Mark Frauenfelder
Editor: Gareth Branwyn
Creative Director: Daniel Carter
Art Director: Brian Scott
Designer: Gerry Arrington
Production Manager: Terry Bronson
Indexer: Patti Schiendelman
Cover Photography: Gerry Arrington, Bill Bumgarner,
Zach DeBord, Roger Ibars, Topher Lucas, Sam Murphy,
Dave Prochnow, Casimir A. Sienkiewicz, Carla Sinclair

Contents

Re: MAKE

Imagine, at your next backyard luau, serenading everyone with a guitar you built from a cigar box, while a friend croons through an amp made from a cracker box (hilarity is sure to ensue). Or how about delighting your friends with a twitchy little "robot" you magically fashioned from a toy motor, a mint tin, and a coat hanger? Or maybe using digital video trickery to "beam" yourself into your favorite TV show?

Welcome to *The Best of MAKE*, a collection of beloved projects from the first ten volumes of MAKE. We've been surprised and delighted by the phenomenal response we've gotten to the magazine (and our blog, podcasts, and Maker Faires). But we're also *not* surprised. There's a movement afoot. Suddenly, all sorts of people, of different ages and walks of life, of varying skill levels, seem itching to head to their basements, garages, and backyards to make cool stuff. Many reasons are given, theories offered, as to why. And why now. A growing environmental awareness, a desire to reuse and recycle? Digital world burnout and backlash fueling interest in the pleasures of "hacking" the real world? A growing strata, generations of techno-junk, filling up our basements, begging to be brought back to life, reanimated in some creative way? It's likely all of this and more. Whatever it is, it's very exciting to watch and to be on its forefront.

For *The Best of MAKE*, we wanted to cover a range of project types, from kitchen table quickies (e.g., *LED Throwies* and *Vibrobot*) to afternoon-long builds (e.g., *Cigar Box Guitar* and *Two-Can Stirling Engine*) to weekend projects (e.g., *Rocket-Launched Camcorder* and *Mousey the Junkbot*). There's something here for every skill level and interest area, from hacking off-the-shelf robots and game consoles to building a wind generator and running your car on fryer grease. One of the things we've tried to do in the magazine is create projects that make the reader want to jump up and proclaim: "I can do this!" We've chosen projects here that most clearly resonate with that enthusiasm.

Reading through these pages, you may be tempted to put down this book, run to your toolbox, and begin before you've even finished reading the article. Don't be *that* enthusiastic. Before starting into anything, finish reading the instructions and then go to the web page for each article (URLs can be found at the end of each piece). Read the discussion and any corrections posted there. If the author has a website listed in the piece, visit that too and look for updates. Also, please read our safety note on page ii.

We're very excited by this collection of outstanding projects and by what we've accomplished with MAKE so far. But we're even more excited by what we plan to do next. Stay tuned...

— The Editors of MAKE

IF YOU CAN'T OPEN IT, YOU DON'T OWN IT

A Maker's Bill of Rights to accessible, extensible, and repairable hardware.

By Mister Jalopy

■ Recently, the gas gauge on my 2000 Chevrolet pickup started acting perquacky and, as I'm a lazy person by nature, I asked the Chevrolet dealership what it would cost to repair. At a staggering $800, I briefly considered living without a gas gauge. Picturing certain roadside disaster, I buckled down and decided to fix the problem myself.

Hopeful that I would be able to buy just the fuel sender, Chevrolet broke the news that I would have to buy the combined $500+ Delco fuel pump and sender assembly. Now, only the fuel sender unit was faulty. The fuel pump still worked like a champion but I had to buy the whole pump/sender assembly. Mercifully, my local auto parts store sold the same exact unit for $259.

After draining and dropping the gas tank, I removed the old assembly; it's clearly designed to have a removable, replaceable fuel sender unit. It's held in place by two plastic tabs and a single wire connector. And to prove my point, I did remove it. It took longer to get the pliers from the toolbox than it did to disassemble.

Sometimes components fail and you have no idea why, but in this case, the cause of failure was obvious. There are two little spring-metal tangs that glide over the PCB resistor contacts, and one

tang had broken off. The metal tangs are fragile, under pressure, and move whenever gas sloshes in the tank, so failure was only a matter of time. After seeing how fine the tangs were, I was surprised that they hadn't broken earlier. A quick Google search proved I was lucky that it had lasted as long as it had — it's a very common problem.

> "When your covered wagon broke a wood spoke, did you throw away the whole wheel? The whole wagon?"

I bet Chevrolet specified to their subcontractor that the fuel sender unit would be removable. Perhaps they were planning to offer it as a separate SKU. Is it a purely financial decision by Chevrolet to not sell the fuel sender independent of the fuel pump? Or how about just selling the tangs for a dollar? When your covered wagon broke a wood spoke, did you throw away the whole wheel? The whole wagon?

In MAKE Volume 03 (*see page 7*), Dale Dougherty wrote an essay on what makes a product maker friendly. And it was an idea that stuck in my head as I was building the retromodern remote-control

Photography by Mister Jalopy

"Buying a piece of equipment that you can't open, repair, or refer to a schematic for means you're setting yourself up to throw it away tomorrow."

LP-to-MP3 converter cabinet (*see page 54*) and was extremely frustrated that the Mac mini is a sealed box. Apple techs open it with some sort of putty-knife-like special tool. Sometimes, smart engineering and new solutions require new tools. The Model T required special tools, but they were included with the car. It's hard to imagine the case for requiring a special tool to open a Mac mini. For all the props that Apple gets for industrial design, would it kill them to put four screws on the bottom? Would that greatly harm the aesthetics?

In the same way digital rights management (DRM) locks up data, buying a piece of equipment that you can't open, repair, or refer to a schematic for means you're setting yourself up to throw it away tomorrow.

You don't own the iTunes songs you buy. Apple does. Granted, I can't get inside to look at the digital rights software and fully understand what I have agreed to, but I know that it has a limited life. I listen to my grandfather's 78 RPM records and will always be able to play an unlocked CD, but will my grandkids get to listen to my iTunes library?

If you can't open it, you don't own it. You bought the hardware but, like DRM, the manufacturer restricts your use by controlling access, replacement parts, and information. It's yours and usable only as long as the manufacturer chooses to support and repair it.

Inside the gas tank of an older car, there's a fuel sender with a float that "sends" a resistance value to the gas gauge to record the tank level. Then, outside the tank, the fuel pump pumps gas from the tank to the engine. Newer cars have combined the fuel pump and sender in a single mechanism (as shown) that is inside the gas tank and can only be replaced as a complete assembly. In 30 seconds, the sender assembly can be removed from the assembly but Chevrolet won't sell you a replacement.

So, what does all this have to do with Chevrolet and the fuel sender unit? Clearly, components should be available at a granular enough level to be able to make repairs at reasonable prices. Ideally, you would be able to buy the little metal tangs, but I would be satisfied to buy the sender unit. Chevrolet's decision to sell assemblies rather than components is unfortunate but understandable. There is a rationale, as they are in business to make money and selling bigger pieces means more money.

After thinking about this, I've come up with a Maker's Bill of Rights. I expect and hope that other makers will add and make changes to this list. Post your suggestions at makezine.com/04/ownyourown.

Mister Jalopy breaks the unbroken, repairs the irreparable, and explores the mechanical world at hooptyrides.com.

THE MAKER'S BILL OF RIGHTS

■ Meaningful and specific parts lists shall be included. ■ Cases shall be easy to open. ■ Batteries shall be replaceable. ■ Special tools are allowed only for darn good reasons. ■ Profiting by selling expensive special tools is wrong and not making special tools available is even worse. ■ Torx is OK; tamperproof is rarely OK. ■ Components, not entire sub-assemblies, shall be replaceable. ■ Consumables, like fuses and filters, shall be easy to access. ■ Circuit boards shall be commented. ■ Power from USB is good; power from proprietary power adapters is bad. ■ Standard connecters shall have pin-outs defined. ■ If it snaps shut, it shall snap open. ■ Screws better than glues. ■ Docs and drivers shall have permalinks and shall reside for all perpetuity at archive.org. ■ Ease of repair shall be a design ideal, not an afterthought. ■ Metric or standard, not both. ■ Schematics shall be included.

Big ups to Phillip Torrone and Simon Hill for their Maker's Bill of Rights ideas.

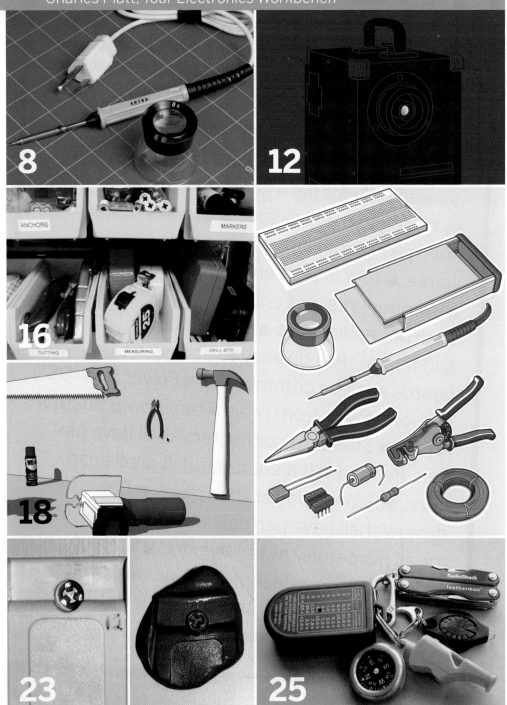

8

12

16

ANCHORS

MARKERS

CUTTING

MEASURING

DRILL BITS

18

23

25

RadioShack

leatherman

THE BEST OF
TOOLS
PROJECTS

from the pages of MAKE

AND YOU SHALL KNOW US BY THE CHARACTER OF OUR TOOLS

They say tools are part of what make us uniquely human. While we're not the only lifeform that uses them, we may be the only species that revels in their use. We venerate our tools, we love them, not only for what they help us do, but for the possibilities of future creations that they encode. We aspire through our tools. We take comfort in the fact that they're always there, at the ready, waiting for a plan, and a few moments of stolen time. Finding "the right tool for the job" can become something of an obsession. What follows are some of our recommendations for right tools, whether for your next Eureka moment, or your next toaster-oven repair.

Your Electronics Workbench

WHAT YOU NEED
TO GET STARTED
IN HOBBY ELECTRONICS.

By Charles Platt

THE BASICS

First, you will need a breadboard. You can, of course, call it a "prototyping board," but this is like calling a battery a "power cell." Search RadioShack online for "breadboard" and you will find more than a dozen products, all of them for electronics hobbyists, and none of them useful for doing anything with bread.

A breadboard is a plastic strip perforated with holes $1/10$" apart, which happens to be the same spacing as the legs on old-style silicon chips — the kind that were endemic in computers before the era of surface-mounted chips with legs so close together only a robot could love them. Fortunately for hobbyists, old-style chips are still in plentiful supply and are simple to play with.

Your breadboard makes this very easy. Behind its holes are copper conductors, arrayed in hidden rows and columns. When you push the wires of components into the holes, the wires engage with the conductors, and the conductors link the components together, with no solder required.

Figure 1 (on page 10) shows a basic breadboard. You insert chips so that their legs straddle the central groove, and you add other components on either side. Figure 1 also shows the bottom of a printed circuit (PC) board that has the same pattern

of copper connectors as the breadboard. First you use the breadboard to make sure everything works, then you transpose the parts to the PC board, pushing their wires through from the top. You immortalize your circuit by soldering the wires to the copper strips.

Soldering, of course, is the tricky part. As always, it pays to get the right tool for the job. I never used to believe this, because I grew up in England, where "making do with less" is somehow seen as a virtue.

When I finally bought a 15-watt pencil-sized soldering iron with a very fine tip (Figure 2), I realized I had spent years punishing myself. You need that very fine-tipped soldering iron, and thin solder to go with it. You also need a loupe — the little magnifier included in Figure 2. A cheap plastic one is quite sufficient. You'll use it to make sure that the solder you apply to the PC board has not run across any of the narrow spaces separating adjacent copper strips, thus creating short circuits.

Short circuits are the #2 cause of frustration when a project that worked perfectly on a breadboard becomes totally uncommunicative on a PC board. The #1 cause of frustration (in my experience, anyway) would be dry joints.

Any soldering guide will tell you to hold two metal parts together while simultaneously applying solder and the tip of the soldering iron. If you can manage this far-fetched anatomical feat, you must

Illustration by Damien Scogin

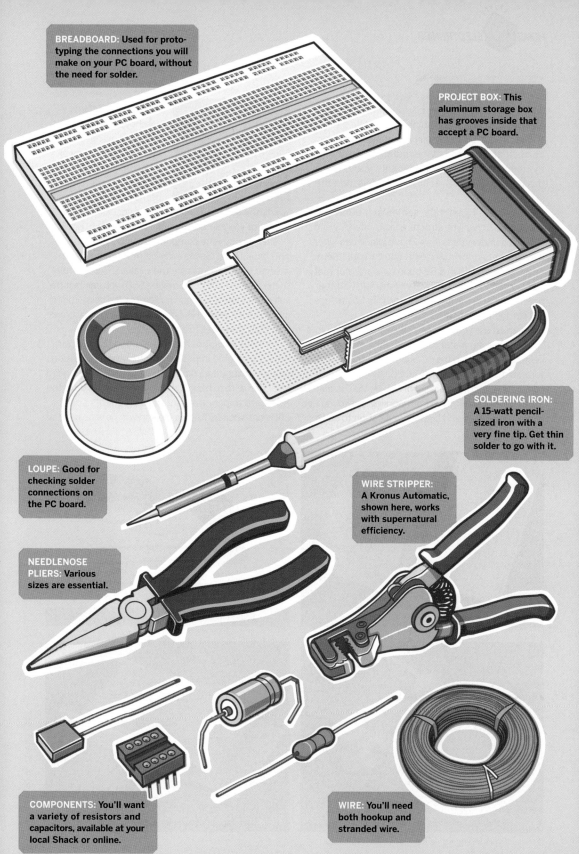

BREADBOARD: Used for prototyping the connections you will make on your PC board, without the need for solder.

PROJECT BOX: This aluminum storage box has grooves inside that accept a PC board.

SOLDERING IRON: A 15-watt pencil-sized iron with a very fine tip. Get thin solder to go with it.

LOUPE: Good for checking solder connections on the PC board.

WIRE STRIPPER: A Kronus Automatic, shown here, works with supernatural efficiency.

NEEDLENOSE PLIERS: Various sizes are essential.

COMPONENTS: You'll want a variety of resistors and capacitors, available at your local Shack or online.

WIRE: You'll need both hookup and stranded wire.

also watch the solder with supernatural close-up vision. You want the solder to run like a tiny stream that clings to the metal, instead of forming beads that sit on top of the metal. At the precise moment when the solder does this, you remove the soldering iron. The solder solidifies, and the joint is complete.

You get a dry joint if the solder isn't quite hot enough. Its crystalline structure lacks integrity and crumbles under stress. If you have joined two wires, it's easy to test for a dry joint: you can pull them apart quite easily. On a PC board, it's another matter. You can't test a chip by trying to pull it off the board, because the good joints on most of its legs will compensate for any bad joints.

You must use your loupe to check for the bad joints. You may see, for instance, a wire-end perfectly centered in a hole in the PC board, with solder on the wire, solder around the hole, but no solder actually connecting the two. This gap of maybe 1/100" is quite enough to stop everything from working, but you'll need a good desk lamp and high magnification to see it.

A FEW COMPONENTS AND TOOLS

Just as a kitchen should contain eggs and orange juice, you'll want a variety of resistors and capacitors (Figure 3). Your neighborhood Shack can sell you prepackaged assortments, or you can shop online at mouser.com or eBay.

After you buy the components, you'll need to sort and label them. Some may be marked only with colored bands to indicate their values. With a multimeter (a good one costs maybe $50) you can test the values instead of trying to remember the color-coding system. For storage I like the kind of little plastic boxes that craft stores sell to store beads.

For your breadboard you will need hookup wire. This is available in precut lengths, with insulation already stripped to expose the ends. You'll also need stranded wire to make flexible connections from the PC board to panel-mounted components such as LEDs or switches. To strip the ends of the wire, nothing

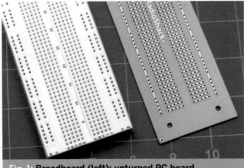
Fig. 1: **Breadboard (left); upturned PC board.**

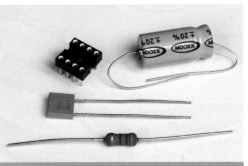
Fig. 3: **Socket, big and small capacitors, resistor (front).**

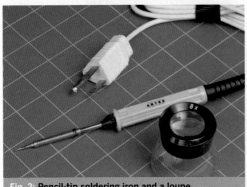
Fig. 2: **Pencil-tip soldering iron and a loupe.**

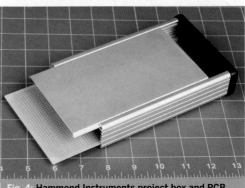
Fig. 4: **Hammond Instruments project box and PCB.**

Photography by Charles Platt

beats the Kronus Automatic Wire Stripper, which looks like a monster but works with supernatural efficiency, letting you do the job with just one hand.

Needlenose pliers and side cutters of various sizes are essential, with perhaps tweezers, a miniature vise to hold your work, alligator clips, and that wonderfully mysterious stuff, heat-shrink tubing (you will never use electrical tape again). To shrink the heat-shrink tube, you'll apply a Black and Decker heat gun.

If this sounds like a substantial investment, it isn't. A basic workbench should entail no more than a $250 expenditure for tools and parts. Electronics is a much cheaper hobby than more venerable crafts such as woodworking, and since all the components are small, it consumes very little space.

For completed projects you need, naturally enough, project boxes. You can settle for simple plastic containers with screw-on lids, but I prefer something a little fancier. Hammond Instruments makes a lovely brushed aluminum box with a lid that slides out to allow access. Grooves inside the box accept a PC board. My preferred box has a pattern of conductors emulating three breadboards put together (Figure 4). This is big enough for ambitious projects involving multiple chips.

LEARN THE RULES

The final and perhaps most important thing you will need is a basic understanding of what you are doing, so that you will not be a mere slave to instructions, unable to fix anything if the project doesn't work. Read a basic electronics guide to learn the relationships between ohms, amperes, volts, and watts, so that you can do the numbers and avoid burning out a resistor with excessive current or an LED with too much voltage. And follow the rules of troubleshooting:

» LOOK FOR DEAD ZONES. This is easy on a breadboard, where you can include extra LEDs to give a visual indication of whether each section is dead or alive. You can use piezo beepers for this purpose, too. And, of course, you can clip the black wire of your meter to the negative source in your circuit, then touch the red probe (carefully, without shorting anything out!) to points of interest. If you get an intermittent reading when you flex the PC board gently, almost certainly you have a dry joint somewhere, making and breaking contact. More than once I have found that a circuit that works fine on a naked PC board stops working when I mount it in a plastic box, because the process of screwing the board into place flexes it just enough to break a connection.

» CHECK FOR SHORT CIRCUITS. If there's a short, current will prefer to flow through it, and other parts of the circuit will be deprived. They will show much less voltage than they should.

Alternatively you can set your meter to measure amperes and then connect the meter between one side of your power source and the input point on your circuit. A zero reading on the meter may mean that you just blew its internal fuse because a short circuit tried to draw too much current.

» CHECK FOR HEAT-DAMAGED COMPONENTS. This is harder, and it's better to avoid damaging the components in the first place. If you use sockets for your chips, solder the empty socket to the PC board, then plug the chip in after everything cools. When soldering delicate diodes (including LEDs), apply an alligator clip between the soldering iron and the component. The clip absorbs the heat.

Tracing faults in circuits is truly an annoying process. On the upside, when you do manage to put together an array of components that works properly, it usually keeps on working cooperatively, without change or complaint, for many decades — unlike automobiles, lawn mowers, power tools, or, for that matter, people.

To me this is the irresistible aspect of hobby electronics. You end up with something that is more than the sum of its parts — and the magic endures.

Charles Platt is a frequent contributor to MAKE, has been a senior writer for *Wired*, and has written science fiction novels, including *The Silicon Man*.

THE MAKER'S
ULTIMATE TOOLS

The tools we use — or wish we could get our hands on. By Saul Griffith

Here's what would go into an extremely expensive ideal toolbox for someone who wants to be able to make pretty much anything, from ultimate fighting robots to hybrid go-karts, and even play around with microelectromechanical systems. You can and will make do without these, but in a perfect world, where the streets are paved with socket wrenches, these five tools would be in your basement. For the complete list, turn to page 14. For an ultimate tools narrative, go to *makezine.com/03/ultimate*.

3D Printer, $25,000
zcorp.com/products/printers.asp

This makes surprisingly beautiful parts; just don't expect them to be robust. It's the fastest way to go from computer model to physical part. My pick of the bunch is Z Corp's printer — it's the cheapest and fastest. Neat fact: They're used to print replacement body parts.

3D Scanner, $30,000
kmpi.konicaminolta.us/vivid/default.asp

These machines are still quite expensive, and accuracy depends on how much you spend and the size of the object you are scanning. They're used a lot these days for restoration of antiquities and sculptures as well as assisting in surgery.

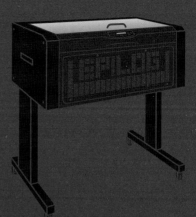

Plasma Cutter, $10,000

toolking.com/hobart/view.asp?id=4276

It's more difficult to use than a laser cutter, but there's a big advantage: it cuts metal or anything that conducts electricity. Think of it as a robotic oxy torch. You can be up and running for about 10K. Make your own parts for that car restoration project or build custom aluminum chandeliers. Poor maker's alternative: An oxy torch and a very steady hand or a high-quality bandsaw and lots of patience.

Laser Cutter, $19,900

epiloglaser.com/mini2412.htm

CAD-driven high-powered lasers cut plastic, paper, and wood in thicknesses up to about ⅜ inch with very high precision. For kicks, you can write your name on toast or etch your face on an eggplant. They're also good for cutting rubber stamps. Poor maker's alternative: Print the patterns with your inkjet printer and cut them out with a scroll saw. Not as accurate or as fast, but a workable workaround.

Water Jet, $100,000

omax.com

This rich man's plasma cutter cuts through 8 inches of granite with a barely subsonic jet of abrasive grit-filled water. It has none of the material restrictions of the laser cutter or the plasma cutter (though it isn't great for wood). The water tank weighs a ton (literally), so you'll need to reinforce your garage floor. Neat fact: Used extensively for cutting up chicken carcasses and chocolate bars (though with water only — no abrasive grit). Poor maker's alternative: Rumor has it you can do something similar with a washing machine pump and a hypodermic needle.

THE ULTIMATE TOOL
BUYING GUIDE

A complete list of tools you need to make almost anything.

If a genie were to grant me my wish for a shed full of tools, this is what I'd ask for. Think of it as an extremely biased guide to outfitting yourself with the ultimate shop for launching your own space program. Tool reference numbers from McMaster-Carr (mcmaster.com)

Legend:
- Necessity
- Priority
- Extremely useful
- Surprisingly useful
- Infrequent but handy
- Can do without, better with
- You didn't know it was so lovely

Tool Name	McMaster #	$ Budget	$ Deluxe
Hand Tools			
Box Knife	3814a11	1	10
Precision Blade	35435a11 38995a71 35515a12	1	10
Claw Hammer	6484a21	10	50
Ball Peen Hammer	6481a31	10	50
Blacksmith's Hammer (Heavy Weight)	6462a24	10	80
Rubber Mallet	5917a8	10	40
Miter Box	4201a11	15	45
Hacksaw	4086a34	5	25
Tight Spot Hacksaw	4060a16	2	5
Hole Punch Tool	3461a22	40	150
Center Punches and Chisel Set	3506a76	25	120
Metric and Imperial Socket Sets	7290a24 5757a35 5582a11	30	1200
Torque Wrench	85555a221	50	300
Hex Key Sets, Imperial and Metric	5541a31 5215a24 7162a13 5215a12	2	80
Torx Key Set	6959a85	2	40
Mini-Hex Drivers	52975a21 7270a59	2	40
Combination Wrenches, Metric and Inch	5314a62 5304a73 5314a25 5772a53	25	800
Vise Grip Long Nose Locking Pliers	7136a19	2	50
Needlenose Pliers, Small and Large	5451a12	2	35
Bull Nose Pliers, Small and Large		2	35
Vise Grip, Large	7136a15	5	60

Tool Name	McMaster #	$ Budget	$ Deluxe
Vise Grip, Med Curved	5172a17	5	45
Adjustable Wrenches	5385a12 5385a15	3	40
Crow Bar / Ripping Bar	5990a2	2	30
Tube Cutter	2706a1	15	80
Glass Cutter	3867a16	2	25
Bolt / Chain Cutter	3771a15	50	150
Sheet Metal Snips	3585a13 3908a11 3902a9	10	40
Finishing Saw	4012a1	10	30
Coping Saw	4099a1 6917a11	4	10
Hole Saw Kit	4008a71	25	120
Pull Saw	4058a52	10	20
Metric / Inch Tap and Die	2726a66	40	1200
Drill Sets	28115a77 31555a55 31555a56 31555a57 8802a11 8802a12 8802a13	5	1200
Deburring Taper	3018a4	5	80
Deburring Tools	4253a16 4289a36	2	25
Drill Stops	8959a16	2	10
Vise	5344a31	10	1500
Clamps	5165a25	2	45
Quick-Grips	51755a7	15	50
Jaw Puller	6293k12	50	180
Files	8176a12 8194a12	2	100
Hydraulic Floor Jack		25	200
Block And Tackle / Lifting Winch		50	500
Screwdrivers, Flat and Phillips	8551a31	1	90

Tool Name	McMaster #	$ Budget	$ Deluxe
Jeweller's Screwdrivers	52985a21 52985a23	10	40
Propane Burner		10	50
Heat Gun		50	250

Power Tools

Tool Name	McMaster #	$ Budget	$ Deluxe
18V Electric Drill	29835a16	25	300
Band Saw	4164a12	250	5000
Reciprocating Saw (Sawzall)	4011a25	120	250
Sliding Compound Miter Saw	3001a21	200	600
Tilting Table Saw	27925a12	300	2000
Drill Press	28865a31	100	2500
Plunge Router	36485a11	100	300
Manual Lathe	8941a12	500	5000
Mig Welder	7899a28	200	1500
Stroboscope	1177t92	25	250
Adjustable Hot Plate	33255k61	50	800
Dremel	4344a42 4370a5	50	150
Angle Grinder	4395a16	50	250
Bench Grinder	20535A654	75	300
Belt Sander	4892a21	100	200
Disc / Belt Sander	46245a49	250	1500
Bridgeport Mill		500	15000
Heisseschneider Hot Knife		50	200
Sewing Machine		25	2500
Air Compressor	4364k3	200	2500
Spot Blaster	31195k11 3210k11	50	500
Vacuum Pump		100	1000+
Oxy / Acetylene Torch	7754a12	250	1500
Plasma Torch		600	3000

Computer Controlled Tools

Tool Name	McMaster #	$ Budget	$ Deluxe
Inkjet Printer		25	250
Large-Format Printer		900	25000
Nc Mill		2500	120000
Nc Lathe		5000	150000
Laser Cutter (Co2)		12000	50000
Plasma Cutter		3000	20000
Wire / Sink EDM		100000	250000
Water Jet		80000	150000
3D Printer (Z Corp, FDM, STL)		25000	250000
Plotter / Cutter (Roland)		1000	25000

Electronics Tools

Tool Name	McMaster #	$ Budget	$ Deluxe
Wire Stripper		2	80
Pliers Set	5323a49	10	120
Work Holder And Magnifier	5007a14	5	100
Multimeter		75	250
Temp-Control Solder Station		150	1000
Hot Air Tool for Point Reflow / Desoldering		30	500
Bench Power Supply, Multi-Output		150	500
Toaster Oven, Adjustable Time / Temp		40	60
Microscope (See Safety / Measurement / Visualization)			
Oscilloscope		500	5000
Micro-tweezer Sets		2	100
Pick-n-Place		3000	25000

Fetish Tools

Tool Name	McMaster #	$ Budget	$ Deluxe
Optics Bench		1000	400000
Mask Writer		50000	1000000
Mini-jector		4000	50000
Thermoformer		1000	20000
ESEM		25000	500000
3D Scanner		5000	100000
Excimer Laser Cutter		100000	1000000
PCR			100000
Micropippettes		20	2000
Spin Coater		500	25000
High Temp / Vacuum Oven		2000	30000
Chemistry Hoods and Glass Equipment		2000	1000000
Ultrasonic Welder		5000	25000
Tube Bender		1000	40000
Tanks for Anodizing, Etching		25	2500
Kiln		500	5000
Anvil		250	1000
Crucible		20	2500
Thin Film Evaporator / Sputterer		5000	100000

Safety, Measurement, and Visualization

Tool Name	McMaster #	$ Budget	$ Deluxe
Safety Goggles	2404t21	1	10
Ear Muffs	9205T6	2	30
Micrometer	2054a75	5	300
Caliper	8647a44	5	500
Head-Mounted Magnifier	1490t3 1509t14	5	120
Feeler Gauges	2070a7	1	25
Spirit Level	2169a4 2169a1	5	50
Tape Measure	19805a74	1	25
Adjustable Stereomicroscope	10705t64	500	25000
Hot Gloves		5	100
Work Gloves		1	40
Welding Mask		15	100

TIGHT-FIT WORKBENCH

Make an inexpensive workspace for crowded quarters. By Todd Lappin

It's hard to be a maker if you don't have a good place to do your making. Yet two things often stand in the way of building out a basic home workbench: high cost and limited space.

Industrial-grade fixtures and spiffy garage storage systems cost a pretty penny. Likewise, domestic real estate is a scarce commodity — garages must still be used for parking cars, basements for storing stuff, and utility rooms must shelter washing machines and assorted whatnot.

I faced those constraints and a little more when I set out to build a simple workbench in my narrow garage. To avoid getting in the way of my car, the bench had to be shallow — no more than 2' deep. I needed lots of storage for tools, small parts, and bulky boxes of big stuff.

And just to make things more challenging, I also had to build the bench around several pre-existing drain and sewer pipes that intruded upon my already-limited workspace. Here's how I built the simple bench setup shown above.

Lighting

Bright, shadow-free light is essential when doing precision work or manipulating small parts. This was one area where I lucked out. We'd recently renovated our house, so my garage started out with brand-new fluorescent light fixtures running along the ceiling. Otherwise, a plug-in overhead fluorescent fixture would have been an inexpensive way to go. I also keep a simple $5 clip lamp on hand for task lighting.

Photography by Todd Lappin

The Workbench

Given my spatial constraints, I was tempted to build my own workbench from scratch, using 2×4's and plywood. Ultimately, however, I decided it was easier (and probably cheaper) to look for something off-the-shelf. The workbenches sold at many of the big chain hardware stores are overpriced and under-built, but Global Industrial (globalindustrial.com) offers several industrial-grade benches for $150 or less. Trouble is, they're also big, typically 60"×30". I didn't have that much room, and because of our intruding drain pipes, I also had to find something that didn't need to sit flush against the back wall.

I found the ideal solution at Ikea, much to my own surprise. Ikea's "Antonius" line is a cantilevered storage system built around upright metal rails that screw into the wall. A compact workbench configuration is offered, with a laminated particleboard top that's just 24" deep and 47" wide. It's sturdy, versatile, and very cheap — less than $50 for all the required parts.

Tool Storage

Ikea offers a pegboard option for the Antonius storage system, but it uses a square hole pattern that's incompatible with standard pegboard fittings. I avoided that problem by simply screwing a half-sheet of standard round-hole pegboard to the back of the workbench.

To store the rest of our tools, my wife donated the red Sears Craftsman tool chest that she'd previously used in her home office (I knew I'd married well). These are also surprisingly affordable, and basic models can be had for around $175.

Parts Storage

Ah, the little stuff: nuts, bolts, screws, nails, tapes, glues, wall anchors, wire, and whatnot. These things should be readily accessible, but storing parts in coffee cans and plastic deli containers quickly grows cumbersome.

A "pick rack" of removable plastic bins — the kind used in factories and warehouses — is a simple and affordable way to get the job done. Global Industrial sells bin unit sets that come with 32 small bins and a 36"×19" wall-mounting panel, all for around $50. Bigger sizes, with many more bins, are also available.

A "pick rack" of removable plastic bins is a simple way to organize the little stuff.

Bulk Storage

A good shelving system is the best way to make efficient use of limited floor space in garages and other mixed-use areas. Again, the temptation here is to simply build a vertical shelving unit from scratch, but in the interests of future-friendly expansion and flexibility, I'm going prefab.

The steel Gorilla Rack shelves sold at Costco or Home Depot are sturdy and cheap, but they're usually sold in just one size and configuration (which may or may not suit your needs). Global Industrial offers a variety of commercial-grade shelving systems in a very wide range of heights, widths, depths, and shelving configurations, at very reasonable prices — all the better to make the most of every square inch of precious space.

Options and Accessories

With my major infrastructure in place, I added a few more bolt-on components to complete the setup. I screwed a 2' power strip into the top-rear edge of the Ikea workbench, to provide plenty of electrical outlets for rechargeable tools and soldering irons. Magnetic strips designed for kitchen knife storage also work well to organize frequently used hand tools. Ikea sells these on the cheap, so I bought one and mounted it to the pegboard. Now I'm well-lit, neatly organized, and fully powered up. Time to get to work!

Todd Lappin (telstar@well.com) moonlights as fleet operations officer for Telstar Logistics, a leading provider of integrated services.

SKETCHUP WORKBENCH

Design your own work area with Google's free drawing application. By John Edgar Park

Google SketchUp is my favorite design tool, and if all goes according to plan, it'll soon be yours, too. Even though I use higher-end 3D software all day at work, SketchUp still blows me away; it enables fast, fun, and accurate 3D sketching unlike any other program (it's free too!).

Makers will find SketchUp useful for all sorts of things, from furniture design to workshop layout, from project enclosures to robotic exoskeletons. It's good for this kind of stuff because you can rough out your designs quickly, using real-world dimensions.

I decided to use SketchUp to design a much-needed workbench. The first phase was to create the conceptual model, which is a rough 3D sketch of the form. The second phase was design engineering, where I figured out the real-world materials list and construction plan for the project.

TOOLS

PC or Mac of somewhat recent vintage
Google SketchUp software (free download)

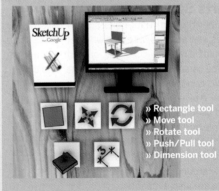

» Rectangle tool
» Move tool
» Rotate tool
» Push/Pull tool
» Dimension tool

Illustrations by John Edgar Park

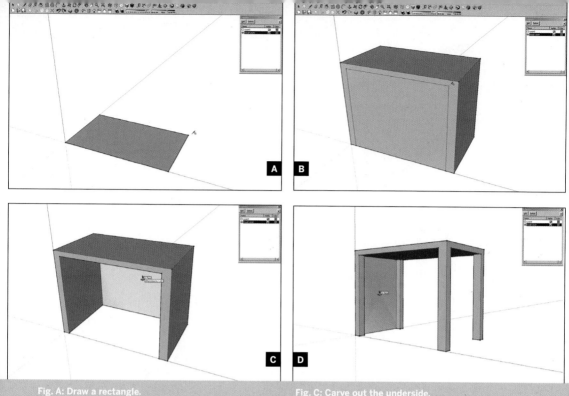

Fig. A: Draw a rectangle.
Fig. B: Mark the side for the cutout.
Fig. C: Carve out the underside.
Fig. D: Make the leg cutouts.

Build a Workbench in SketchUp

Phase I. Conceptual Design

1. Get SketchUp running.

1a. Download and install SketchUp from sketchup.google.com. It's available for OS X and Windows XP (please join me in begging for a Linux version).

1b. Launch SketchUp and do the introductory tutorial listed under Help→Self-Paced Tutorials →Intro to get a feel for viewport navigation and the basic drawing tools.

2. Prepare your project.

2a. Create a new project by clicking File→New. Set the units to fractional inches by going to Window→ Model Info, choosing the Units category on the left, and then picking Fractional from the Format list. This means that measurements in this project will be listed in inches only, instead of feet and inches. Also, go to Window→Styles, click the Edit tab, and turn on Endpoints. This makes vertices easier to see.

2b. Use the Select tool (found under Tools→Select) to click on the 2D man living in your scene. His name

is Bryce. Click Edit→Hide and wave goodbye to Bryce.

2c. I like to organize the models within each project on their own layers to control visibility and interaction between parts. Add a new layer for the conceptual phase by choosing Window→Layers and then clicking the Add Layer button in the Layers window. Name the layer Rough Layer and make it active by clicking the Active radio button. The active layer is where all new objects will go.

3. Rough out the form.

3a. It all begins with a rectangle. Choose Draw→ Rectangle. Now, click the left mouse button on the origin (the center of the scene where all axes cross) and drag toward an opposite corner, paying atten- tion to the measurements in the lower right corner of the interface. Release the mouse button to finish. Immediately after you draw a shape, you can type in dimensions to set an exact size; type 48", 28" and press Enter on the keyboard. (No need to click anywhere, just start typing.) A shaded rectangle appears (Figure A).

3b. Extrude the tabletop upward to give the model height. Choose Tools→Push/Pull. This tool is fun to use; put the cursor over the tabletop face, then click

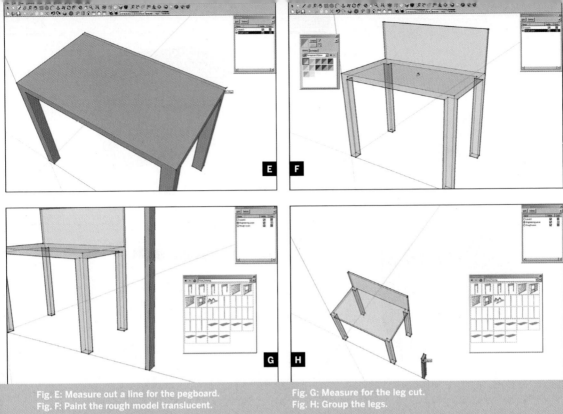

Fig. E: Measure out a line for the pegboard.
Fig. F: Paint the rough model translucent.

Fig. G: Measure for the leg cut.
Fig. H: Group the legs.

and drag upward. Release the mouse button, and key in an exact height of 36".

3c. Time to cut out the underside. Create a measurement guide 3" from the bottom left corner with the Tape Measure tool by clicking once on the corner point and a second time anywhere along the bottom edge. Type 3" and press Enter to set the exact measurement. Use the Rectangle tool to draw on the front face of the model. Start the rectangle at the measurement guide you just made. End the rectangle at around 42", 34" — again, you can type these dimensions to be precise. Although this is a rough model, some of the following steps work best if the rectangles you draw are of a consistent height (Figure B).

3d. Use the Push/Pull tool to push this new face all the way to the back of the model. You'll see an inference pop-up declare "On Face" when your cursor is aligned with the back face. Release the mouse button and you'll have carved a large chunk out of the model (Figure C).

3e. Repeat this procedure twice more on the inner sides of the workbench to leave the tabletop standing on 4 legs. Start each rectangle at the bottom edge, 2½" from the side, measuring this off with the Tape Measure tool first. The dimensions should be 23", 34" (Figure D).

4. Add details.

4a. Next, add a pegboard for tool storage. Choose the Line tool (the pencil) from the Draw→Line menu item. Click a point on the left edge of the tabletop near the back edge of the table. Begin moving the cursor to the right side to draw your line — a red inference line appears when your line is parallel to the x-axis. Continue until you reach the right edge and a message pops up to let you know you've intersected the edge. Click to lay down the second point, which will complete your line (Figure E).

4b. Using your Push/Pull tool, pull up the small face at the rear of the tabletop to an appropriate height, around 16".

4c. In the next phase, you'll use this rough model as a template for your design engineering model. To make that easier, paint a semitransparent material on the rough model. Go to Tools→Paint Bucket, and choose Blue Glass from the Transparent palette. Shift-click your model to paint it (Figure F).

4d. Save your scene by clicking File→Save As and type in the filename *workbench.skp*. Click the Save button.

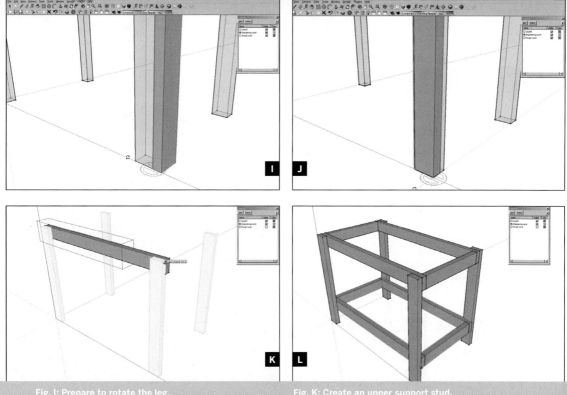

Fig. I: Prepare to rotate the leg.
Fig. J: Rotate the leg 90°.

Fig. K: Create an upper support stud.
Fig. L: Duplicate the upper parts for the lower frame.

Phase II. Design Engineering

5. Choose components.

5a. Create a new layer (Window→Layers). Name it Construction Layer, and make the layer active.
5b. Open the Component window by clicking Window →Components, and then choose Construction from the drop-down menu. Here you'll find the 2×4 we'll need. Click-drag the 12'-long 2×4 stud from the Component window to your scene.

6. Cut lumber.

6a. Right-click on the stud and choose Explode from the menu. Sorry, nothing dramatic happens, but this does let you edit the stud.
6b. You'll measure the cut with the Tape Measure tool. Create a measurement guide by clicking on a corner point at the bottom of the stud and then click again partway up the same edge. Type in the height of your leg cut, 34", and press Enter (Figure G).
6c. Use the Push/Pull tool to drag the top face of the stud down until it snaps to the Guide Point you measured, thus cutting the leg down to 34".
6d. With the Select tool, triple-click the stud to select all connected faces, and then group them by clicking Edit→Make Group (Figure H).

7. Make copies.

7a. Select the leg, then choose Tools→Rotate and rotate the leg 90° on the z-axis (the protractor should be blue). Then, click Tools→Move and move the leg into position over one of the legs of the rough model. Do so by clicking once on one of its bottom points, and then a second time on the equivalent point on the rough model (Figures I and J).
7b. With the leg still selected, duplicate it by clicking on Edit→Copy, and then Edit→Paste. Position the new leg and then repeat for the remaining legs.
7c. Clean up the screen by turning off the Rough Layer visibility in the Layers window.

8. Support frame.

8a. Paste, move, rotate, and resize (phew!) one of the legs to create the front upper support. In order to edit the length of the stud, double-click the group with the Select tool, then use the Push/Pull tool on one end face. Repeat this step 3 times to complete the frame (Figure K).
8b. Duplicate the frame parts downward to support the lower legs, about 6" up from the bottom. Do this by shift-selecting the parts and using Edit→Copy, then Edit→Paste (Figure L).

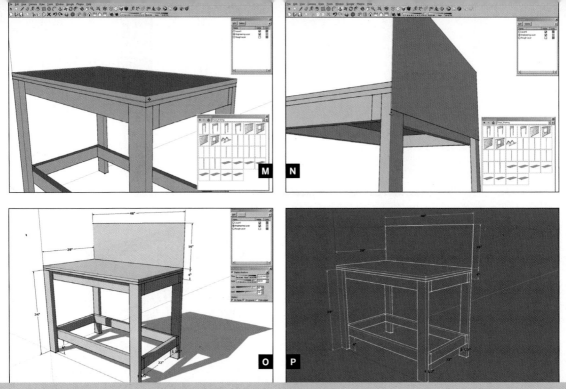

Fig. M: Cut two plywood sheets for the top.
Fig. N: Resize the pegboard.

Fig. O: View the dimensions and shadows.
Fig. P: Impress your friends with the blueprint render style.

9. Top it off.

9a. Grab a standard sheet of ¾" plywood for the top of the workbench from 3D Warehouse, the user-supported model repository. Click File→3D Ware-house→Get Models. In the search field, type "¾ thick plywood". When you find it, click Download Model, then click Yes to load it into your scene. Snap it to the top of the bench, then cut it down to size the same way you did the 2×4s. For a sturdy work surface, lay a second piece of plywood on top of the first (Figure M).

9b. You can now add the pegboard to the back. Use the 3D Warehouse to import and place a sheet of ⅛" plywood. Then, use the Push/Pull tool to extrude the pegboard piece 16" above and 4" below the bench top (Figure N). Save your model by clicking File→Save.

10. Note dimensions.

10a. To add dimension annotations to the model, choose Tools→Dimensions, then click on 2 points you need dimensions for, dragging outward to place the text. Repeat this for any unique cuts or measurements.

10b. Choose a flattering camera view and then click File→Print. Now that you're done, show off your new workbench in its best light; turn on shadow rendering by clicking View→Shadows. You can adjust lighting

in Window→Shadow Settings (Figure O).

By playing with the scene's color setting found in Window→Styles, you can create the clean look of a blueprint, the loose lines of a charcoal rendering, the paranoia of a watermarked painting, and more (Figure P). Snazzy!

RESOURCES:

3D Construction Modeling by Dennis Fukai from insitebuilders.com

SketchUp Level 1 training DVD from go-2-school.com

Official SketchUp forum: groups.google.com/group/sketchup

Live by the golden rule; share your models via the 3D Warehouse. Be sure to tag them with the word *MakeMagazine*.

Looking for more models? Search the 3D Ware-house, the user repository of SketchUp models. Go there by clicking File→3D Warehouse→Get Models.

John Edgar Park rigs CG characters at Walt Disney Animation Studios. He is the author of *Understanding 3D Animation Using Maya*. Read about his house addition at parkhaus.blogspot.com.

No exotic screw head is a match for someone wielding some Silly Putty and a Dremel tool.

FREEDOM TO UNSCREW

Make tamperproof driver bits by molding the screw heads. By Johnathan Nightingale

Photograph by Amy Nightingale

When a friend asked for my help removing some nonstandard screws from his doorframe without damaging them, I expected a little resistance. Many manufacturers use so-called tamperproof or security screw heads to prevent casual would-be hardware hackers; tamperproof Torx, spanner bits, and Tri-Wing being some of the more popular types. This security-by-obscurity approach can usually be foiled with a security bit set available in most hardware and electronics stores, though, and I assured him we'd have that panel off in no time.

The screws in question, however, were not of the standard varieties. Rather than having a bit pattern cut into the center of the screw head, this was basically a round head with three notches removed from the edges to form an equilateral triangle. Some research online revealed that the screw was

a Tri-Groove design (for an excellent reference on standard security screw types, see *lara.com/reviews/screwtypes.htm*). What's more, individual driver bits for this head type can cost up to $10 a piece. When our attempts to use pliers and brute force failed, I decided to make a bit myself. I would need to get a cast of the screw head for reference, find a suitable source material for the driver, and then use a Dremel to handle the metalwork.

TOOLS:	
One "egg" of Silly Putty	Dremel
One set of dollar	Polyfilla drywall
store hex keys	compound, or some
(a.k.a. Allen keys)	other quick-drying
	spackle

There are better molding compounds than Silly Putty, but few are as cheap, and its weaknesses,

First impressions: Make a mold out of the tamperproof screw head with Silly Putty. From this, you'll make a cast out of ordinary wall spackle.

particularly its tendency to "flow" and lose definition, shouldn't matter for the short timespan required. Work the putty to soften it, and then press it onto the panel. One advantage of such a soft casting material is that it easily fills countersunk holes. Once removed, I covered the reverse mold of the screw head with spackle and let it dry. It gets fragile once dried, so handle it carefully; it should be mostly an eyeball reference anyhow.

Grinding Away

I took the cast back to my shop. The decision to use Allen keys was automatic — they are excellent, self-contained tools, they're cheap, and are made of solid, uniform metal. Since the Tri-Groove screw head is basically just three notches, I had to remove metal from the head of the hex key until only three "posts" remained, to match the notches of the head. The fact that the keys are hexagonal in shape helped here with the equilateral spacing of the posts.

Using the Dremel with a cut-off wheel, I began the process of removing material from the hex head to create my bit. It is essential whenever doing metal work like this to have a glass of water handy, so you can quench the key every few seconds. If the metal

starts heating, it gets harder to hold, of course, but the greater problem is that you can overheat a section and temper the metal. This will make it very hard but also brittle, which is not a desirable feature in a driver bit. Make a cut, quench, make another cut, quench. Of course, eye protection is essential as well.

Throughout the process, I referred to the cast I had made to ensure that the posts were positioned and shaped correctly. Several times, I thought I was finished only to find that when I tried to match the key to the spackle cast, the posts were too fat. I was leaving too much material from the center of the key on the posts, so they would have impacted the screw head instead of sliding into the notches. Eventually though, the driver matched the cast, and after some sanding with medium grit sanding cloth to remove any burrs (a tstep altogether), I had a functional, if unbeautiful, Tri-Groove driver.

Johnathan Nightingale is an IBM software maker by day, tinkerer by night.

Photography by Johnathan Nightingale

For more info, corrections, and discussion on this piece, please visit makezine.com/03/diy_driverbits

Whistle while you work: this set of tiny tools will get you out of many a tiny jam.

KEYCHAIN SURVIVAL TOOLS

Whether you're facing a parachute drop into the High Sierras or a jammed button on your mobile, some handy keychain gizmo can be there for you. By Bob Scott

Although my daily routine doesn't include as many parachute drops as it probably should, I still like to be prepared. Here's what's keeping my keys company.

Lighting

Whether reading a menu at your local diner or coping with a blackout in a high-rise office, a reliable light is a must.

LRI's Photon (*photonlight.com*) series are probably the best-known keychain lights and for good reason. They're reasonably rugged, light, and dependable. I've been using the white LED version of their latest Freedom light, which features easily adjustable brightness, extended run time, and

doubles the light output of earlier models, all for about 20 bucks.

If you want something really tiny, check out the hearing-aid-battery-powered Firefli. Barely big enough to find, it features a clever valve arrangement in the on/off switch that extends the normally short "use it or lose it" life of the zinc air batteries.

Signaling

Unless you're an opera star or door-to-door cymbal salesman, you can't bet on being able to signal for help in a crisis. If you've ever enjoyed an evening stuck in an elevator, you know you can shout yourself hoarse in a few minutes. A good whistle,

Photograph by Bob Scott

on the other hand, can attract attention over a wide area and weighs next to nothing.

My old standby is a $5 Fox 40 Mini (*fox40whistle. com*) with the cosmetic side plates dremeled off to reduce its size. Fox has recently released a new Micro model that features a flatter profile than my hacked version, and it's reportedly just as loud. Both have no moving parts and work even after being submerged.

Tools

The Micra and Squirt from Leatherman (*leatherman.com*) are pleasant standouts in an otherwise bleak sea of cast metal junk. Both pack a good selection of tools centered around a clever set of spring-action scissors or needlenose pliers. Also check out RadioShack's version that replaces the standard pliers with a wire stripper. $30 to $40.

Compass

I've used this more than I care to admit. My favorite is the liquid-damped Pocket Compass manufactured by old-school knife maker Marbles (*marblesknives.com*). About $15. If you insist on spending more, the $50 Traildrop II Digital Compass & Temperature Keychain (*www.highgearusa.com*) offers a backlight and all the functions you'd expect from a gizmo with a microchip and an LCD.

Test Gear

How about a $160 nuclear radiation detector? Looking vaguely like a car alarm remote, the NukAlert (*nukalert.com*) operates continuously, sounding an alarm when it detects a life-endangering amount of gamma or X radiation. By listening to the ten distinct alarm levels, you can plot a quick course out of a danger area or, better yet, avoid entering one.

Does it work? Beats me. My lease is vague about storage of high-level radiation on the premises, so I wasn't able to evaluate the manufacturer's claims. Their status as a state-licensed nuclear calibration facility is reassuring, though.

Bob Scott is a statistical construct of various consumer electronics marketing departments.

USB Thumbdrive Fill Up

Rather than haul around a bunch of wasted space on your keychain drive, why not keep some useful data on there between big file transfers? For instance:

Browser: Either a standalone installation of Firefox (see Volume 01 of MAKE) or at least a current copy of your bookmarks, exported from your browser as an HTML file. Add a copy of your RSS news and podcast feeds for access on the road, or to share with friends.

Data: Besides the current project information that you're sure you'll need, grab a copy of all documents less than 90 days old from the "My Documents" folder on your computer. Add a PDF version of your contact list in case your PDA packs up.

Email: If you haven't converted to a web-based email service, you may want a copy of your relevant mailboxes or a critical subset of your Outlook .pst file.

Photos & Music: Interesting photos you've shot in the last few weeks, a couple from the last vacation, and some sentimental favorites can all be big hits at the office or when visiting friends. Throw in your top 20 MP3s as a boredom antidote.

Manuals: Having a PDF copy of the manuals for your cellphone, camera, and car can come in very handy on trips. (Check the relevant OEM's website for these gems.)

Software: You've probably got a list of your "go to" programs, but before dragging all those zip files over, see what you can get from the web (e.g., online virus checkers like Trend Micro's Housecall). You may be better off with just a bookmark.

ID: Put a "Please Return Me To.txt" file containing your contact information in the root directory. You may get lucky.

If you've got any particularly sensitive data, consider encrypting it and keeping a copy of the decryption program (but not the password!) on the drive as well.

Once you've got the drive set up to your satisfaction, copy the files back to a dedicated file folder on your PC. Then you can erase the thumbdrive if you need the space for a big file transfer, and quickly restore it when you get back to your PC. — *BS*

  For more info, corrections, and discussion on this piece, please visit makezine.com/03/diy_keychain

> ≫ "Clock radios are hated devices that designers seem to ignore, judging by how little their features and interface have changed. To add a new capability for would-be sleepy heads, we're going to shoot the alarm. With a gun."
> Roger Ibars, *Gun-Operated Alarm Clock*

30

40

50

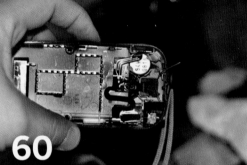

60

63

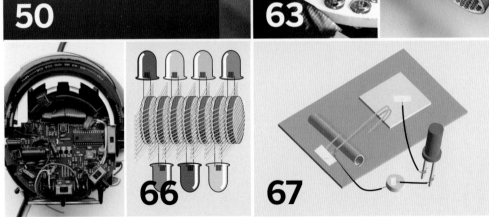

66

67

THE BEST OF ELECTRONICS PROJECTS

from the pages of MAKE

ELECTRONICS: NOT JUST FOR DEEP GEEKS ANYMORE

Far more people are intimidated by electronics than need to be. Partly, this is the fault of a frequently insular priesthood of electronics engineers who struggle to explain the mysteries and movements of electrons to Muggles like you and me. At MAKE, we try to give everyone the tools, the intel, and the friendly encouragement needed to have fun wrangling electrons through the fascinating miniature machinery known as circuits. Properly brought up to speed, anyone can be a solder iron samurai in no time. Here are some innovative, useful, and educational projects for Makers of all skill levels.

Five cables, bad.   One cable, good.

THE 5-IN-1 NETWORK CABLE

By Michael Ossmann

Nothing's worse for a network administrator than being without a needed cable. So I made a single cable to replace the five I used to carry. The result: no more tangles and no more scrounging for a missing link. »

Set up: p.34   Make it: p.36   Use it: p.39

WHY I MADE A 5-IN-1 CABLE

Do you find yourself toting several of these cables everywhere you go? Do you often wish you'd brought a different cable with you after you've arrived onsite? Are you as geeky as me and think that the idea of a 5-in-1 is just plain cool even if you never expect to configure a router in your lifetime? Then I'll show you how I made one.

The 5-in-1 cable consists of a CAT5 Ethernet cable along with four simple custom adapters, giving me an Ethernet cable, a crossover Ethernet cable, a modem cable, a null modem cable, and a Cisco console cable. An added benefit is that I can always extend my cable by finding a longer Ethernet cable than the one I carry in my bag. (It's usually pretty easy to locate a long Ethernet cable, but not so easy to locate a long null modem cable.)

Michael Ossmann (*ossmann.com/mike*) is a Senior Security Engineer for Alternative Technology in Colorado. He can't think of a second sentence that doesn't sound pompous or stupid.

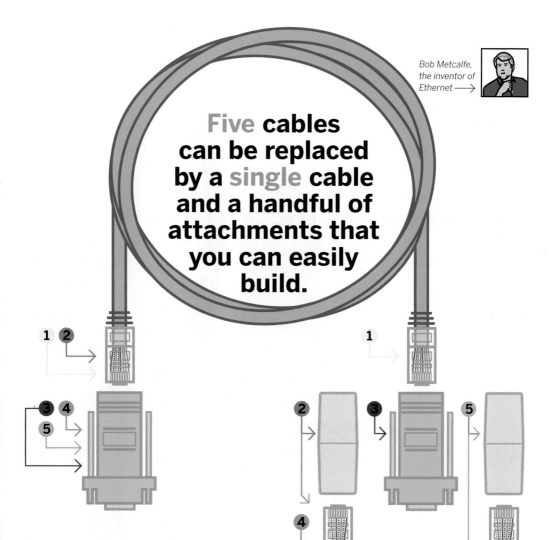

Bob Metcalfe, the inventor of Ethernet ⟶

Five cables can be replaced by a single cable and a handful of attachments that you can easily build.

1 Ethernet cable: Ethernet is the standard way to connect computers that are relatively close to each other. Modern CAT5 Ethernet cables carry data over twisted pairs of wires in order to reduce interference, enabling longer cables and faster data rates.

2 Crossover cable: Connecting computers over Ethernet requires a hub or switch to connect the "send" wire on one machine's Ethernet cable to the "receive" wire on another machine's cable. With just two computers, you can ditch the hub and use a crossover cable. Its send and receive wires cross over from one end to the other.

3 Modem cable: A straight-through RS232 serial cable is called a modem cable because it is used to connect Data Communications Equipment (DCE), such as a modem, to Data Terminal Equipment (DTE), such as a dumb terminal or computer. DTE devices are pinned differently than DCE devices so that they can be connected with a straight-through cable.

4 Null modem: The null modem cable is to a modem cable what the crossover cable is to the Ethernet cable. It allows direct serial communication between two nearby devices such as two computers or two modems.

5 Cisco console: This cable is used on certain kinds of Cisco equipment. It is an RS232 null modem cable with a 9-pin plug on one end and an RJ45 plug on the other.

Illustration by Nik Schulz

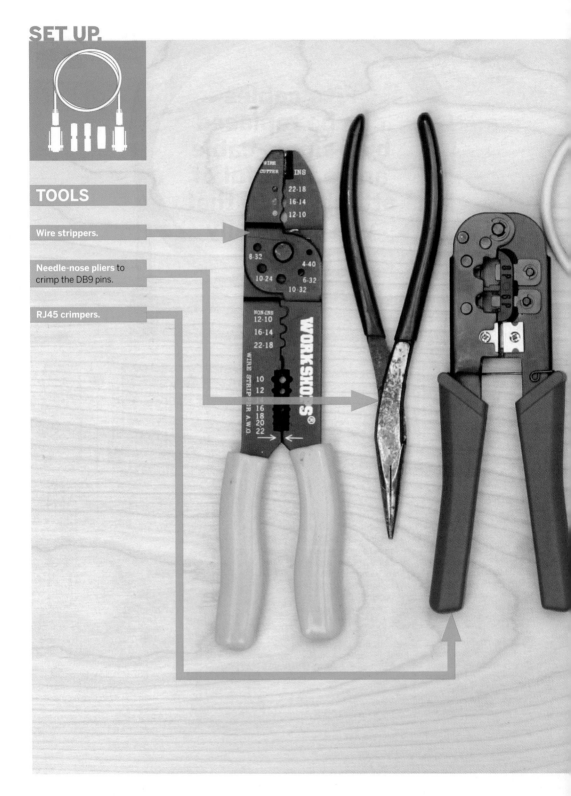

SET UP.

TOOLS

Wire strippers.

Needle-nose pliers to crimp the DB9 pins.

RJ45 crimpers.

Two short lengths of CAT5 cable, preferably of different colors, about 2 inches long . You can cut up existing Ethernet cables.

Four RJ45 modular plugs. Have extras on hand just in case.

Eight female pins for the DB9 connectors. Have a few extra pins on hand unless you are much more dexterous than me.

Two DB9 female to RJ45 female modular adapters. These are the kind of adapters that let you configure your own pinouts.

One RJ45 coupler. The coupler must have all eight conductors. Be aware that many Ethernet couplers only have four.

One straight-through Ethernet cable. This must be an eight-conductor cable, not a four-conductor cable.

MAKE IT.

BUILD YOUR 5-IN-1 CABLE

START ››   **Time: 30 min. Complexity: Low**

1. MAKE "THE WORLD'S SHORTEST CROSSOVER CABLE"

You can actually make the crossover cable as long as you want, but the longer you make it, the more you have to carry around.

This cable must cross the pairs that are not used by Ethernet in addition to the pairs that are. This is why you must perform this critical step and cannot use a standard crossover cable you may already own.

1a. Crimp one of the RJ45 plugs on each end. Order the wires on one end according to the following 568B standard (with the clip facing down):

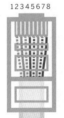

12345678

pin 1: white/orange
pin 2: orange
pin 3: white/green
pin 4: blue
pin 5: white/blue
pin 6: green
pin 7: white/brown
pin 8: brown

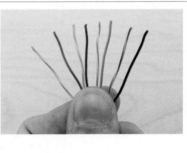

This is the most common order for Ethernet cables, so you could get a head start by snipping off the end of an existing cable; then you only have to do the other end.

1b. Order the wires on the other end this way:

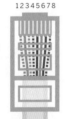

12345678

pin 1: white/green
pin 2: green
pin 3: white/orange
pin 4: white/brown
pin 5: brown
pin 6: orange
pin 7: blue
pin 8: white/blue

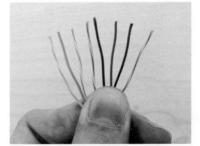

Make sure each pair has at least one twist. Then you can prove people wrong if they scoff, "That's not the world's shortest crossover cable; that's just an adapter!"

2. MAKE THE CISCO CONSOLE ADAPTER

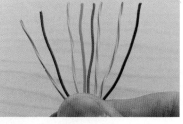

This adapter works with the RJ45 serial port found on most Cisco routers. It also works on some Sun servers. It is important to note that this is not a symmetric adapter. The 568B end will point away from the router and the other end will be inserted into the router. I marked one end with a Sharpie so I wouldn't forget which end was which.

Cisco's (otherwise very helpful) cabling page (*cisco.com/warp/public/701/14.html*) has RTS and CTS reversed on the DB9/RJ45 console cable. I verified this by inspecting an actual Cisco cable. They don't really care because their console ports do not use flow control, but doing it the right way enables interoperability with Sun servers and perhaps some other things.

This is like the crossover cable but with a different pinout.

2a. Make the first end according to 568B again (clip facing down):

12345678

pin 1: white/orange
pin 2: orange
pin 3: white/green
pin 4: blue
pin 5: white/blue
pin 6: green
pin 7: white/brown
pin 8: brown

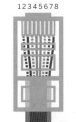

You can read about the 568B standard at *www.utm. edu/~leeb/568/568.htm*.

2b. And the other end:

12345678

pin 1: white/brown
pin 2: brown
pin 3: white/green
pin 4: green
pin 5: orange
pin 6: white/orange
pin 7: white/blue
pin 8: blue

Since this is an asymmetric adapter, mark one end of the adapter with a Sharpie so you know which end is which.

3. MAKE TWO DB9/RJ45 ADAPTERS

This is the trickiest part. In order to make your cable compatible with the largest number of serial devices possible, you need to combine a couple of pins and split another one. Both of the DB9/RJ45 adapters should be wired exactly the same way, regardless of whether they will be used for DTE or DCE devices. Here is the pinout:

DB9 pin	signal	RJ45 pin	color *
1	DCD	8	white
2	RxD	3	black
3	TxD	1	blue
4	DTR	5	green
5	SG	2 & 6	orange and yellow
6	DSR	8	white
7	RTS	4	red
8	CTS	7	brown
9	R		none

* My DB9F/RJ45F modular adapters are colored blue, orange, black, red, green, yellow, brown, white (RJ45 1-8). If yours are different, ignore the colors in the above pinout.

3a. DB9 pins 2, 3, 4, 7, and 8 are easy. Just push the appropriate pin in the back of the DB9 connector until it snaps.

Be careful not to get any of the pins mixed up because errors are a bit difficult to fix unless you have the right tool to pop the pins back out again. Pin extractors are available at *svc.com/mole extractor.html*.

3b. DB9 pin 5 needs two wires connected to it. Snip the pins off of the wires coming from RJ45 pins 2 and 6 (orange and yellow on mine), strip about 3mm off the end of each, and crimp them together onto one of your spare pins.

Use a spare pin to crimp the wires together.

3c. RJ45 pin 8 has to connect to both 1 and 6 on the DB9 connector. Snip the pin off of the white wire, strip the end, cut about an inch of scrap CAT5 and pull out two of the white wires, strip both ends off of them, crimp a pin on each one, and splice all three loose ends together.

You can break the end off of a spare pin and use it to crimp the three wires together. You can use the other two white wires from the inch of CAT5 for the second adapter.

When you are finished, your Ethernet cable will be pinned like so:

1	TxD	pair one
2	SG	pair one
3	RxD	pair two
4	RTS	pair three
5	DTR	pair three
6	SG	pair two
7	CTS	pair four
8	DSR/DCD	pair four

FINISH ☒

NOW GO USE IT ≫

PUTTING YOUR 5-IN-1 CABLE TO WORK

THE FIVE DIFFERENT SETUPS

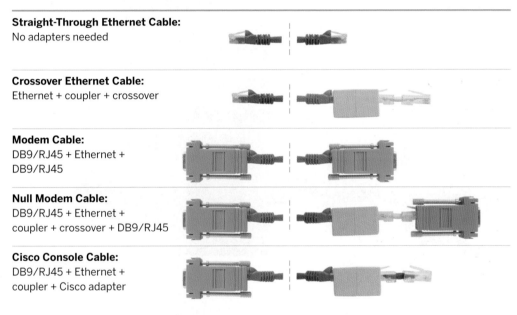

Straight-Through Ethernet Cable:
No adapters needed

Crossover Ethernet Cable:
Ethernet + coupler + crossover

Modem Cable:
DB9/RJ45 + Ethernet + DB9/RJ45

Null Modem Cable:
DB9/RJ45 + Ethernet + coupler + crossover + DB9/RJ45

Cisco Console Cable:
DB9/RJ45 + Ethernet + coupler + Cisco adapter

Remember that the Cisco adapter is not reversible.

NEED A DIFFERENT PINOUT?

If you need a DB9/RJ45 serial cable with different pinouts than the Cisco one, all you have to do is make another little CAT5 adapter. Cable ends are cheap and plentiful.

FAKING FLOW CONTROL

When used as a serial cable, this is a hardware flow control (CTS/RTS) cable. If you are using devices that both require hardware flow control, it should work. If neither of your devices requires hardware flow control, it should still work. However, if one of your devices requires hardware flow control and the other does not support hardware flow control, then you need a cable that fakes flow control. This could be done with an additional DB9/RJ45 adapter or with another CAT5 adapter and some creative crimping (my preference), both of which are left as exercises for the reader.

EXTRA ADAPTERS

Many additional adapters could easily be added to this set. A few that leap to mind would be for other kinds of serial ports such as DB25 and various DIN and miniDIN ports for Macintoshes and other things. For the pinouts for these adapters, please visit my website: *www.ossmann.com/5-in-1.html*.

CHECK YOUR CONDUCTORS

If you grab a random Ethernet cable to use with your serial adapters, remember to make sure that it has all eight conductors. Also keep in mind that most Ethernet cables have only been tested for connectivity on four conductors (1, 2, 3, and 6) if they have been tested at all. I used those conductors for the most important serial signals (Transmit, Receive, and Ground) just in case, but some serial devices won't talk without all eight working.

VCR CAT FEEDER

By James Larsson

Liberate a motor from an old VHS deck, attach it to a food chopper, and program the deck's recording timer to fill Fluffy's bowl on schedule. >>

Set up: p.44   **Make it: p.45**   **Use it: p.49**

A VCR TO FEED YOUR PET

Any old VCR has a programmable timer that connects to motors for recording TV shows. This is analogous to feeding a cat, and following this principle, you can convert a VCR into a weekend pet feeder. You set the VCR's timer, and when feeding time comes, the motor that would ordinarily spin the video head operates a food delivery mechanism instead. You can even program different size portions for different days, for times when you plan on returning midday.

Pet feeders are sold commercially, but few match the versatility of a modified VCR (no matter how silly this project might sound). My feeder is based on an auger mechanism, like some vending machines. A helical shaft propels food from a hopper into the pet's bowl. You can use the same basic mechanism to drop food into a fish tank.

James Larsson is an electronics engineer and IT historian from London, for whom hardware hacking is both work and play. In addition to designing electronic equipment, he lectures and broadcasts about computer history. He also regularly performs comedy science shows, where advanced scientific principles are used to do ridiculous things.

PROGRAMMABOWL VCR: HOW IT WORKS

Cats need a regularly controlled food supply, but can otherwise look after themselves for a few days. Hijack a motor from inside an old VCR, and you can use its timer-recording system to dispense scheduled meals.

A gearbox converts the fast-spinning VCR motor to a much slower rotation with correspondingly higher torque — enough strength to turn the crank of a food chopper or other auger mechanism. Some gearboxes, like the ones shown here, reduce the rotation step-by-step with a series of planetary gears, where the box's total ratio equals the product of the gear ratios of each element. Other gearboxes, like traditional music boxes, use a single screw-shaped worm gear. You can also reduce rotation with different width spools and rubber-band pulleys.

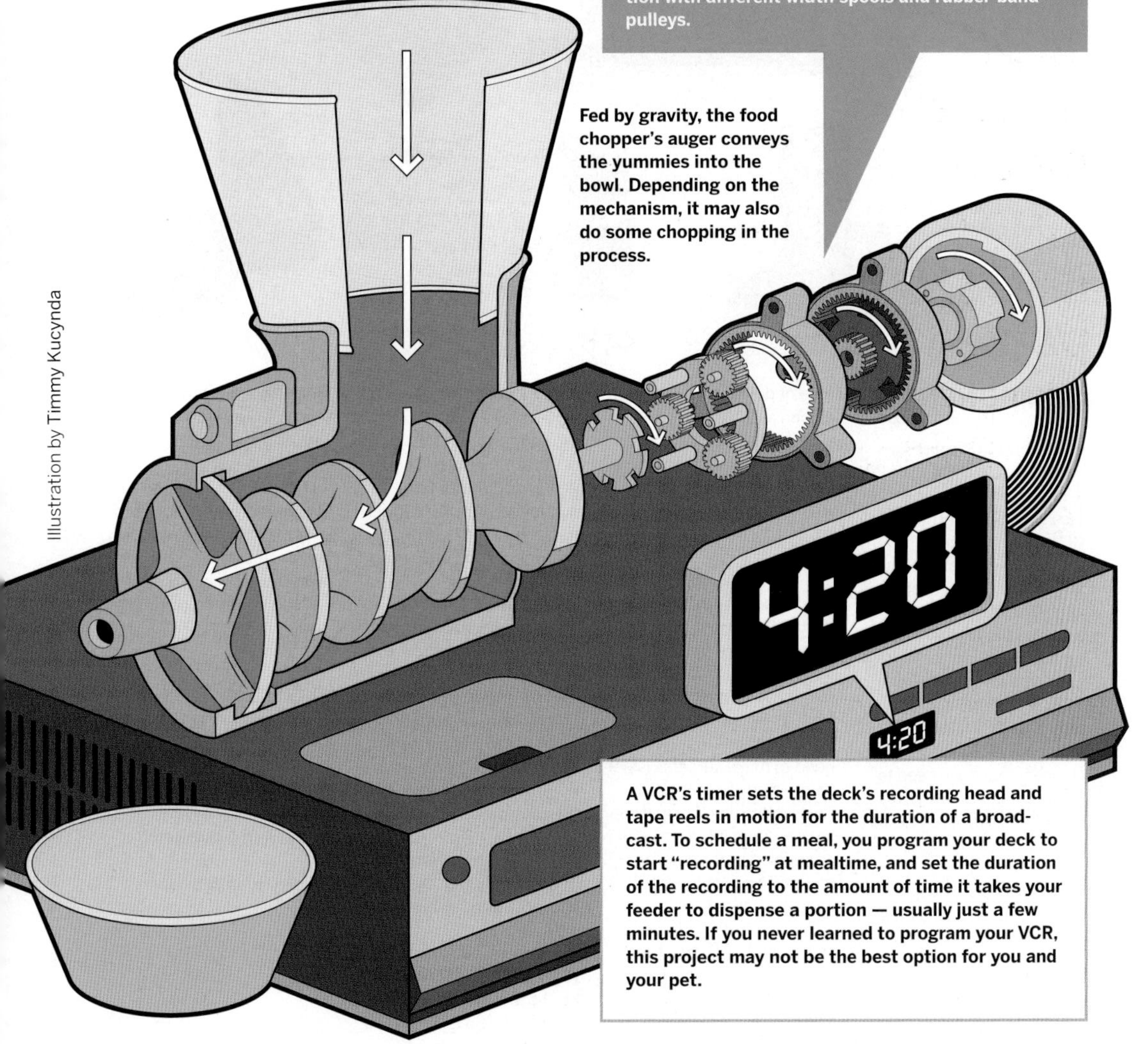

Fed by gravity, the food chopper's auger conveys the yummies into the bowl. Depending on the mechanism, it may also do some chopping in the process.

A VCR's timer sets the deck's recording head and tape reels in motion for the duration of a broadcast. To schedule a meal, you program your deck to start "recording" at mealtime, and set the duration of the recording to the amount of time it takes your feeder to dispense a portion — usually just a few minutes. If you never learned to program your VCR, this project may not be the best option for you and your pet.

SET UP.

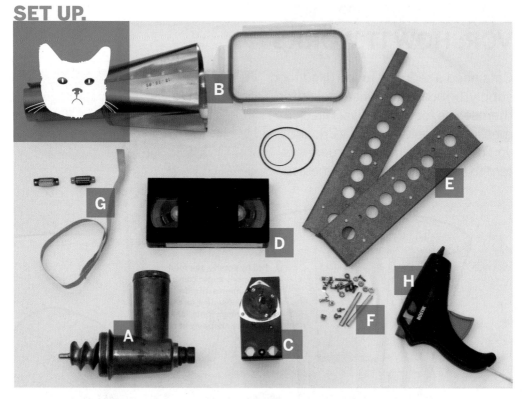

COMPONENTS

A VCR that still more-or-less works. Test the VCR first to make sure its timer and tape transport mechanisms still function, even if it doesn't produce a watchable picture. The VCR should activate its mechanism at the set time, run the tape for the set period, then stop.

If you have a choice of VCRs, go for one that you can program entirely from the front panel — this project gets cumbersome if you need to program the VCR with a remote control or via on-screen menus.

[A] Some kind of auger system. I used an old meat grinder with a helical shaft, with the cutting blades removed.

This propels the food from a container into the bowl. Make sure it works with the type of pet food you want to dispense. When choosing your auger, bear in mind that your hungry pet might try to eat, paw, or lick the system while it's in motion, and, if so, you don't want it to cause any injuries.

[B] Some kind of container that will connect with the auger system. I fashioned a sort of metal hopper head out of the magnetic shielding from an oscilloscope tube.

As with the auger, you'll need to make sure this is pet-safe — and also pet-proof. The system won't help if your cat can jump into the hopper or knock it over. I covered mine with a plastic lid, to keep the cat out.

[C] A small gearbox (or pulley system) which can reduce the standard 1800 RPM counterclockwise rotation of a video head motor into rotation suit-able to drive the auger. A turn ratio of about 600:1 typically produces the right speed and torque. I used a gearbox from a defunct cam sequencer, but you can also get these from hobby retailers. Some good ones are made by Tamiya.

The "fast side" of the box must be able to couple to a shaft of approximately ½ cm in diameter. Also, make sure that the "slow side" of the gearbox can connect solidly to your auger and rotate it in the correct direction (remember the "fast side" will be going counterclockwise).

[D] A videotape you don't mind sacrificing. Make sure the recording-enable tab is unbroken.

[E] Something that can hold the assembly of auger, gearbox, and video head motor all together. I used a metal card frame.

[F] Electrical tape, nuts, bolts, screws, strong glue, etc.

[G] You might also need a multiway electrical connector and some hook-up wire. This will depend on what you find inside your VCR and how you arrange things.

TOOLS

[H] Hot glue gun

Screwdrivers

Soldering equipment

Ability to improvise

Photography by James Larsson

MAKE IT.

BUILD YOUR VCR CAT FEEDER

1. **UNPLUG THE VCR AND OPEN IT UP.** To do this, you'll probably just need a Phillips screwdriver. WARNING: As with all 110V AC-powered equipment, once you open the cover of your VCR, you are exposing yourself to the risk of serious and possibly fatal electric shock. Generally speaking, this risk is confined to the power supply and any associated switches, cables, or connectors. This article only involves the safe, low-voltage sections of a VCR. Nevertheless, it is crucial that you know WHERE NOT TO TOUCH, especially since some of the experiments involve switching on the VCR while the cover is off. It is a good idea to place some sort of insulating shield (e.g., a piece of plastic) on top of the power supply area. Hacking a VCR is only to be attempted by people with a good knowledge of electricity and its risks.

2. **IDENTIFY THE VIDEO HEAD DRUM.** Find the motor that drives the rotating video head drum. This motor works independently from all the other mechanical systems in a VCR, so you can disconnect it with impunity, without affecting the VCR's control systems. VCRs contain several timer-controlled motors you could use, but these other ones are often linked to sensors and interdependent systems, and their absence or misuse might stop the VCR from doing what you want. That's why I chose to use the video head motor.

Video head drum

3. **REMOVE THE VIDEO HEAD DRUM ASSEMBLY.** Remove the screws that hold the video head drum assembly in place, but don't disconnect any of the wires leading to it. You'll be pulling the video head motor outside of the VCR and using it to power the auger via the gearbox.

4. **DETACH THE DRUM FROM THE MOTOR.** To reduce the load on the motor, remove the actual video head drum. These are usually attached to the motor by screws on top, but you might have to unsolder connections to the heads.

Video head motor assemblies carry a drive system, a feedback system, and more, so they need a lot of wires. Newer models have small controllers on an integral PCB, but even with these, numerous wires still lead out to other parts of the VCR. Since all you care about is the motor, you can cut away any other wires you identify as unnecessary. They are usually the cables nearest the top of the head assembly, a short distance away from the motor connection.

5. **EXTEND THE MOTOR CON-NECTIONS.** You need to get the motor out of the VCR and into a location where it can drive the gearbox, and, in turn, the auger. To do this, you may have to splice some additional length into the wires that feed the motor. If so, keep the length as short as possible; I sited the whole motor/gearbox/auger/catfood assembly directly on top of my old VCR for this reason. It's also nice to solder in a multiway connector, so that you can unplug the feeder assembly from the VCR, thus making it easier to clean.

6. ASSEMBLE THE FEEDER.

Connect the motor to the "fast side" of the gearbox. How you do this will depend on the gearbox you have chosen and the length of shaft available from your VCR's motor. For my feeder, I cut off the cog from my gearbox's original, attached the motor, and, making sure it was dead central, simply glued the VCR motor on with strong glue. Similarly, connect the "slow side" of the gearbox to the shaft of the meat grinder (or other auger mechanism). I attached the two using a cog from an old lawn mower and more glue. Finally, attach the motor/gearbox/auger assembly to whatever you're using to hold it in place. I secured it to my metal frame with a combination of bolts and glue.

7. TEST.

First, make sure everything is aligned and that the couplings on each side of the gearbox turn smoothly. Power up the VCR, insert the sacrificial video tape, and press Record. Ideally, the video head motor will rotate and drive the auger. If so, you're lucky; your feeder is ready to roll. Just be sure the tape is sufficiently rewound before each use; if it reaches the end, your pet will go hungry!

Don't worry if the motor slows a bit under load, but if the motor stalls completely, the VCR's microcontroller will sense this and shut the system down, probably forcing you to switch the VCR off and then on again. If you have persistent problems with overloading, you might need to swap in a gearbox with a greater reduction ratio. Alternately, you could try using one of the other motors in the VCR. If you do this, you'll have to take into account the motor's original role, and arrange a kludge for any sensors associated with it, as discussed later.

VCRS AND TRASH TECH

VCRs have been around for about 30 years, and in that time they have gone from being suitcase-sized machines stuffed with motors, belts, and PCBs to small boxes that seem relatively empty. What you see when you take the cover off your VCR will have more to do with its age than with its brand or model. As a general rule, older machines are better for hacking. Their designs are less integrated; fewer systems are locked away in chips, and there's simply more stuff to alter and adapt.

You might simply scavenge these junked machines for individual components, but it's more interesting to use whole, functioning sections for some entirely new purpose. If you wanted to build a pet feeder from individual pieces, you'd have to assemble a power supply, a timing system, and a mechanical control system. In a VCR, not only are all of these subsystems ready-made, but they already work together. Sure, you could rip the timer out of an old VCR and use it to trigger any electrical device, but it's connected to tape transport and read-head motors — so why not base a project on more of the original machine? This high-trash approach saves effort and minimizes the number of new components you need to buy, adding to the project's trash-tech value.

Note that trash-tech projects like this one require more improvisation than ordinary construction projects, because you probably won't be using the same old VCR model that I used. You'll have to find your own way with your trash, and in some places, I can only describe the principles, theories, examples, and pitfalls, rather than give a step-by-step. Working with junk technology is rarely going to produce a device of great engineering elegance or optimal performance. Nevertheless, it's fun, interesting, and inexpensive — and it works!

8.

TRICK THE SENSORS AS NECESSARY. Your cat feeder still may not work due to its sensors reading abnormal conditions. Or you might want it to operate continuously, with no tape to rewind. The following tricks might make it work the way you want. See next page for explanations and other strategies.

Trick #1. Disassemble your sacrificial videocassette and remove the tape reels. Reassemble, cover the holes on each side with opaque tape, and load it in. If your VCR accepts this empty shell and still "records" without stopping, consider it conveniently gullible!

Cover holes on each side of cassette

Trick #2. With your sacrificial videocassette partly rewound, remove its screws and reassemble it with adhesive tape. Load it into the VCR. Once it's happily loaded, unstick the tape, take the top half off, and remove the reels. If your VCR precludes you from disassembling the videocassette *in situ,* remove its windows so you can get your fingers inside.

Stretch a rubber band from the right-hand spindle to the capstan. Then press Record and see what happens. If all goes well, the spindle will turn at the correct speed and your VCR will continue "recording." If it does not, try placing another rubber band between the left-hand and right-hand spindles.

This is a fiddly operation that you won't want to repeat very often. Moreover, your VCR might need to stay powered up afterwards (so that it remembers that it's been through the tape loading procedure). Thus, you might want to finish all other aspects of your pet-feeding system and treat this as the last stage before use.

Capstan

Connect the right spindle to the capstan, and it rotates as if a tape is loaded.

FINISH ⊠

NOW GO USE IT »

USE IT.

FEED ME.

TROUBLESHOOTING.

A typical VCR presents several obstacles to hacking attempts. Here are the most common problem sources, and ways to get around them:

Hack-resistant circuitry. Many VCR subsystems are surprisingly distributed, and some microcontrollers sense the absence of any circuitry. Don't disconnect or remove any PCBs or other systems, even if they appear to play no part in your project.

Weak signal. Some VCRs won't record a show if the signal is too weak. To avoid having to connect your pet feeder to a TV aerial, set it up to record from a (nonexistent) camera or other external line source.

Various optical sensors. These can be sensitive to ambient light, and will trigger the VCR into doing spurious things when the case is open or there's no tape inside. You may have to work in subdued light, or locate and shield all of the offending sensors.

Tape-loaded sensors. These sense the presence of the videocassette, and are usually linked with the mechanism that loads and ejects it. The easiest kludge is to load a tape or modified tape case.

Tape-end sensors. These detect the start and end of the tape using light. Put opaque adhesive tape over the two sensors that flank the cassette, or cover the corresponding holes on a cassette itself.

Recording tab sensor. This detects whether a videocassette's record-enable tab is present. It's usually a little leaf switch. Use adhesive tape to hold it in the "pressed in" position, or else connect or break the switch's contacts as appropriate.

Spindle motion sensors. These sense whether the cassette's reels are moving at a normal speed, triggering shutoff if the tape jams or breaks. The right-hand spindle always has one of these, but the left may not. One workaround is to drive spindles from the capstan by using rubber bands as pulleys.

Mode switch. This usually looks like a cog with electrical connections, and it tells the VCR's microcontroller the device's current state (Play, FF, REW, etc.). For this switch, as well as some tape-loaded and spindle motion sensors, there's too much variation among VCR models to permit any sure advice. Different models of VCR exhibit huge variations in system design and in what sorts of misuse they will tolerate. You'll just have to experiment both electrically and mechanically to get around these.

If nothing else works, try to determine what happens when an ordinary videocassette is loaded, and re-create these events by manually twiddling the spindles with your fingers, simulating the tugging that a tape would do. You'll need to observe your VCR operating, and identify which bits of the mechanism are in what position, and which internal switches, sensors or optical systems are in use. I like to think of it as a puzzle which gradually teaches you how your VCR works. And once it's done, your pet can look forward to happy days of automatic feeding!

FEEDER OPERATION.

You'll schedule feedings as timer recordings on the VCR, but first you will need to figure out how long each "recording" should last. After filling the hopper with food, put the VCR into the Record state and time how many minutes it takes for it to dispense a single portion. This is your program time. With my meat grinder auger, it takes only two minutes to fill the bowl.

GUN-OPERATED ALARM CLOCK

By Roger Ibars

KILLING TIME

Hack a retro gaming light gun with some tilt switches to control a vintage digital clock radio. After the alarm wakes you up, you can grab the gun and kill it off. Isn't that what you've always dreamed of doing?

Clock radios are everyday hated devices that designers seem to ignore, judging by how little their features and user interfaces have changed. This project adds a new capability, letting would-be sleepyheads enjoy a human-machine interaction of a different sort. Don't worry, we're not going to connect your clock to an MP3 player and play mellow New Age sounds. We're going to shoot the alarm off. With a gun. Wake-up time is now payback time.

We'll base the project around a digital clock radio and a light gun for gaming; huge selections of both of these are available inexpensively second-hand, with many beautiful and well-designed examples. To enable our FPSI (First Person Shooter Interface), we'll outfit the gun with five tilt sensors, arranged at different angles on a small circuit board. A cable tethers the gun to the clock and carries your tilt and trigger signals to the clock's time and alarm control button contacts.

Set up: p.53　Make it: p.54　Use it: p.59

Roger Ibars lives and works as an Interaction Designer in London. He is interested in how people understand technology and how technology understands people. See more of his work at **selfmadeobjects.net**.

IMPLEMENTING GUN CONTROL

Our control circuit uses tilt switches to detect the gun's position.

Tilt switches contain a metal ball or a conductive liquid (such as mercury) that rolls inside a small capsule. When the switch's contacts point downward, the conductor bridges the contacts, closing the switch. Turn the switch upside down, and the conductor falls away from the contacts, opening the connection.

The tilt switches are arranged in a plane perpendicular to the gun's barrel, to detect the approximate rotation of the handle when the gun is aimed forward (its "roll" in aeronautical terms). The gun doesn't need to point at the clock to work (but it's more satisfying that way).

Tilt switch: A rolling ball (or some mercury) makes or breaks the connection.

The tilt circuit is wired into the clock radio to spoof three buttons: Alarm, Hour, and Minute. (We ignore the Snooze button.) Tilting the gun at different angles lines up connections through the tilt switches that bridge different buttons' contacts. Pulling the trigger closes a shared ground connection, completing the connection as if you pressed the button (or button combination).

⚠ **WARNING: Do not work on the alarm clock while it is plugged in, or you risk serious and possibly fatal electric shock.**

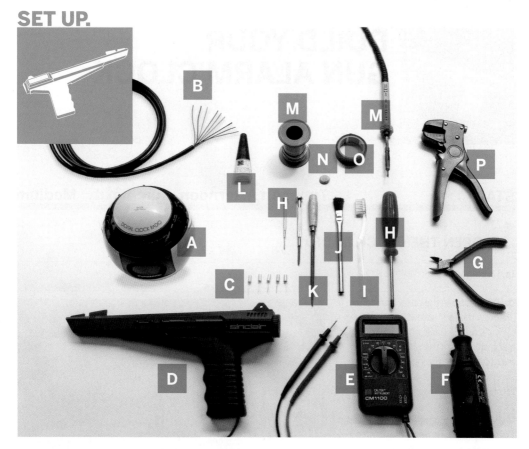

MATERIALS

[A] Digital clock radio
Almost all of these operate the same way and will work. If you go retro, don't go too far back; 15 years old is a good limit. Make sure all the buttons, the alarm, and the screen still function. Check for cracks in the case, and inspect the underside to see how much the color of the plastic has yellowed or faded. Finally, play with the radio's volume to make sure it doesn't sound like frying eggs. I chose a vintage, spherical Panasonic RC-70.

[B] Multi-way cable with at least four conductors, about 5' long I used a fancy SCART cable consisting of 9 color-coded, stranded wires wrapped in a grounded, metallic Mylar screen, all sheathed in black PVC to an overall diameter of 6mm. This top-quality choice can be found in hi-fi stores and will give you a nice curvature of the cable.

[C] Small tilt switches (5) I recommend non-mercury switches for environmental reasons. These cost about $1 each, and are available from electronics suppliers such as Farnell (farnell.com), Newark InOne (newark.com), Rapid (rapidonline.com), and RS (rswww.com). For models with just one lead, the case works as the other contact.

[D] Light gun Many are available secondhand.

I particularly like the Nintendo Zapper, the SEGA Light Phaser, the Atari G1, the Konami Justifier, and the Sinclair Magnum (my choice). Make sure there's enough space inside the grip to fit the circuit that we are going to build.

Small perfboard Available from electronics suppliers, including RadioShack (not shown).

TOOLS

[E] Multimeter

[F] Rotary tool and bits

[G] Side cutter

[H] Screwdrivers

[I] Toothbrush

[J] Artist paintbrush

[K] Round metal file, no more than 5mm thick

[L] Strong glue (or glue gun and hot glue)

[M] Soldering equipment

[N] Poster putty

[O] Electrical tape

[P] Wire stripper

[NOT SHOWN]
Spray plastic polish
The kind used to shine your car dashboard works perfectly.

Kitchen soap

Cotton cloth

MAKE IT.

BUILD YOUR
GUN ALARM CLOCK

START ⠿   **Time: An Afternoon**   **Complexity: Medium**

1. OPEN THE CLOCK RADIO

1a. Unplug the clock radio.

1b. Disassemble the case. Carefully unscrew the four screws concealed by wells on the back, and separate the electronics from the plastic parts. To avoid damaging the plastic, choose a screwdriver that fits well, press down firmly, and turn slowly.

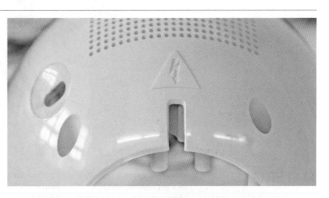

Marvel at the design details and quality! The shell is extremely well-crafted, with neat cavities for the screws and an elegant, raised icon for the power cord. The controls even resemble a face. This clock is full of design generosity, which is quite rare nowadays.

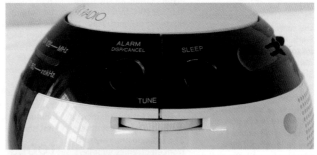

1c. Untie the electric cable, and separate the electronic parts from the plastic shell.

Remember exactly how you took apart the electronics block and untied the electric cable, so you can put it all back together later.

2. CLEAN AND SHINE IT

2a. Use the paintbrush to clean the dust accumulated on the electronics.

2b. Clean the plastic case with the toothbrush and kitchen soap. The bristles won't damage or scratch the surface. Take your time to enjoy cleaning every corner! Let all pieces dry thoroughly.

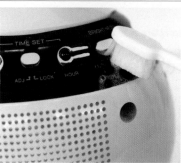

NOTE: After the pieces dry, spray them with plastic polish, following the instructions on the can.

Wait a few minutes after spraying, then polish the surfaces with a clean cotton cloth.

You'll be amazed how new they'll look! Now your alarm clock is ready for more serious work.

3. DRILL THE HOLES

3a. On the front half of the plastic shell, measure and mark a point to drill near the base and centered below the display. Protect the surface around the hole area with electrical tape, in case your drill skips away from its proper destination.

3b. Find a drill bit that's a bit thinner than your multi-conductor cable and drill the hole.

NOTE: This is the riskiest part of the work: drilling holes for the cable through the clock's case. (I see some vintage collectors out there raising their hands in objection.)

3c. Widen the hole with the round file. Keep widening slowly until the cable can just go through the hole with a bit of pressure. We don't want a hole bigger than the cable.

3d. Drill and widen a second hole in the plastic shell about 1" to the left and slightly back from the first hole, in the interior wall that holds the battery case (see picture below right). The cable will enter the first hole and make a sharp left turn into this hidden hole as it routes its wires around to the back of the clock.

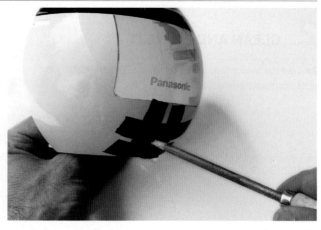

4. ATTACH THE CABLE

4a. Strip about 16" of sheathing off of each end of the cable, to reveal the color-coded wires inside. Pass one end of the cable through the holes you just drilled.

4b. Use hot glue or another strong adhesive to fix the cable firmly in place, keeping the individual wires free inside the main compartment. Leave it neat and strong since this cable is going to be used.

4c. Put the electronics block back into the front half of the case. Now we are ready to hard-wire the color cables to the clock switches.

5. HARD-WIRE THE CLOCK

If you aren't modding a Panasonic RC-70, don't worry, because almost 99% of digital clocks work the same way. Just remember that each button has 2 connections, which come into contact with each other when you press the button. You want to extend these connections so that, instead of closing the circuit with the button, you'll be using the light gun.

5a. Find the buttons that control the essential functions: set time and alarm (hours and minutes), and alarm off. The RC-70 uses just 3 buttons for these: Alarm, Hour, and Minute. The Alarm button does double duty, shutting off the alarm and switching the Hour and Minute buttons from "time-set mode" (the default) to "alarm-set." Some clock radios use a switch instead of a button to change between time-set and alarm-set modes.

5b. Locate the essential buttons' contacts, 2 per button, and use a multimeter to follow each of them out to a solderable connection point. With the RC-70, all 3 buttons shared a common ground, so I needed to find a total of 4 connection points.

5c. Solder wires from your cable to the contact points, using the color coding to track what goes where. Following convention, I connected the cable's black wire to the shared ground, and designated colors for the 3 button-specific connections.

5d. Organize and fix all the color cables between the free spaces of the clock. Trim extraneous cables to get them out of the way. Avoid placing cables near parts that get warm, such as the power converter.

5e. Reassemble the clock and screw it back together.

5f. Test the clock by plugging it in and touching together the other ends of the cable wires you just soldered. Confirm that bridging the wires mimics the functions of the buttons you connected them to.

If you didn't make any big mistakes, everything should work! Don't worry about touching the wires going to the buttons with your fingers, because the current going through them is very low. Once it's working, unplug it again.

6. HARD-WIRE THE LIGHT GUN

6a. Take the light gun apart, and clean and shine it up the way you did with the clock radio in Step 2.

6b. Cut a rectangle of perfboard that's big enough to carry the 5 tilt switches flat, but narrow enough to fit facing backward inside the handle. For my Sinclair Magnum, my board was about ⅝"×2⅜".

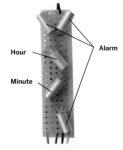

Hour   Alarm

Minute

6c. If your clock radio is an RC-70 or work-alike, arrange the tilt switches on the perfboard as shown at right. Leave some extra room around each switch to let you bend and fine-tune their positions later.

6d. Following the wiring diagram online (makezine.com/08/alarmgun), solder the tilt switches into place and connect them to the cable wires. Don't worry about which sides you connect; the switches are functionally symmetrical.

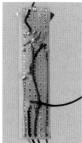

6e. Connect the ground to each sensor section and route it through the trigger switch contacts in the gun. To make my circuit more readable, I routed all wiring from the cable and trigger through the front of the board, at the edges. I also split 1 ground coming from the trigger and connected it in 2 places.

6f. Use poster putty to hold the tilt circuit in position inside the gun-half with the trigger. Plug in the clock radio, and test the interface by tilting and firing. Experiment with different firing angles, and bend the tilt switches around to refine their operation.

6g. When the tilt switches work together properly, glue them in place, and then glue the circuit board into the gun handle. Re-assemble the gun.

6h. Set the time, set the alarm to +1 minute, wait a minute, and FIRE!

Now send a photo of your design to roger.ibars@gmail.com, and I'll send you back an exclusive preview of my latest projects.

FINISH ✕

DO YOU FEEL LUCKY?
WELL, DO YA CLOCK?

OPERATION

The Panasonic RC-70 has a switch that locks the time set and disables the Hour and Minute buttons from changing the time. For our gun to work, this switch needs to be in the unlocked position.

To use this new hard-wired device, imagine that the clock is a circle of degrees (not difficult with a spherical clock). A straight up-and-down shot kills the alarm. Tilting left sets hours, and tilting right sets minutes. Small tilts set the time, and big tilts, below the horizon, set the alarm.

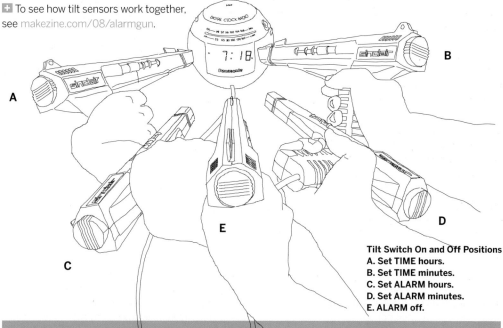

To see how tilt sensors work together, see makezine.com/08/alarmgun.

Tilt Switch On and Off Positions
A. Set TIME hours.
B. Set TIME minutes.
C. Set ALARM hours.
D. Set ALARM minutes.
E. ALARM off.

SOURCE

HARD-WIRED DEVICES BY ROGER IBARS

This project comes from my Hard-wired Devices collection, which pays tribute to great consumer electronics designs from the 70s and 80s. They are remanufactured vintage devices that blend two cultures of interface: computer games and household appliance design.

In these pieces joysticks, light guns, and controller pads from the classic age of computer games connect actively via cable to digital alarm clocks and other iconic devices.

You can see the full collection of hard-wired devices at selfmadeobjects.net.

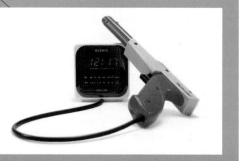

Illustration by Roger Ibars

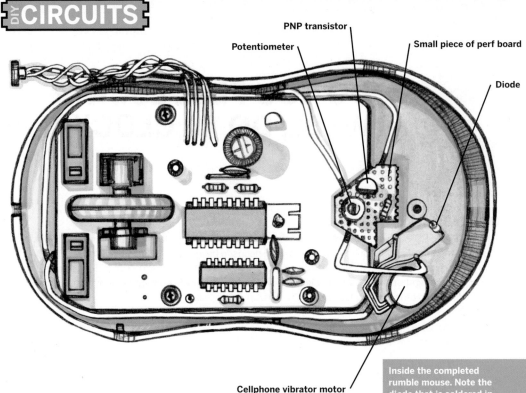

PNP transistor

Potentiometer

Small piece of perf board

Diode

Cellphone vibrator motor

RUMBLE MOUSE

For FPS gaming, a cellphone vibrator gives kick to your clicks. By Greg Lipscomb

Have you ever been playing your favorite first-person shooter with someone's rumble controller and thought, "I would love to have that capability while playing on my computer"? A friend of mine had a spare rumble pack lying around, so he decided to stick it in a mouse. This inspired me to create my own version.

I determined that my rumble mouse should meet certain specifications. I wanted it to be fully enclosed, with no parts sticking out of the case. It should be an optical mouse, rather than a roller-ball one, connected and powered by USB. In play, the mouse would give a satisfying rumble-recoil when you click the left button — the trigger in most FPS (first-person shooter) games.

I found a cheap suitable mouse from a local

surplus store. It had a scroll wheel that I liked, and was large enough to fit the extra components inside. For the rumble motor, I wanted something small, and my fiancée suggested that I use a cell-phone vibrator. I had several old cellphones lying around, so I cut one open and located its small motor near the top left corner, which looks like a watch battery with two wires coming out of it.

Naturally, different phones are different. Nokias I've cracked open use small cylindrical motors, not as flat, and in an old flip phone I took apart, the motor was in the same piece as the speaker, close to the LCD screen.

I also needed a switch to connect the USB power to the motor when the left mouse button clicks. I chose a PNP transistor for this. The final

Illustration by Damien Scogin

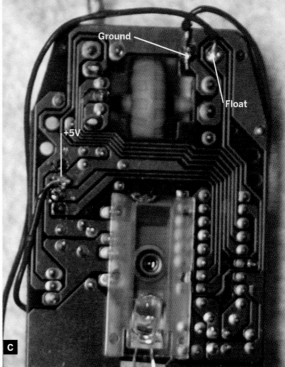

For the rumble motor, I opened up an old cellphone and dug out the vibrator, which looks like a thick watch battery (Figure A). Bench test for the rumble mouse circuit on a solderless breadboard (Figure B).

The three pins used in the circuit: +5V is direct from the USB cable, Ground is common ground for the circuit, and Float is the pin that momentarily grounds when you click the left mouse button (Figure C).

Photography and illustration by Greg Lipscomb

MATERIALS
Optical USB mouse I used an off-brand, model HTM-67WT
PNP transistor, 2N3906 RadioShack #276-1604
Electric vibrating motor Salvage from cellphone, or use Sanko Electric #1E120, available at alleelectronics.com, catalog #VB-1
5kΩ or 1kΩ micro trim potentiometer A 10kΩ pot, like RadioShack #271-282, works, but lower values fit the adjusting range better
Diode, 1N4001
1kΩ resistor
Mini perfboard RadioShack #276-0148

mandatory part was a 10kΩ potentiometer, which would be used to regulate the current going to the motor.

Find and Adjust the Power

The rumble circuit takes power from the USB port and directs it to the motor, so I needed to find +5V power and ground contacts both on the USB cable and inside the mouse. For the cable, I found a USB pin-out diagram of a USB cable online, and learned that the two outer pins of the USB cable are the +5V and the ground. Then I opened up the mouse and used a digital multimeter to probe for terminals on the board that connected to the USB's power and ground pins. On my mouse, I found that the green wire from the USB was ground, and the blue wire was the +5V. Then I connected the cellphone vibrator directly to the USB power, just to make sure that it could pull enough power to run. It worked perfectly.

I soldered jumper wires to the +5V and ground contacts on the phone, and started putting my circuit together on a solderless breadboard. The vibrator motor had a labeled rating of 3.6V, so I needed a resistor in series with the motor to lower the incoming voltage from 5V. I chose to use a 10k potentiometer, which would let me adjust the voltage and therefore the speed of the motor. You can use your multimeter to make sure that the voltage across the motor is less than 3.6V.

An alternative to the potentiometer is to use two diodes in series. Each diode drops the voltage 0.7V, so if you have two, you would drop 1.4V to give 3.6V over to the motor. Either way, it's important to have a separate diode in parallel with the motor. When the transistor is first activated, the motor voltage can spike really high, and a diode will protect the transistor from this high-voltage spike.

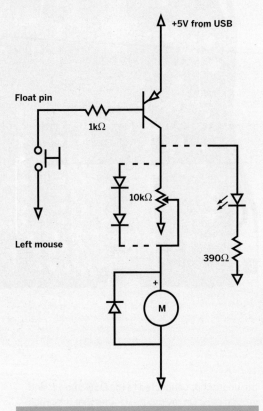

+5V from USB

Float pin

1kΩ

10kΩ

390Ω

Left mouse

M

+

Rumble mouse schematic (optional circuitry shown by dotted lines). Note that the potentiometer can be replaced with two diodes in series, as shown. There is also an optional LED-resistor subcircuit that lights when the left mouse button is clicked.

Intercept the Left-Click

The next step was to locate the board contacts for the mouse's left-click button. It was a single-pole momentary switch, and using my multimeter, I found which pin was ground and which pin was floating. I soldered a jumper wire to the floating pin contact, and continued breadboarding. Since the mouse button operates by grounding the floating pin, I needed an electrical switch that would be activated by a low voltage signal. A PNP transistor was perfect for this requirement. Connect the float pin contact to the base of the transistor, and low voltage from a click action completes the circuit between the emitter (connected to power) and the collector (connected to the motor). Note that this is opposite from an NPN transistor, which is open with a high base current. (Another approach would be to forget

the transistor and just hook the motor up between +5V and the floating pin. Then, while the floating pin is grounded, 5V runs through the motor to ground.)

To limit the current flowing through the transistor base, I put a 1k resistor between the floating pin and the transistor base. Then I connected the emitter of the transistor to my +5V from the USB port, and connected the collector of my transistor to the motor through the 10k potentiometer. The other pole of the potentiometer was connected to ground. (When wiring a potentiometer, it is important to attach one wire to the middle pin, and place the other wire on either of the two outer pins. If you connect to both outer pins, you will always get the total 10kΩ.) This completed my circuit on the solderless test board, so I tested it and fixed the bugs.

Put It Together

The final step of my project was to solder everything together. I soldered my components to a small piece of perfboard, using small wires to connect the appropriate leads. Then I used double-sided tape to connect the motor to the casing at the back of the mouse. I also routed the wires to fit nicely into the body of the mouse.

When I completed the assembly, I had a fully functioning rumble mouse that would vibrate on every left mouse click. It is perfect for any first-person shooter game.

Some ideas for improvement would be to add a simple toggle switch between the +5V from the USB and the emitter of the transistor so that the rumble part can be turned off. All in all, this is a straightforward DIY project that is sure to enhance your gaming experience.

Greg Lipscomb is an electrical engineer (Auburn University) who is in his second year of medical school at the University of South Alabama in Mobile. You can see his work at diylive.net.

MY LOVE AFFAIR WITH LEDS

Build a bright, low-powered desk lamp.
By Charles Platt

Photography by Charles Platt

Quartz halogen lamps are more efficient than old-fashioned incandescent bulbs; fluorescents are more efficient than quartz halogens; and white light-emitting diodes (LEDs) are the most efficient of all (if you consider the ballast that is needed by fluorescents). So why haven't white LEDs become as fashionable in homes as a Prius in the driveway?

One reason may be that a white LED isn't quite white. It has an eerie purplish tint. This doesn't seem to bother people when the product is a flashlight, but it has been a barrier to the application of LEDs in domestic settings. Personally, I like freakish technology that looks as if it came straight out of *The Jetsons*, so a purplish tint is fine by me. With this in mind, I set out to be the first on my block to have an LED desk lamp.

After an unproductive attempt to find ultra-high-power LEDs using Froogle, I tried eBay and hit pay dirt. Hong Kong manufacturers are now accepting PayPal for state-of-the-art LEDs ordered by the hundred, and are charging less than 25¢ per diode. Diodes are 1cm in diameter and have a claimed light output of at least 100,000mcd.

Now, you may be wondering what this "mcd" measurement is. It stands for "millicandles," which is confusing, since old-fashioned light bulbs are rated in lumens. A lumen measures the total amount of light emitted in all directions, whereas a millicandle is a measurement of intensity per unit of area — more appropriate for an LED since it has a lens that emits a tightly focused beam, often as narrow as 5 or 10 degrees. It can best be compared with a reflector bulb, but since those aren't rated in millicandles, the only way to evaluate LED brightness

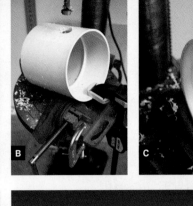

A B C

D E

Fig. A: The body consists of PVC pipe pieces glued together, sanded, and spray-painted silver. Fig. B: The base is drilled at an angle to accommodate the stem and filled with sand. Fig. C: The stem's other end is glued into a 1" hole in the head. Fig. D: The mounting plate is ABS plastic drilled with 72 holes. Fig. E: The finished lamp, with the head tilted to show the LEDs. Total cost: about $40; total power consumption: about 5W.

conclusively is by buying some and trying them for yourself.

When my Hong Kong package arrived, I opened it and found not only LEDs but also little brown blobs on wires — resistors to be put in series with the lights. Evidently, a common application of raw LEDs is to trick out cars with underbody lights and other effects. So, manufacturers helpfully deliver LEDs with resistors that reduce the 12V from a car battery to the 3.5V preferred by a typical diode.

This was not so helpful from my point of view, since a resistor dissipates excess power as heat. Fortunately, there are a couple of ways to avoid using this wasteful device in your ultra-efficient lighting system.

If you want to use an LED light source at home, you can simply buy an AC power adapter that converts house current directly to the 3.5V DC that you need. However, I wanted my lamp to work equally well in an RV, so I wired my LEDs in groups of three. Because each LED has its own internal resistance, you no longer need a wasteful series resistor. When you put three together in series and attach a 12V supply to the ends of the chain, each LED requires

4V. You can add more chains like rungs of a ladder, as shown in Figure B on page 65.

Because LEDs are fussy about getting precisely the correct amount of power, you do still need a very small amount of resistance in order to optimize the voltage. For this purpose, I placed a wire-wound potentiometer (variable resistor) at the bottom of the ladder, to adjust the supply for all of the LEDs simultaneously.

I used a multimeter (available for about $20 from RadioShack) to check the voltage across a few sample LEDs while adjusting the potentiometer to the correct setting. By turning the potentiometer back from this point, I could make it do dual duty as a lamp dimmer.

The natural next question is "How do I wire everything together?" You will need a soldering iron — the low-wattage type designed for electronic components (15W is typical) — which you can find at RadioShack. To avoid roasting the diode, you must apply this to your LED wires very briefly. I place an alligator clip on the wire just below the soldering iron as a heat sink.

(Soldering guides will tell you that "real solderers don't use heat sinks," but since the difference

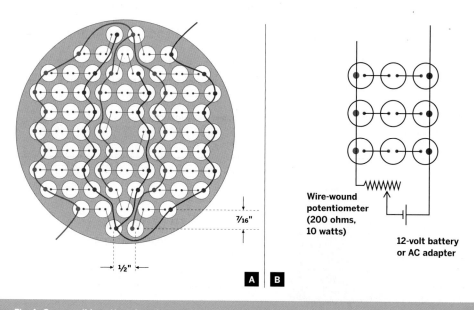

7/16"

1/2"

Wire-wound
potentiometer
(200 ohms,
10 watts)

12-volt battery
or AC adapter

A | B

Fig. A: One possible pattern for a lamp using LEDs wired in threes. The result is 24 chains, with a total of 72 LEDs powered by 12V.

Fig. B: Basic schematic for wiring LEDs in chains of three, driven by a 12-volt DC power source and with a potentiometer to fine-tune the voltage (usually 3.5 volts across each LED).

between a diode that works and a diode that doesn't work can be 6 seconds of soldering time instead of 5, why not take an elementary precaution?)

Most importantly, you will need to remember that the long wire of the LED goes to the positive side of your DC supply, whereas the short wire goes to the negative side. If you wire the LED the other way around, you may burn it out, depending on how much voltage you apply.

With all of these details, you may be wondering whether LEDs are just too much trouble. They're really not. I composed a kind of sunflower pattern for my 72-LED lamp using Adobe Illustrator, then printed it and used it as a template for drilling holes into a piece of ¼"-thick, white ABS plastic.

If you don't happen to have any metric drill bits in the house (I certainly don't), you can use a 25/64" bit, which is almost exactly equivalent to the 1cm diameter of the LED. After you drill the holes, poke the LEDs into them, glue them in place (I smothered them with silicone caulking), and solder them together. The whole project took me an afternoon.

If you have a recreational vehicle, you may like the idea of ultra-efficient lighting that won't add to the power drain on your electrical system. In fact, I'm expecting RV owners to go for white LEDs long before homeowners.

Once your system is working, you will enjoy its peculiar economic benefits: a chain of three 100,000mcd LEDs lasts about 100,000 hours and uses a mere 20mA (0.02A) of electricity. If you follow my lamp design, using 24 chains (a total of 72 LEDs) powered by 12V, the whole assembly will consume approximately 5W, while emitting about as much light as a 60W incandescent bulb.

Of course, you will have to get used to that freaky purplish glow, but early adopters should be willing to embrace that kind of thing.

Charles Platt is a freelance journalist who wrote the profile on cold fusion researcher Ed Storms (MAKE, Volume 03).

LED Throwies

By Graffiti Research Lab

Make and toss a bunch of these inexpensive little lights to add color to any ferromagnetic surface in your neighborhood.

You will need:

10mm diffused LED, any color(s) 20 cents each from HB Electronic Components (hebeiltd.com.cn).

1" strapping tape
One roll will make many throwies.

CR2032 3V lithium batteries
25 cents each from CheapBatteries.com.

½"×⅛" NdFeB disc magnet, Ni-Cu-Ni plated
25 count for $13 from Amazing Magnets (amazingmagnets.com).

Conductive epoxy (optional)
Weather-resistant alternative to tape. Available from Newark InOne (newark.com).

1. Test the LED.

Pinch the LED's leads to the battery terminals, with the longer lead (the anode) touching the positive terminal (+) of the battery, and the shorter lead (the cathode) touching negative (-). Confirm that the LED lights up.

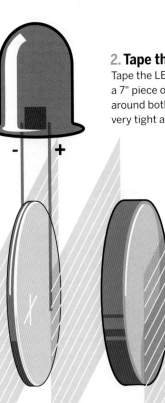

– +

2. Tape the LED to the battery.

Tape the LED leads to the battery by cutting off a 7" piece of strapping tape and wrapping it once around both sides of the battery. Keep the tape very tight as you wrap. The LED should not flicker.

NOTE: The battery's positive contact surface extends around the sides of the battery. Don't let the LED's cathode touch the positive terminal, or you'll short the circuit.

3. Tape the magnet to the battery.

Place the magnet on the positive terminal of the battery, and continue to wrap the tape tightly until it's all done. The magnet should hold firmly to the battery. That's it — you're ready to throw (or make a few dozen more). Throw it up high and in quantity to impress your friends and city officials.

Throwies naturally chain together in your pocket, making multi-segmented throwie bugs, which will also stick to metal surfaces if they aren't too long.

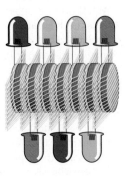

A throwie will shine for about 1-2 weeks, depending on the weather and the LED color. To get one off a ferromagnetic surface, don't pull it, or it may come apart. Instead, apply a lateral force to the magnet base, and slide it off the surface while lifting it with a fingernail or tool.

Graffiti Research Lab (graffitiresearchlab.com) is dedicated to outfitting graffiti artists with open source technologies for urban communication.

Illustrations by Kirk von Rohr

For more info, corrections, and discussion on this piece, please visit makezine.com/06/123_led

Magnetic Switches from Everyday Things
By Cy Tymony

Control many devices from afar with the magnetically sensitive Sneaky Switch.

You will need: Magnet, paper clip, aluminum foil, tape or foam, cardboard, wire, LED or buzzer, 3-volt watch battery or equivalent

Optional: Ring, battery-operated toy, X-10 universal interface and appliance module

1. Make a magnetic activator.

You'll want a strong magnet to activate devices from at least an inch away. Tiny rare earth magnets can be found in most micro radio-controlled cars, and in the packaging of some hearing aid batteries. Glue a magnet to the face of a ring or a wand, or affix it to some object so that when it's near the switch, or moved away, it will cause the desired effect.

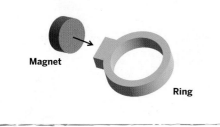

Magnet

Ring

2. Make a Sneaky Switch.

In this magnetic switch, the paper clip lies across a "spring" of rolled tape, one end hovering just above the aluminum foil and the other end taped down. (A small piece of foam can also be used as the spring.) When a magnet passes over the switch, it tugs the clip to touch the foil, completing the circuit. Connect the switch to a 3V watch battery to light an LED, buzzer, or other low-current devices and toys.

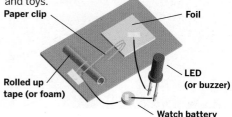

Paper clip

Foil

Rolled up tape (or foam)

LED (or buzzer)

Watch battery

3. Connect switch to a relay.

Your magnetic switch can be attached to a relay to control devices that need higher current. Mount your switch and relay behind the dashboard to secretly activate a cut-off switch, alarm, or other car accessories. Or hook your switch to an X-10 controller and universal interface module to control a variety of appliances. Pretty sneaky!

Bonus: Detect counterfeit money.

Legitimate currency has iron particles in the ink. Fold a bill so half of it stands up vertically — if the top edge moves toward your magnet, it's the real deal. If not, phone the Secret Service!

Cy Tymony (sneakyuses.com) is a Los Angeles-based writer and is the author of *Sneakier Uses for Everyday Things*.

70

78

80

86

93

96

THE BEST OF MICRO-CONTROLLERS PROJECTS

from the pages of MAKE

>>

YOUR WORLD IS FILLED WITH COMPUTERS THAT ARE TALKING TO YOU. TALK BACK!

Ask anyone how many computers they own and they'll likely say one, or two, or maybe a few more (if they count their TiVo and their cellphone). The actual answer is hundreds. Your house, your car, nearly every electronic gizmo you own has one or more computers-on-a-chip, called a microcontroller. There's a quiet revolution going on of precocious users messing around with these micro-controllers, programming them for fun and profit. The chips are cheap, and the programming software and other tools get easier to use by the day. Here's some of what you need to know to become your own digital controller.

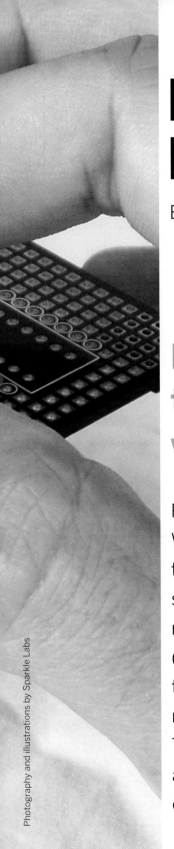

Microcontroller Programming

By Sparkle Labs

Easy-to-program chips tell circuitry to do what you want.

Press a button and a light flashes a pattern. What makes it flash? It seems like there's a tiny monkey in there flipping the switch. If so, many household items contain these tiny monkeys. They're what send the infrared (IR) codes out of our remote controls and then decode them in our televisions. They run our washing machines and toasters. These tiny monkeys are microcontrollers, and you can train them to help you with your own projects.

Microcontrollers are small computers, all on one chip. The chip carries a central processing unit (CPU), program memory, data memory, and input/output (I/O) pins that can connect to various devices. The chip works by the CPU following the instructions in the program memory, which tell it what to read and write to data memory, and what to input and output to the pins.

Programming the microcontroller means writing and storing these instructions in the chip's program memory. The microcontroller speaks assembly language, which consists of binary instructions (ones and zeros). You can program directly in assembly language, but most people prefer to use a higher-level language like C or BASIC because they're easier to understand. When you do it this way, programming a microcontroller is a four-step process.

This article explains the process. In our example, the chip will simply make an LED flash. This may not seem like much, but the hard part is setting up your programming environment and making all the pieces work together. After you get the light blinking, you can take over the world!

PROGRAMMING A MICROCONTROLLER

write code → compile code

download code

1. **Write the code.**
2. **Compile the code.**
3. **Download the assembler code onto the chip.**
4. **Build and test your target circuit.**

Low-Level Microcontrollers vs. BASIC Modules

First, decide whether you want to use a low-level microcontroller or a BASIC module. Inexpensive low-level chips, from Microchip (PICs) or Atmel (AVRs), require additional hardware, software and effort. BASIC modules, from Parallax (Stamp) and NetMedia (BasicX), are easier for beginners, but pricier.

BASIC modules combine their microcontrollers with with oscillators and other components into one, pluggable package, simplifying circuit design. They also include all development hardware and software.

For our example circuit, we're using the low-level, two-dollar Microchip PIC 12F675.

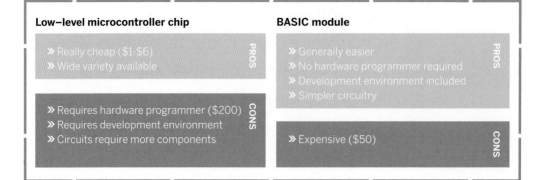

Low–level microcontroller chip

PROS
» Really cheap ($1-$6)
» Wide variety available

CONS
» Requires hardware programmer ($200)
» Requires development environment
» Circuits require more components

BASIC module

PROS
» Generally easier
» No hardware programmer required
» Development environment included
» Simpler circuitry

CONS
» Expensive ($50)

Microcontroller Programming Hardware and Software

To program a low-level chip, you need three pieces of hardware: a PC, a hardware programmer with compatible cable, and the target circuit that you'll plug the micro into. You'll also need some software.

1. PC: The vast majority of software tools for programming microcontrollers run on Windows machines (*but see sidebar, page 74*). You don't need a fast machine; any PC with the proper port for your hardware programmer cable (serial, USB, etc.) will do.

2. Hardware Programmer: This is what you plug your chip into in order to transfer your program from the PC. Traditionally, you then remove the chip and place it in your circuit, but some hardware programmers support in-circuit programming, which lets you burn the chip in place, within the circuit, making it easier to debug and re-run the software.

For our example here, we used Microchip's PICkit 1 Flash Starter Kit, a $35 USB programmer that contains a small demo board and can program some but not all of Microchip's 8- and 14-pin micros.

3. Target Circuit: If you're just starting out with microcontroller programming, you can experiment with a demo board, like the one included with the PICkit 1. These are printed circuit boards with a space to plug in your chip and various input and output devices such as buttons, LEDs, and potentiometers. Using one of these boards, you can explore the features of your chip and run different programs without worrying about wiring.

If you have a standalone circuit idea in mind, the next step is to build your own. There are plenty of circuits published online, and you may be able to find one that's close to what you need. But schematics often contain errors, so you need to be careful. If you're up to it, you can also design your own circuit from scratch, as discussed below.

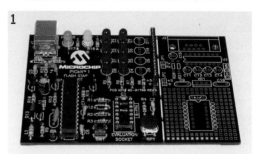

Software: In addition to the hardware, you need to put together your software development environment. This will include the text editor where you write your code, a compiler, the software that drives your hardware programmer (which probably came included with the hardware), and microprocessor simulation and debugging tools.

You can buy most of this software grouped together into an integrated development environment (IDE) from companies like Microchip and Micro-Engineering Labs. We used Proton Lite, a free trial-version IDE that restricts you to 50 lines of BASIC code — which is plenty for our simple blink program.

Designing a Circuit

First, think of what inputs and outputs your circuit will have: switches, sensors, lights, motors, etc. Then you can determine the power requirements. For simplicity, our circuit uses batteries, but a wall-wart with a voltage regulator is more reliable.

Next, determine how your inputs will interface with the microcontroller. Some pins take only digital inputs, a.k.a. logic inputs, where 5V means 1 and 0V means 0. The general rule for these is that power brings the voltage up and ground draws it down. To make a button that changes an input pin from 0 to 1, for example, connect the pin to ground, through a resistor, and also connect it to a button that, when pressed, completes a connection to power.

Some micro pins take analog as well as digital inputs; you can feed these from analog sensors that produce a range of electrical values. For example, a potentiometer's knob changes its resistance, which changes a voltage fed through it. Connect a pot to a pin that works as an analog-digital (ADC) converter, and the micro will convert the current position of the knob into a number you can program with.

An LED will light up directly from a micro's output pins, but things like motors require more current. You can supply this by connecting an output pin to the base of a transistor that has higher current running through it. Motors may generate voltage spikes that can damage your chip, but a diode running in reverse across the transistor will protect the circuit.

Once you know which sensors connect to which types of pins, you need to study your microcontroller's datasheet. As with most micros, pins on the PIC12F675 perform multiple tasks, and you set registers in your software to tell the pins how to behave. In the registers table from the PIC12F675 datasheet on Microchip's website, we see that the TRISIO register tells a pin to be input or output, and the ANSEL register determines which pins connect to the ADC. These registers have eight bits, one for each pin. So, to connect a simple binary button to a pin, set its corresponding bit in the TRISIO register to 1 (input) and in the ANSEL register to 0 (disconnect). To connect an LED, set the TRISIO to 0 (output). To read a potentiometer value, set the pin's TRISIO to input (1) and set the ANSEL to connect the ADC (0).

WRITING THE CODE

There are many general references for programming in BASIC, but here are two handy sample code segments for microcontrollers.

Button on pin GPIO.0 lights an LED on pin GPIO.1:

```
if GPIO.0 = 1 then
  GPIO.1 = 1
else
  GPIO.1 = 0
endif
```

Turn an LED on for 1,000 program cycles:

The main program runs the startlight subroutine when a button is pressed, turning the LED on, and calls the endlight subroutine inside a loop, to turn it back off.

```
startlight:
GPIO.1 = 1
counterVar = 1000
return


endlight:
counterVar = counterVar -1
if counterVar = 0 then
  GPIO.1 = 0
endif
return
```

Open Source PIC Programming

Contrary to Microchip documentation, PIC development does not require a Windows PC. I use free Unix tools on Mac OS X, plus a USB-to-serial adapter to connect my Mac to my hardware programmer. Here's all the software you'll need.

gputils: Package includes gpasm assembler, which translates compiled source code into the hexified format suitable for burning onto a PIC.

gpsim: PIC simulator steps through code and indicates pin status, for wiring-free debugging.

picp: Utility for PICSTART Plus and Warp-13 hardware programmers, writes hex code to the PIC.

Fink: Unix package manager for Mac OS X lets you install the utilities above, and other software.

X11 for Mac OS X, and Xcode Developer Tools (with X11 SDK): Available from Apple, these let you install Fink.
—Mikey Sklar

Make an LED Blinky

Here's how we built and programmed our microcontroller-based LED blinky circuit.

START »

1. SET UP YOUR DEVELOPMENT ENVIRONMENT.

Install the IDE and the PICkit 1 software and hardware. Adjust the IDE settings to make sure it knows where to find the programmer and that it's set up for our controller, the Microchip PIC 12F675.

Adjusting IDE settings.

2. ASSEMBLE THE CIRCUIT.

Before getting down to building and coding specifics, take a look at the PIC 12F675's pinouts, from its datasheet on the Microchip website. You can see here how our circuitry's wiring diagram follows the chip's pinouts.

Our circuit includes the microcontroller, a power source, a timing crystal, support components for the microcontroller, and the output LED.

Power comes from three AA cells that provide 4.5V, which is close enough to digital logic's 5V "high" voltage. A 10K-ohm pull-up resistor maintains a high voltage level to the chip's master clear pin, Pin 4; sending low voltage (logical 0) to this pin would make the PIC reset. A 4KHz oscillator supplies the source blink pulse, which is slowed down by two capacitors. Voltage to the LED comes from the microcontroller's output Pin 7, connected to ground in series with a 220-ohm resistor.

Follow the wiring diagram at right. We soldered our circuit together using a small piece of perfboard.

PIC 12F675 Pinout

1. VDD
2. GP5/T1CKI/OSC1/CLKIN
3. GP4/AN3/T1G/OSC2/CLKOUT
4. GP3/MCLR/VPP
5. GP2/AN2/T0CKI/INT/COUT
6. GP1/AN1/CIN-/VREF/ICSPCLK
7. GP0/AN0/CIN+/ICSPDAT
8. VSS

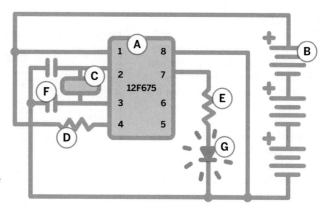

A. PIC 12F675	**D. 10KΩ resistor**
B. Three AA cells	**E. 220Ω resistor**
C. 4KHz crystal oscillator	**F. 22pF capacitor**
	G. LED

»

3. WRITE THE BASIC CODE.

We wrote ours in the text editor window of Proton Lite. Here it is in full, with comments included to explain what's going on:

Device = 12F675	Tell the compiler what kind of chip we are using.
DelayMS 500	Wait 500 milliseconds for things to settle down.
XTAL = 4	What speed of crystal or oscillator.
ALL_DIGITAL = True	Turn off all analog-to-digital converters (ADC); some of the chip's pins can be used to detect analog voltage levels, which we don't want to do.
TRISIO = %00000000	Write the byte "%000000000" to the register.
"TRISIO"	TRISIO decides which pins are input and which are output.
GPIO = %00000000	This line sets all of the I/O pins to be output. The GPIO register tells the status of our output pins; all output pins are now set to low (0V).
While 1 = 1	1 is always equal to 1, so this creates a loop forever.
GPIO.0 = 1	Here we make the "0" bit of GPIO equal to 1. GPIO is now %00000001, with Pin 7 high (+5V). This makes Pin 7 output light up the LED.
DelayMS 500	Wait .5 seconds.
GPIO.0 = 0	Set Pin 7 to low again, switching LED off.
DelayMS 500	Wait .5 seconds.
Wend	Go back to the beginning of the While loop.
End	This marks the end of the code.

Proton Lite code window.

4. COMPILE THE CODE.

Save your code (which is a plain text file) to your hard drive, and then click Proton Lite's "Compile and Download" button. This creates a hex version of the file and launches the PICkit 1 software.

Within the PICkit 1 software interface, click "Import HEX" to open your newly created hex file, which was written to the same directory as your original code file. This hex file contains the microcontroller assembly code.

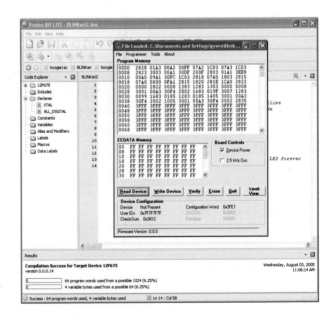

5. BURN THE ASSEMBLER CODE ONTO THE CHIP.

Insert the microcontroller chip into the PICkit hardware programmer and click "Write Device." It's that simple; your microcontroller is now programmed!

6. BUILD AND TEST YOUR TARGET CIRCUIT.

One common method is to prototype it on a breadboard. This is what we did for our example.

Plug it into your target circuit, connect up the power, and bask in the blinking glory!

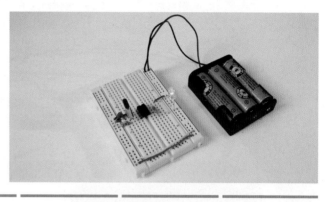

NEXT STEPS

Now you can experiment with other input and output devices. Resistive sensors are fun; these change their resistance in response to things like light, flex, and temperature. You can also use gravity sensors, compasses, accelerometers, and rangefinders.

For outputs, buzzers and small speakers make sounds, and motors give your project legs or wheels. LCD panels show information, and LED matrices make colorful displays.

Mount a rangefinder on a servomotor, and your micro can point it around and read its physical surroundings. Have your micro output serial data to a PC, and you can make a sensor control a game character. We've created a lot of fun projects with our little micro monkeys, and we invite you to do the same.

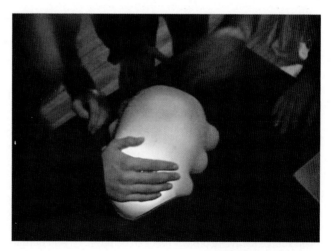

The Platypus Amoeba has touch sensors on its back, and responds to petting with lights and sounds.

FINISH ☒

Sparkle Labs (sparklelabs.com) is a product development firm in NYC. They build "hi-tech, hi-touch" environments and products, using new technologies to create soft and playful interactions.

Arduino Fever

The tale of a cute, blue microcontroller that fits nicely in the palm of your hand, and the expanding community of developers who love and support it. By Daniel Jolliffe

ARDUINO IS SPREADING RAPIDLY across the globe. But before you reach for the Merck Manual to find the symptoms you're sure to have, check this out. Arduino is actually an open source hardware project that can be programmed to read temperatures, control a motor, and sense touch, and gets its name from an ancient Italian king. And it's fun!

Named after the 11th-century king of Ivrea in northern Italy, the Arduino is both a cute, blue microcontroller platform that fits nicely in the palm of your hand and an expanding community of developers who support it, distributed across two dozen countries, four continents, and counting.

Decidedly 21st-century in its design and construction, the Arduino board is for anyone who wants to build a basic level of intelligence into an object. Once programmed, it can read sensors, make simple decisions, and control myriad devices in the real world.

Using it is a snap: first, hook up a few sensors and output devices to the Arduino, then program it using the free developer's software. Next, debug your code and disconnect the Arduino. Presto — the little blue Arduino becomes a standalone computer.

The original intention of the Arduino project was to see what would happen if community support were substituted for the corporate support that is usually required for electronics development. The first developers — Massimo Banzi, David Cuartielles, David Mellis, and Nicholas Zambetti — ran a series of workshops on assembling the Arduino, giving away the board to stimulate development.

Photography by David Cuartielles (workshop), Ren Wang (robot), and Nicholas Zambetti (chip)

A scant year later, the project has spread far and wide the message that electronic design doesn't have to be a solitary, complex, and painful process, and that it needn't cost much if you have a little help along the way. Says Ren Wang, a student at China's Xiamen University who used the Arduino to power Eye, a walking robot: "Arduino is open and friendly. To make a cool robot was always my dream, and the Arduino made it come true."

Today there's a thriving website with sample code, tutorials, and a forum that serves as the meeting point for Arduino developers. While the original developers still give workshops, the project is increasingly a standalone endeavor, with newcomers taking up the idea that electronics can be a community effort.

Back in Ivrea, a friendly Italian manufacturer, who was courageous enough to support the project from its inception, still provides low-cost Arduinos, in both assembled and kit form. In Europe, the price is €20; Sparkfun (sparkfun.com) is the United States distributor and sells the USB version assembled for $30. And since the project is open source, all the plans, code, and instructions are available online free for those who prefer to roll their own.

Asked what's next, Cuartielles says: "Arduino for kids! We have been asked to evaluate the use of Arduino for technology classes in secondary schools

Arduino assembly workshop (previous page) at Malmo University in Sweden.

"Eye" (this page top) is a robot based on Arduino, designed and built by Ren Wang, a student at Xiamen University in China.

Arduino (bottom) in the palm of a hand.

in Madrid, Spain. Can you imagine one million kids a year making experiments in electronics based on this open hardware platform? It would be massive!"

To get in on the massiveness, and to become a contributor yourself, check out arduino.cc.

Daniel Jolliffe is the designer of *One Free Minute*, an anonymous public speech project. He wrote "Throw Your Voice!" in MAKE, Volume 04.

The MAKE Controller

Announcing a just-maybe-revolutionary microcontroller for all things DIY.

By David Williams and Liam Staskawicz

Microcontroller chips and the tools required to use them were once so specialized that few people could explore their potential without a degree in electronics and software engineering.

From BASIC Stamp kits to Wiring and Arduino boards, manufacturers and open source communities now offer somewhat easier ways to program microcontrollers, enabling cross-pollination between the engineering world and hobbyists, do-it-yourselfers, artists, and students.

But the effort needed to program a chip and design a circuit is still a barrier to countless creative people, even as the latest 32-bit chip designs offer new features like built-in networking.

All the elements are in place for the next revolution in DIY microcontrollers: a friendly, open platform that can do it all, launch a thousand projects, and empower both novices and pros.

Open Source Hardware

It's always exciting when powerful technologies become accessible to ordinary people. This is happening now with microcontrollers. After revolutionizing many industries (and requiring industrial-sized development budgets), now these do-anything chips are becoming easier to program and use with free, open source tools.

We realized, as did MAKE, that a general-purpose controller kit served by a free and open development environment would offer limitless potential to makers. A smart building block like this would short-cut (and even enable) diverse projects, collapsing pages of complicated instructions into, "connect these components to the controller board, upload this code, and *voilá* — you've got an amazing new device for your kitchen" (or garden, or living room, or backyard, or ... *see sidebar, page 85*). This platform would also inspire more advanced makers to develop and share their own original projects. Ideally, such a controller kit would be visible enough to attract a healthy developer community, and rewarding enough to see it grow.

Last year, MAKE magazine approached MakingThings to create this platform, the MAKE Controller. We were delighted, and we're even more delighted to announce that it is now available. In this article, we explain how the kit's two PC boards work, and why we designed them the way we did.

Basic Principles

Right from the outset, it was imperative that this be an open project. We knew from our experience with similar products, and from our customers (MakingThings has been producing and selling a range of controller boards since 1998), that anything short of freely available hardware and software would frustrate the creativity of the experimenter and the hacker — precisely the audience we're aiming for. We will publish all of our schematics, to encourage modifications and improvements to the hardware. The software environment(s) had to be free and easily accessible as well, to promote development. By supporting all aspects of the project with cross-platform, free, and open tools,

we created an open hardware platform that many disparate communities can support, contribute to, and enjoy. For further accessibility, we also wanted the kit to be relatively inexpensive, and we managed to keep it down to less than $150.

To serve both novice users and experienced engineers, we decided to base our design around two boards rather than one. The Controller Board contains the microcontroller itself, plus the delicate circuitry needed to run it it, and all essential peripherals. This Controller Board can then plug into a larger Application Board that provides circuitry to

> We decided on the newer, faster class of chip, to create a new generation of controller boards.

interface with real-world electronics like sensors and actuators. Users of an Application Board can just plug a stepper motor or a pneumatic valve right in, rather than having to design the circuitry to drive them from the chip. We designed one general-purpose Application Board with a nice mix of control and communication capabilities, and over time we will add others with different features, such as CMOS cameras and audio input and output.

With the two-board approach, novices can experiment with an Application Board with little fear of breaking anything, while experts can either use an Application Board or plug the Controller Board into a board of their own design.

For the most advanced programmers, the Controller Board makes it easier to get at the microcontroller's densely packed pins. Everything is accessible. They can read the 700-page processor manual, dig deep into the chip's architecture, choose or write their own software and device drivers, and have full control over every other aspect of the board's operation.

Novices also have options. The Application Board includes onboard switches that select and control useful and fun software that's pre-installed on the Controller Board. This makes the kit enjoyable and configurable right out of the box, without any programming. After exhausting the switches, a novice's next step can be downloading different code and uploading it into the boards using USB.

Photograph by Susan Williams

stats (a 32-bit 55MHz ARM core, 256KB of flash memory, and 32KB of RAM), the part has Ethernet addressability, a USB port, a CAN controller (CAN = Controller Area Network, a simple, flexible, two-wire protocol originally developed for cars), several serial ports, and more, making it one of the most versatile microcontrollers on the market.

The SAM7X's Ethernet port opens up entire realms of possible applications — for example, it can host a web server or telnet session, or retrieve data from the internet all by itself. And its USB port lets the system interface with a computer, to offload any computational heavy lifting. With either port, you can have a computer perform specialized processing while the connected microcontroller exchanges information with the rest of the world.

Finally, because of early excitement around the introduction of the SAM7X, many software providers have adapted their development environments and other tools to this chip, resulting in strong software support even before we got started.

Liam and David with the final MAKE Controller Kit and several prototypes.

Choosing the Microcontroller

The first and most difficult task was choosing the microcontroller. Most microcontrollers that address the hobby/artist/DIY markets are 8-bit systems, in which most operations in the processor work on 8 or 16 bits at a time. All of the predecessors of the MAKE Controller are 8-bit, including our own products and the Wiring and Arduino boards.

32-bit microcontrollers have recently become available, some of which are ten times faster than 8-bit chips, with far more onboard memory and sophisticated on-chip resources. They can comfortably multitask and communicate over bridges and networks, in addition to running simple control operations. On the downside, they're also a bit more expensive (about $9 rather than $7 or less) and they use more power.

We decided on the newer, faster class of chip, to create a new generation of controller boards. Then we had to choose a specific one, since Philips, STMicroelectronics, Texas Instruments, and Atmel all have excellent 32-bit products. Several meetings and a masochistic one-day trip from San Francisco to Boston later, we had enough information to select our chip: the Atmel AT91SAM7X256 — or "SAM7X." This is a brand-new part with a staggering list of features that hobbyists could only dream of before its release. In addition to quite exciting basic

Selecting the Development Environment

The next thing we did was confirm that there was a full set of free tools that would allow people to develop with the SAM7X. Fortunately, GNU's incredible open source compiler, gcc, will compile code for most ARM processors. With every release, gcc becomes more capable and more efficient, although at the time of writing, the debugging environment that gcc provides for ARM parts is somewhat limited.

A few expensive professional tools also target the SAM7X processor and use gcc, including Rowley and Associates' CrossWorks for ARM. This means you can use either a high-end environment or open source tools to develop for the MAKE Controller.

Designing the Hardware

By far, the hardest part of the process is designing the hardware. Here are the steps we followed:

1. **Decide on basic functionality.** What will each board include? One such decision was to squeeze all Ethernet support (except for the connector itself) onto the Controller Board, making it available to all users, instead of putting it on the Application Board.

2. **Decide which microcontroller pins do what.** The SAM7X has 100 pins, but that's not much, considering what the chip can do. Manufacturers rely on

MAKE CONTROLLER KIT: CONTROLLER AND APPLICATION BOARD LAYOUTS

Layout of Controller Board plugged into Application Board.   **Dashed connectors are detachable.**

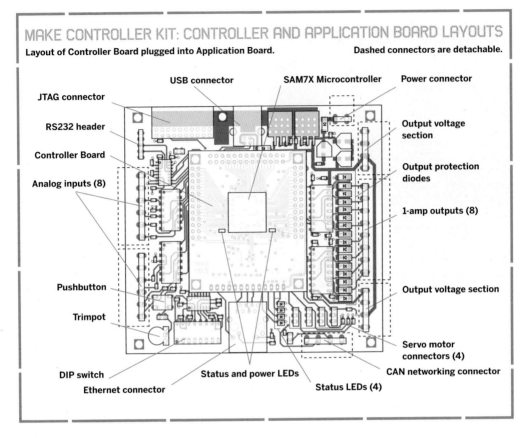

JTAG connector

RS232 header

Controller Board

Analog inputs (8)

Pushbutton

Trimpot

DIP switch

Ethernet connector

USB connector

SAM7X Microcontroller

Power connector

Output voltage section

Output protection diodes

1-amp outputs (8)

Output voltage section

Servo motor connectors (4)

CAN networking connector

Status and power LEDs

Status LEDs (4)

---

pages worth of usage modes and dependencies in order to squeeze more functionality onto their chips, so the task of allocating all the signals becomes a monumental logic puzzle.

3. **Work out power issues.** What power will be needed by what parts? What is the maximum voltage the board should withstand?

4. **Draw schematic.** Wire all components per manufacturer's notes and common design practices.

5. **Select packages for all devices.** We favored surface mount (SMT) components since they're smaller, cheaper, and cheaper to assemble — although you can't solder them easily by hand.

6. **Map schematic onto a PCB board.** Each trace connects two points, and they can't overlap or run too close. Ground and power can go on the internal layers of a four-layer board, as a wiring convenience and to reduce interference from electrical noise.

7. **Check and double-check everything.**

8. **Generate Gerber files.** PCB fabricating machines use these to manufacture boards. Most PCB layout software has a Gerber export option.

9. **Send Gerber files to a board shop that can do the job.** The traces on the Controller Board, and even the pads of the SAM7X itself, are very close together. With a design this fine, many cheaper shops will send boards back with shorts between contacts.

10. **Assemble the boards.** Order all the parts, and send them with the boards to a manufacturing facility.

11. **Test.** With our first boards, we found mistakes in the complex Ethernet area. Correcting these required some real white-knuckle soldering tricks.

12. **Celebrate!** Designing a new board is tense, and failure means a loss of money as well as pride. When it all finally works, everyone breathes a sigh of relief.

One of the difficulties of programming a micro-controller directly is having to write for the device's specific inner workings. This means structuring code to permit other functions to operate, and deciphering arcane header files and the bit-by-bit operation of complex peripherals. We wanted none of that.

To liberate our users (and ourselves), we wrote a series of code layers that access all the devices on the boards at a more abstract level. Instead of needing to understand every aspect of each device, developers can include these files with their projects, and then call simple methods that encapsulate lots of fiddly hardware settings. For example, with one simple call, our AdcInit() routine initializes the SAM7X's Analog to Digital Converter, a process which otherwise requires a hideous chunk of code.

Another difficulty with bare-bones microcontroller programming is having to explicitly manage all control-passing, sharing the processor among all running tasks. The cure for this pain is to use a Real Time Operating System (RTOS). With complex projects where functions run simultaneously or timing needs to be managed closely, an RTOS makes coding applications quicker, simpler, and more elegant.

RTOSs have traditionally been expensive, and they consume too much overhead to be practical with 8-bit microprocessors, but there is an excellent, free RTOS, helpfully called FreeRTOS.org, that runs on the SAM7X. When connected to lwIP, a small library of code designed to provide internet capabilities, we have a full RTOS with internet capability.

Programming with the benefit of an RTOS means you can start tasks upon initialization, and run them concurrently as permanent loops that poll for input and communicate with each other. There's no need for spaghetti code that breaks out of each ongoing task to check on everything else.

Conclusion and Future

We are very pleased with the MAKE Controller Kit. It is an absolute delight to program, and connecting real devices to it is very simple. We look forward to adding more and more functionality, and seeing what others think of doing with it. Try one and tell us! Email us at info@makingthings.com.

Project Possibilities

Some applications of the MAKE Controller that you may be seeing in the near future.

Drink-O-Mat
Timed control of valves and motorized swizzle stick can dispense and mix measured amounts of different beverages from inverted bottles.

Plant Cyborg
Light sensor and motor controllers position houseplant for optimal growth, while moisture sensor triggers valve to ensure perfect watering.

Glue Gun 3D Imaging
3-axis stepper motors and trigger controller render stable 3D forms by building them up layer-by-layer using hot glue.

Automated Pet Feeder
Motor device feeds pet remotely via web-based interface or timer program, and dispenses treats when pet stands in a specific position.

Liquid Temperature Controller
Thermostat feedback loop keeps liquid within a specified temperature range — for brewing, aquariums, candy-making, photo developing, etc.

Global Sensor Network
Large-scale survey system registers sensor input from hundreds of physical nodes connected to a common server, which publishes sensor info.

Get Your MAKE Controller Kit!
On sale now at makezine.com/controller, where you can also learn more about the kit, including full schematics, and share your experiences in our online forum.

David Williams enjoys the connection between machines and intelligent software, and has designed and built professional tools for software, electronic, and mechanical engineering. Liam Staskawicz has explored sensor-driven hardware and software systems from the perspectives of both engineer and musician. See makingthings.com.

PROPELLER CHIP

BASIC STAMP'S CHIP GRACEY PUTS A NEW SPIN ON MICROCONTROLLERS.

By Dale Dougherty

The head guy at Parallax, Chip Gracey, is truly self-taught, which means that he has had to find his own way. Twenty years after teaching himself to program the first generation of personal computers, the creator of the new Propeller microcontroller still speaks with the enthusiasm and amazement of a bright teenager: "The tools are out there. These days with the internet, it is so easy; you can learn anything. What used to be obscure stuff that only a few people were interested in — well, today those people put it on the net to share among themselves, and the rest of us have access to it."

Lately he's been "playing around" with speech synthesis. "I've put about two steady years of work into it. Just now, I'm on the cusp of having a working voice synthesizer." He immediately starts walking down the idiosyncratic path he took to get where he is today, and I do my best to follow along. He conveys all the details as though they are keys to finding the levels of an adventure game he created.

He explains how the key to speech synthesis is to reproduce vowel sounds, which are vocal resonators. The long sounds, the *oohs* and *ahhs*, resonate inside the hollow spaces in our skulls when we say words like "food" or "bath." If he's able to replicate vocal resonance in software, the result will be something much closer to normal speech.

"I used to look at the human voice on a scope in high school and you can see the waveforms of speech. But you can't look at speech in terms of

waveforms because speech is really a spectral phenomenon." So he believed the key to his approach was to write programs that allowed him to visualize speech patterns in more detail. "If you can visualize it accurately then you know what you have to do to re-create it."

"I kept reading about digital resonators. Everybody talked about them but no one explained how they work. I couldn't find a recipe. Then I finally came upon CORDIC math, which means coordinate rotation. [CORDIC is an acronym for Coordinate Rotation Digital Computer.] It was developed in the 50s. Real interesting stuff and very simple."

He points to the computer screen where there are lines of code that don't explain themselves. "I realized that a CORDIC rotation algorithm could be used to create a resonator. You just add in the stimulus vector (add x's and y's) and then rotate the

Photography by Robyn Twomey

resulting point. Repeat. Repeat. Repeat. Any in-band energy will have the effect of growing the distance of the point from (0,0) as it rotates at the resonant frequency. This is key to making vowel sounds."

I probably look more than a little confused. He adds: "You've got to excite that resonator. It's like a bell. *Bing-gggg*. That took me forever to figure out but it was very simple." I'm not sure I follow everything he says, but I get that here is a guy who works and works an idea and pushes it forward, bit by bit, and eventually gets somewhere. He works at a level of electronics that is completely ethereal to most of us, but not to him.

"Ultimately, everything has to be physical so we can perceive it with our senses," Gracey says. "The hardware is like the body and the software is like the spirit. The whole point of the Propeller chip is to make an able body for the kind of spirits you want to create. Like speech synthesis."

Gracey's immediate goal is to write a state-of-the-art sound synthesis library for the new Propeller chip. He explains that existing speech synthesis applications "almost always require a dedicated speech chip." By adding sound synthesis as a software capability with the Propeller, he can make it affordable to create applications that generate a vocal track. He adds: "The whole point is to enable inventors to make stuff they couldn't make before."

From the beginning, Parallax has offered an embedded platform, distinct from the PC and Windows, for hobbyist and educational projects, not to mention professional uses. Why not just use a PC?

Gracey says a PC can get expensive for a number of applications, especially when compared to using the BASIC Stamp controller. However, he says the "big sleeper issue is reliability because Windows seems to blow itself up in time."

Neither BASIC Stamp nor the Propeller have an operating system. Why? "The systems are so simple that the code you write is the operating system." He sees the microcontrollers as deterministic systems that do what you tell them to do; put an operating system onboard and you will have something with a mind of its own. PCs are interrupt-driven, which means the operating system is designed to interrupt what it's doing if it thinks it should be doing something else; a PC usually hosts multiple applications. Embedded applications typically are designed to do one thing — such as read an RFID tag and unlock a door, or monitor and store GPS coordinates of a rising weather balloon — very well.

The landmark BASIC Stamp controller is a custom circuit based around an 8-bit Microchip PIC chip, running Parallax's now famous BASIC interpreter, plus a high-level development environment. Other models in the BASIC Stamp family are based on a Scenix SX processor or may come with more I/O

pins, faster speeds, or other enhanced features.

The new Propeller chip has eight separate 32-bit processors called "cogs" that can independently process information, with a shared memory space. The Propeller is an entirely custom integrated circuit design of Gracey's own making, and a completely new development environment and programming language (called Spin). It's a completely new path that he set out on, and he's got to convince customers to go along with him.

On the Parallax user forums, (forums.parallax. com) one user wrote that the new Propeller chip was untested and unproven and that he'd only consider using it in a hobby application. Gracey wrote back:

"I know that the Propeller is solid because I designed, debugged, tuned, and tested it myself. The only other person involved in the silicon design was one other layout engineer. This took eight years of my time, and two years of the layout engineer's time. An excruciating amount of attention went into every aspect of the Propeller's design and testing, and I allowed no compromises."

As Gracey sees it, he's always been ahead of his customers. Ever since he learned to program, he knew he was doing something that other people found valuable. "I would think of something that might be neat and I'd make it, and then I could sell it to people."

Today, Parallax employs 40 people, most of them located in an industrial park in Rocklin, Calif., about a half-hour from Sacramento. None of them are salespeople. "We have nobody calling anybody trying to make them buy anything," Gracey says proudly.

Gracey's business career started in his bedroom when he was in high school. He created a software utility for disk duplication on the Commodore 64. In 1987, he made $130,000 in royalties; he was only 17. Next he made development tools for the Apple II. He was so into his programming that he barely finished high school. He tried a year of college to please his

mother, but quickly dropped out.

Then something happened to him and he dropped out of business as well. Today, he calls it a "malaise," a term Jimmy Carter used to characterize America's "crisis of confidence" in 1979. Gracey says that he thought the reason he liked to program was to buy cool stuff. He spent some of his money on a stereo, a car, remote control toys, video games, and the like.

Even though at 19 he was still living with his parents, he had everything he could ever want, but he discovered none of it made him happy. He got to the highest level of his own adventure game, only to find out he didn't care anymore about the achievement. Game over.

He recalls: "You're trained to think that you do things for money, that success equals money." Gracey was all mixed up and it took him a while to separate the work he enjoyed from the rewards it had gained for him. "Money is not the object. What I liked about what I was doing, was doing it," he said. "What was making me happy was the learning and the creativity."

So he came back and started Parallax with a friend, Lance Walley, and they eventually shared an apartment where they also worked. They began making microcontroller development tools for the Intel 8051, an all-in-one computer chip, meaning it had a CPU, RAM, ROM, and I/O on board. (Intel's 8086, the heart of the first IBM PC motherboard, is just a CPU.) Why the 8051? "I thought it would be neat to build development tools for a single chip."

Then in 1993 he developed the BASIC Stamp product. "I loved all these little microcontrollers because they are so much fun. The trouble was that you had all this arcane setup to do to program them. I wondered: what if you could make a little computer that you could program in a high-level language and that was about as cheap as a microcontroller?"

He saw that Microchip had PIC chips, a popular family of 8-bit microprocessors notably used in Microsoft PS/2 mice, and 8-pin EEPROMs (Electronically Erasable Programmable Read-Only

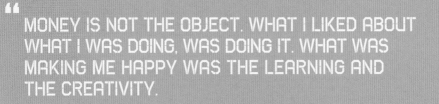

> MONEY IS NOT THE OBJECT. WHAT I LIKED ABOUT WHAT I WAS DOING, WAS DOING IT. WHAT WAS MAKING ME HAPPY WAS THE LEARNING AND THE CREATIVITY.

Memory), simple serial-based devices that store small amounts of data. He realized he could develop a BASIC interpreter that he could program into the Microchip PIC device, and then use the EEPROM to hold the user-generated code that the PIC would pull out and execute. In addition, he created a development tool on the PC that's a high-level compiler. "You write code on the PC, which compiles it down and then you hit a button and send the code down to the chip and it starts running."

That's the essence of the BASIC Stamp line of products, the bread-and-butter of Parallax, which made it easier for more people to program microcontrollers. For several years, Gracey and Walley were getting orders, as many as 20 a day, and assembling the circuit boards and kits themselves, in their apartment. "We worked really hard," recalls Gracey. Now Parallax has machines that automate the assembly of the circuit boards.

Walley and Gracey parted ways in 1996, and Chip's younger brother, Ken, came on in 1997 to help manage operations as the company began to grow. I wondered if Ken, who had finished college, was considered the good brother in the family. "For a long time, my parents really felt like their hopes

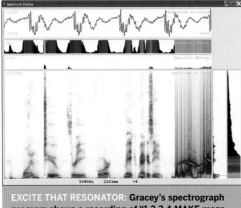

EXCITE THAT RESONATOR: Gracey's spectrograph program shows a recording of "1 2 3 4 MAKE magazine" followed by some whistling.

had been lost in me," Chip recalls. "Now they think of us both as good brothers."

An environmental studies major, Ken has picked up quite a bit of the technical side of the business as well as handling day-to-day operations. "He's not as enthused about the technical side the way that I am, but he can always learn as much as he needs to know." Ken is the multitasking manager handling sundry activities and interfacing with customers while Chip remains singularly focused on development. Parallax has grown up to become a satisfying family business, with their father, Chuck, coming in most days to work as well.

The decision to build the Propeller chip was driven not only by Gracey's developing interest in chip design. (On his bookshelf, he points out *Principles of CMOS VLSI Design* by Weste and Eshraghian, which he says taught him all he needed to know.) He felt that large chip manufacturers were not willing to experiment with new design ideas. Moreover, he felt the chip manufacturers had too much control in the relationship. "We had to have our own silicon," he says. He had to go his own way.

There's a rock-star-sized poster of the Propeller chip's schematic on the wall in Gracey's office, and he gives me a tour of all its functional areas, which he knows so well. "The Propeller has eight capable processors and a shared memory. Any one of those can synthesize speech, generate a VGA signal, talk to a mouse or a keyboard, or digitize something."

He's not the first to put multiple processors on a chip. However, previous approaches required parallel processing to use the chip; the challenge is to figure out how to split up a program so that parts of it can run in parallel. "That's not what you do with multiple processors," says Gracey defiantly. "Instead you let them each do something different."

The chief design goal of the Propeller was to allow multiple processes to run concurrently without interrupting each other. For the new speech synthesis library, this means each processor can produce a voice independently of the others. "I've added vibrato,"

Joe Grand on the Propeller chip.

Joe Grand, a member of MAKE's Technical Advisory Board, has worked closely with Parallax and has followed the development of the Propeller.

Chip Gracey designed the Propeller chip (parallax.com/propeller) completely from the ground up, a feat rarely attempted in a world where most products are based on some existing prior art, reference design, or chipset.

Even the most fundamental base of the Propeller, having multiple "cog" processors, each active for a given time "slice" and sharing a common memory space, is sufficiently unique and different from existing microprocessor technologies.

I remember visiting Parallax for the first time a few years ago; Chip had just got his code working to output graphics to a TV monitor using an FPGA development system — the precursor used to test code before moving to actual silicon.

He showed me the code in his excited fashion, and we spent the next few hours hacking away at different parts of it, trying to display multiple Parallax logos on the screen. His enthusiasm was unparalleled and contagious. I knew he was on to something big. To see the fruits of his labor come together at an early, pre-release Propeller training session was just phenomenal.

Even with the release of the Propeller, the BASIC Stamp family is still a hugely popular device for the hobbyist and electronics community, and I personally don't see that going away. Much to the chagrin of Chip, who I think wants

BIRTHPLACE OF CHAMPIONS: A Parallax solder stencil for the now-classic Basic Stamp II.

to move everyone over to the Propeller, the two product lines seem to attract different audiences (at least for now).

On one hand, using Stamps for simple, low-cost tasks is just so easy. I absolutely love using them for quick prototypes that aren't worth the effort or energy to deal with a more complicated microprocessor development environment. On the other, I've never seen such a unique and interesting device as the Propeller, and the Spin code people have written for it already is amazing, even in its very short life.

Gracey says. "You can have all eight voices singing, just like a choir." (Months later, he sent me a cool sample called "Propeller Monks" generated by his code; hear it at makezine.com/10/propeller.)

It took years to develop the Propeller chip. The path was not straight, and there were many obstacles. It required new machines, new investment, and new knowledge of manufacturing techniques. It also took longer than Ken hoped it would. Perhaps recalling something that had worked so well before Chip had his midlife crisis at 19, Ken promised to buy him a very high-end stereo system, an audiophile's dream, if he could complete the design on time. And that's where we stop in an otherwise empty

office, and sit down, listening to a jazz standard on the turntable. "You can hear so much more with this system," says Gracey, with the glee of a kid having friends over to listen to his record collection.

One thing Gracey said earlier in the day stands out. "If you have a sense of mastery, even if it's not complete, then it goes a long way toward helping you get to where you want to go." It's a statement as much about himself as about what Parallax hopes its products do: help others to find their own way.

Dale Dougherty is editor and publisher of MAKE.

THE MACHINE ROOM

We built up our own lab that gave us the "hands" and "eyes" that we needed to work on our chip. First, we invested in a Micrion FIB (focused ion beam) machine (shown at right) that allowed us to perform microscopic surgery, so that we could check failure hypotheses and make experimental modifications. Think "wire cutters," "soldering iron," and "solder" for the sub-micrometer wiring inside the chip.

The other big thing we acquired was a Schlumberger e-beam prober — essentially a scanning electron microscope that can use its electron beam to measure voltages on those same tiny wires while the chip runs at full speed. Think "7GHz, non-loading, 10nm-tip, contactless oscilloscope."

These machines are almost *Star Trek* in their technology, and they get you all the way down to where you need to be in order to see and fix problems.

We were able to purchase these machines, used, for only 0.5% of what they cost new. The real investment, though, turned out to be the six months it took to get them running, and to learn how to use them.

Now, we can even do our own maintenance on them, which is not trivial. All this was a huge adventure in itself, but invaluable in getting the Propeller's silicon perfected.

—Chip Gracey

TURBO SALAD: Chip offers this backstory: "This is a turbo molecular pump [used to pump stray molecules of air from the vacuum chamber where chips are fabricated] that crashed. It weighs about 45 pounds and has about 10 pounds of stacked turbine blades on ceramic bearings that spin at 40,000 rpm. The bearings wore down critically over a weekend and the pump seized within a few revolutions, resulting in 'turbo salad.' A tremendous amount of kinetic energy was let loose and the 9/16" stainless steel bolts were all bent 10 degrees in a circular pattern as a result of the sudden breaking. We usually get our turbo pumps rebuilt when they are starting to show wear through either excess whine or noticeable heat. Getting such a pump rebuilt is about $1,500. Buying a rebuilt used one costs about $4,000. Buying a new one can cost over $20,000. So, our tactic is to rebuild, and if we blow that, get a used rebuilt one."

Turbine photograph by Dale Dougherty

IT'S EMAIL TIME

Innocent-looking "clock" monitors the unread-message pileup in your inbox.
By Tom Igoe

I have a lot of anxiety about email. Every kilobyte in my Inbox destroys another minute of my life, but I can't stop checking it. So I decided to embody my anxiety in a device that would worry about my incoming mail for me. I've always liked clockwork mechanisms, so I made my email fetish object in the form of a clock. For each kilobyte of new mail I receive, the clock ticks relentlessly forward.

Here's the basic design I came up with. The clock itself is driven by a microcontroller, which connects to the internet and queries a program that checks my email accounts. The program reports back the number of kilobytes, and the microcontroller moves the clock forward a tick for each kilobyte. Simple!

To build this yourself, you need to know how to program a microcontroller and how to do some

basic web programming. For microcontroller programming, see the Primer article (page 70), or my book *Physical Computing: Sensing and Controlling the Physical World with Computers*. For the web component, I used a Common Gateway Interface (CGI) script written in Perl, but you can also write CGI scripts in PHP, Python, Ruby, and other languages.

Building It

I started by looking for the simplest way to drive the clock. I carefully took the clock apart and examined the circuit board inside. There was a mystery chip in the middle, sealed in plastic, but I found that the main gear of the clock, which drives all the other gears, is controlled by a solenoid.

Photography by Tom Igoe

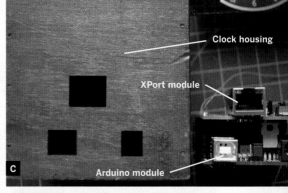

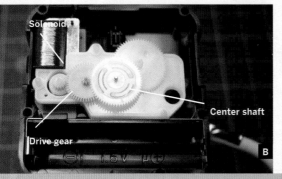

A

B

C

D

Clock housing

XPort module

Arduino module

XPort module

Arduino module

Solenoid

Center shaft

Drive gear

UNITS
600 ms MAX.

Wires soldered to solenoid contacts. Note crystal at bottom and "mystery chip" under blob in center (Figure A). Cheap battery clocks all generally work the same way (Figure B). Rear view of clock, showing holes drilled in housing for Ethernet, power, and the USB-B port for programming the Arduino microcontroller board (Figure C). Front view of clock, face removed (Figure D).

MATERIALS

Here's what I chose for the basic building blocks of my email clock.

Microcontroller I used the Arduino module (arduino. berlios.de), a small I/O board and development environment that's built around Atmel's ATMega8 microcontroller. Arduino is based on Wiring, another open and easy development environment, which uses the ATMega128 chip. I opted for Arduino because I knew I wanted to eventually make my own custom circuit board, and the ATMega8 is easier to solder than the ATMega128.

Serial-to-Ethernet Converter Instead of bothering to write my own TCP/IP stack for the Arduino board, I used some handy hardware: Lantronix's XPort serial-to-Ethernet module. These modules can route data between an Ethernet connection and a serial port that feeds a microcontroller. To avoid some soldering, I also recommend the Cobox Micro, which has the same programming interface as the XPort (you simply telnet in) but a simpler physical interface.

Clock I would have loved to build a mechanical clock but didn't have time, so I took the guts out of a cheap battery clock and connected its drive shaft into the hub of a clock-like antique test gauge.

| Mail Server | Mail Server | Mail Server | Mail Server |

Web Server

Internet

Microcontroller

Clock

System diagram. A CGI script queries my mail accounts and returns new mail volume. A microcontroller calls the script, and moves the clock's hands forward accordingly.

The two solenoid terminals were pretty easy to spot, and I figured that all I'd have to do was send a pulse through these connections, and the clock would tick happily away.

I soldered a couple of leads onto the solenoid contacts and put the clock back together. Then I pulsed the solenoid directly with 5VDC power, bypassing the mystery chip. The motor jumped, but the clock didn't tick. Clearly, there was more to learn, so I put the clock's battery back in and connected the leads to an oscilloscope to see how the voltage changed as it ticked. The pattern was more complex than I expected. Each second, the pulse would alternate: high-low-zero one second, and low-high-zero the next.

I programmed the microcontroller to duplicate this pattern, and fed its output into the clock. This took some experimentation, but eventually I got the microcontroller to control the clock pretty well. Each tick it generated moved the clock forward about two seconds. Since I wasn't concerned with keeping actual time, this was fine.

The next step was to get the microcontroller to check mail. The XPort needed only three wires connected from the microcontroller side: serial receive, serial transmit, and a reset connection to allow the microprocessor to restart the port. Making a circuit board for the XPort was a challenge because its pins don't follow a 1/10" grid, the hobbyist perfboard standard. So I used CadSoft's Eagle software to design a custom board to mount the module into. You can find the layout file at tigoe.net/emailclock.

Then I configured the XPort from my laptop through a USB-to-serial converter cable. Following Lantronix' instructions, I gave it an IP address, gateway address, and subnet mask. I also configured the serial port settings.

I opened a terminal window, telnetted into the XPort, and entered "Hello World!" to confirm that messages were passing through. Then I quit telnet and tried connecting to my web server by entering its numerical address (port 80) in the serial window: C82.165.199.37/80.

The XPort confirmed by returning a "C". I responded with an HTTP request for a web page on my server, http://tigoe.net/pcomp/index.shtml:
GET /pcomp/index.shtml HTTP/1.1
HOST: tigoe.net

The server returned the HTTP header and contents of the requested page:

HTTP/1.1 200 OK
Date: Tue, 13 Dec 2005 20:50:27 GMT
Server: Apache/1.3.33 (Unix)
Transfer-Encoding: chunked
Content-Type: text/html
<html>
  <head>
... and so forth.

Success! Seeing this exchange of HTTP (HyperText Transport Protocol), the normally hidden language of web browsers and servers, meant that I had gotten the XPort to perform as a browser. Now I just had to get the microcontroller to do the same: open a net connection, request a page, and read the results. You can find my code to do this at tigoe.net/emailclock. I uploaded the compiled firmware from my laptop via the Arduino's onboard serial-to-USB converter.

The microcontroller isn't requesting an HTML page — it's calling a common gateway interface (CGI) script, also available at the link above, that checks all of my mail accounts, queries for the volume of new messages, adds the numbers up, and sends the total back to whoever's asking. Since this script isn't expecting to be called from a browser, it doesn't format its results as HTML. To minimize the programming needed on the microcontroller side, it keeps things simple, returning just the HTTP header and one line of text: <KB: 1234>.

Once the whole system was working, I found a housing for the clock: a nice, antique piece of electronic test equipment with a hole behind its clock-like face that could accommodate a driveshaft. The clock shaft screwed in easily, and the Arduino module and the XPort fit snugly inside. I drilled a couple of holes in the back for the power and Ethernet cables, and the clock was done.

How well does it work? I still check my email compulsively, but for the couple of hours it took to build this clock, I didn't open my mail program at all!

Tom Igoe heads the physical computing area at the Interactive Telecommunications Program at New York University.

Insert firefly in plastic bubble and a $5 photodiode chip converts light to frequency.

FIREFLY METER

Bioluminescence detector lights way toward insect-cyborg pollution sensor.
By Christopher Holt

For me, the memory of catching fireflies in a jar brings me back to the dog days of my childhood summers. Now I'm a part-time biology graduate student, and one of my research projects focuses on the impact of air pollution on firefly flash duration, intensity, and period. Understanding this link could enable us to use fireflies as environmental contaminant sensors, providing an indirect method of monitoring air quality.

If you couple the fireflies with a low-cost way of measuring their bioluminescent properties, you have a sensor. Fusing whole organisms with hardware in this fashion is an exciting new research area, which has produced sensors capable of measuring biological and chemical agents at part-per-trillion concentrations.

Fireflies, or lightning bugs as they are commonly called, are technically neither flies nor bugs, but beetles of the order *Coleoptera*. There are more than 2,000 species worldwide, with 170 species in the United States alone. Fireflies use their light to find mates, which makes them among the few nocturnal insect types that discriminate mates visually. Numerous other insects are luminescent, but fireflies have the rare ability to turn their bioluminescence on and off. Typically, the males will signal with a specific flash pattern, and then wait for a female to respond with another flash pattern.

Most instruments for measuring bioluminescence are bulky, require advanced signal conditioning circuitry, and are far too expensive for a

Photograph by Kirk von Rohr

self-funded grad student. I needed an easy-to-use instrument that could measure firefly flash duration, intensity, and period. I surveyed my electronics workbench (yes, I'm probably the only biology grad student with one of these), and my eyes wandered to a BASIC Stamp 2 (BS2) Board of Education (parallax.com). I pondered the possibilities of using this simple, easy-to-use microcontroller kit for my bioluminescence meter.

On a few robotics projects in the past, I had already interfaced cadmium sulfide (CdS) photoresistors with the BS2 to measure light intensity. So first I thought I would just plug a photoresistor into a breadboard, build a chamber to house the firefly, and collect data until the cows came calling.

I soon discovered a big problem with my CdS photoresistors: long response and recovery times. In some cases, it took the photoresistor up to half a second to recover its value after an LED flash.

I figured out that Texas Advanced Optoelectronic Solutions (TAOS, www.taosinc.com) had basically solved my problem with their $5 TSL230 light-to-frequency converter. This 8-pin chip has a photodiode and built-in signal conditioning circuitry that does all of the analog-to-digital conversion on board. Also, the chip can be programmed to measure light for a specified duration, and transmit serially to a microcontroller or PC. With this key piece, I had what I needed.

Let's Build It

The hardware components needed for making the firefly flash meter are simple: a BS2 microcontroller, 0.1µF capacitor (to reduce noise), the TSL230 light-to-frequency converter chip, firefly test chamber, and a computer.

For software, I used StampDAQ (Stamp data acquisition) to translate the data coming via serial cable from the BASIC Stamp microcontroller into a format that's usable by Microsoft Excel on the computer. StampDAQ software is available as a free download from parallax.com.

For my delicate measurements, I set the TSL230 pins S0, S1, S2, and S3 high (+5V), which tells the chip to run at maximum sensitivity and resolution. I assembled the circuitry right on the Board of Education, which is an easy way to prototype.

I built the firefly test chamber out of a plastic pipette (VWR #14670-149). I found that cutting

```
Code to interface the BS2 to StampDAQ:
'{$STAMP BS2}
'Define variabless
light   var  word          'stores light intensity
sPin    con  16            'serial transmit pin
Baud    con  84            '9600 baud, 8-bit
S0      con  0             'S0 pin
S1      con  1             'S1 pin

'-------------Main routine--------------------------

Initialize:
pause 1000
serout sPin,Baud,[cr]              'Prepares StampDaq
                                   buffer
serout sPin,Baud,[cr,"label",light",cr]   'label Excel column
                                   with light
serout sPin,Baud,["cleardata",cr]  'Clear all data
dirs = %00000011                   'make pins 0 and 1
                                   outputs
High S0                            'Set S0 and S1 to 5V to
                                   divide frequency by 100
High S1

again:
count 2, 10, light                 'Count on Pin 2 for 10
ms
serout sPin,Baud,["DATA,light,", DEC light,cr]
                                   'Send data to
                                   StampDaq
goto again
```

the bulb of the pipette approximately ½" from the top provided adequate space for the firefly.

To make sure the firefly had adequate air supply, I cut an X at the top of the pipette bulb chamber. To maximize system sensitivity, I placed the firefly chamber directly on top of the TSL230 chip.

I used a styrofoam cup to remove residual background light sources. The hardest part of this project was collecting the fireflies. I enlisted the help of my 4-year-old daughter, Jordan. We learned that it's easiest to collect fireflies right at dusk, when it's still light enough to resolve the flying males from the background during periods of no flashing.

I used the BASIC command COUNT to count the number of pulses from the TSL230 (see code above). I set the COUNT command to measure for 10ms, but I determined that the actual resolution was approximately 31ms. The delay comes from the time it takes to execute the rest of the program loop and transmit the data into Excel.

Christopher Holt is a research scientist at NexTech Materials in Columbus, Ohio.

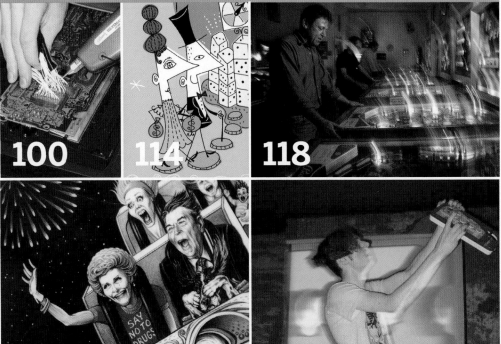

> "My games sell on story, so the first thing I do is try to come up with a good one: zombie fast-food workers fight over a single brain (*Give Me the Brain*) or infected mad cows graze fields to discover unexploded landmines (*Unexploded Cow*)." James Ernest, Homebrew Game Design

100

114

118

127

133

135

THE BEST OF
TOYS & GAMES
PROJECTS

from the pages of MAKE

SMART PEOPLE NEVER GROW UP

Don't believe the pop psych drivel about the nature of maturity, the so-called Peter Pan Syndrome, which tells us that we all need to grow up and act our age. Bollocks. If you want to maintain your sense of wonder about the world, keep your creativity intact and your mind active, connect with your kids, stay physically active, and all sorts of other things that are inherently good for you. Play! Go ahead. You have our permission.

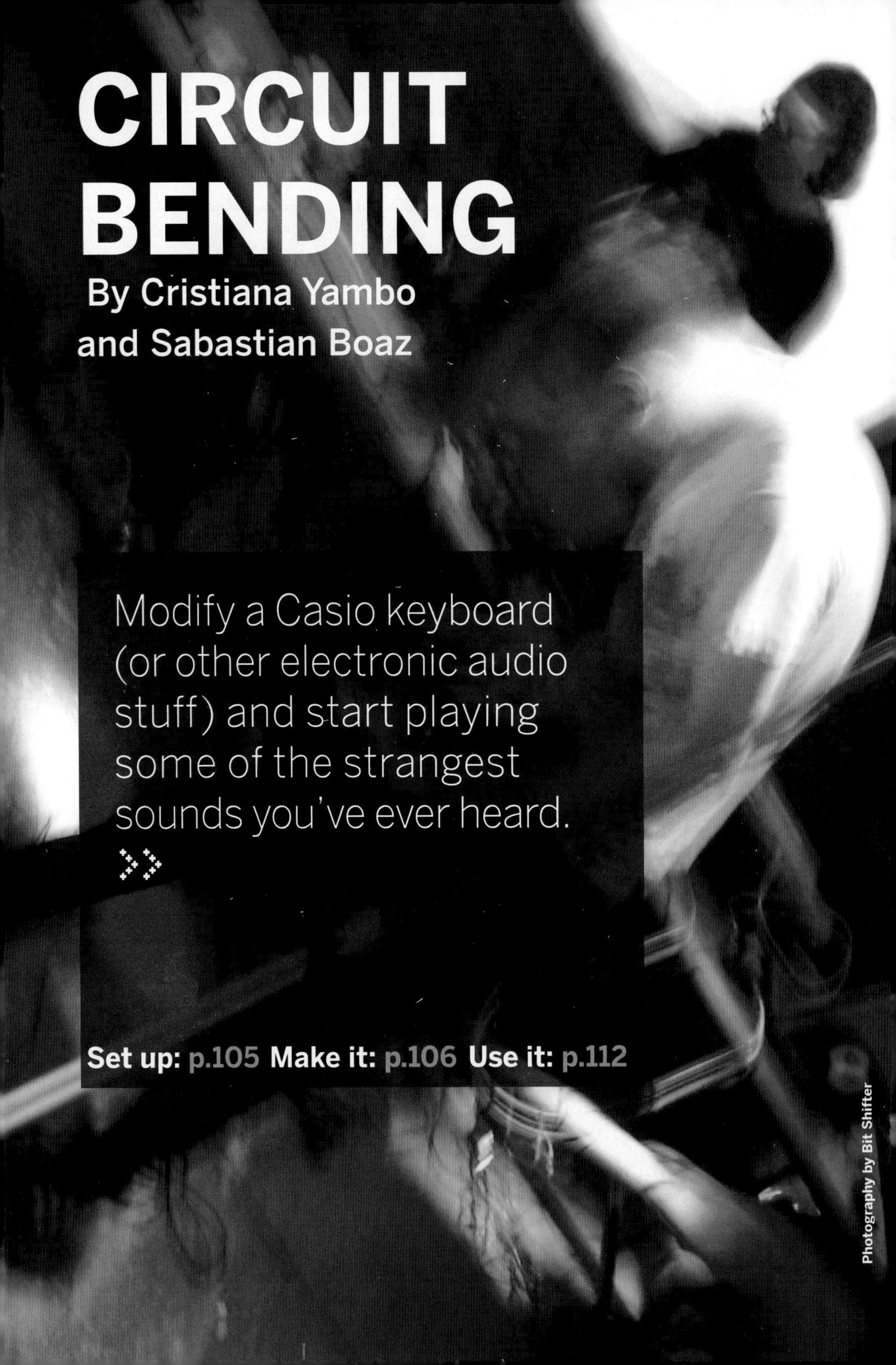

CIRCUIT BENDING

By Cristiana Yambo and Sabastian Boaz

Modify a Casio keyboard (or other electronic audio stuff) and start playing some of the strangest sounds you've ever heard.

Set up: p.105 **Make it:** p.106 **Use it:** p.112

Photography by Bit Shifter

SOUNDS OF CIRCUITS

The easiest way to start circuit bending is "playing open circuits." That's where you open up an audio device and use your hands or alligator clips to mess with the board inside and see what it sounds like. But it's almost as easy to permanently "bend" any suitable device by soldering on a few wires and switches. We'll explain how, and then show you how we transformed a Casio SK-5, a common, 80s-era sampling keyboard, into an unstoppably flexible sound organ and sonic effects generator.

A word of caution: Do not attempt to circuit bend anything that needs to be plugged into a wall, such as a VCR or a television. These devices use high voltages, and playing with the circuitry inside might injure or kill you. Circuit bending is for battery-powered toys and instruments only.

When not sticking her hand in electronic equipment, **Cristiana Yambo** spins industrial beats on the DJ table and hides by day in the guise of a computer programmer. **Sabastian Boaz** doesn't just circuit bend as a hobby but also uses his creations in recording and performing synthpop and industrial music.

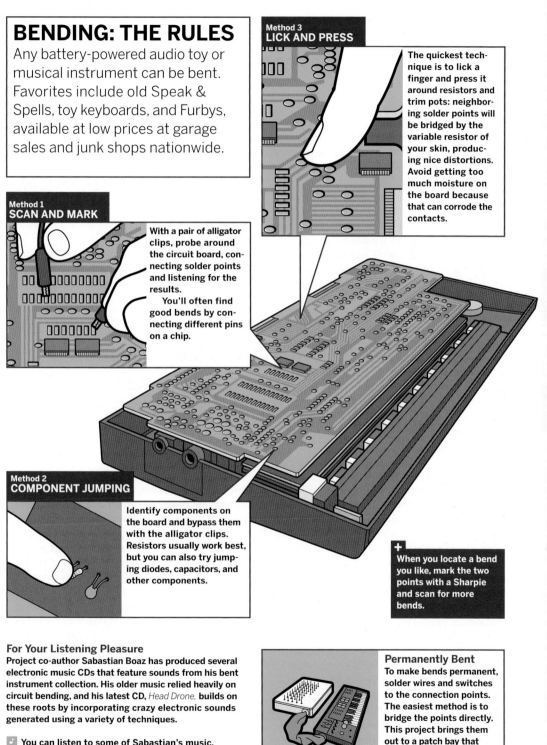

BENDING: THE RULES

Any battery-powered audio toy or musical instrument can be bent. Favorites include old Speak & Spells, toy keyboards, and Furbys, available at low prices at garage sales and junk shops nationwide.

Method 3
LICK AND PRESS

The quickest technique is to lick a finger and press it around resistors and trim pots: neighboring solder points will be bridged by the variable resistor of your skin, producing nice distortions. Avoid getting too much moisture on the board because that can corrode the contacts.

Method 1
SCAN AND MARK

With a pair of alligator clips, probe around the circuit board, connecting solder points and listening for the results.

You'll often find good bends by connecting different pins on a chip.

Method 2
COMPONENT JUMPING

Identify components on the board and bypass them with the alligator clips. Resistors usually work best, but you can also try jumping diodes, capacitors, and other components.

+
When you locate a bend you like, mark the two points with a Sharpie and scan for more bends.

For Your Listening Pleasure

Project co-author Sabastian Boaz has produced several electronic music CDs that feature sounds from his bent instrument collection. His older music relied heavily on circuit bending, and his latest CD, *Head Drone*, builds on these roots by incorporating crazy electronic sounds generated using a variety of techniques.

♪ You can listen to some of Sabastian's music, along with other bent instrument recordings, at makezine.com/04/circuitbending

Permanently Bent

To make bends permanent, solder wires and switches to the connection points. The easiest method is to bridge the points directly. This project brings them out to a patch bay that allows arbitrary connections, for greater flexibility.

Illustrations by Mark Watkinson

ALIEN LIFE FORMS
The father of circuit bending confronts the normals.

It's hard to know who invented electronic music synthesis, but you could say that Reed Ghazala uninvented it. He coined the term "circuit bending" and has been the technique's spiritual father and first major practitioner.

Ghazala stumbled on the idea in 1967. "I was 14 going on 15, in junior high and craving my own synth, but penniless," he recalls. "In my desk drawer was a junked mini amplifier made by RadioShack, battery installed and back off, exposing the circuitry. Closing the drawer after a search for who-knows-what, my room was suddenly filled with weird electronic music: strange 'flanged' oscillator sweeps rising in pitch, repeating over and over again." The amplifier had shorted out against something metal, and that led to a revelation: what if you shorted out circuitry intentionally?

Reed's first instrument took advantage of this idea, but audiences didn't immediately take to the avant-garde sounds. "Rowdy Elvis fans" at an early gig physically attacked Ghazala, hoping to "send me and the band to the emergency room," he says. "But their main target was my instrument. It symbolized a notion of music beyond verbal ballad, telling stories in a language they couldn't understand." The audience succeeded in destroying the instrument.

Despite the lack of early audience support, Ghazala was hooked on creating something new and alien, and his audience-obliterated instrument was followed by decades of prolific instrument building. "When I was first body-contacting bent instruments, I realized neither the circuits nor I stopped at our ends anymore," says Ghazala. "I was extended into the circuit and the circuit was extended into me. We shared the vital electricity that we each depended upon to function, to live. We were no longer two separate entities, like guitar and guitarist. We, literally, were one." He calls the combination a "Bio-Electronic Audiosapian," or BEAsape.

Circuit bending, Ghazala explains, means dropping your intentions. "When I hack a radio receiver, I may clip some diodes in it so that I can tune into a wider frequency range. But that's not circuit bending, because I know where things are going to go. With bending, you cannot presume what you're going to find — just like sending a probe into deep space." And outside presumption lies an alien world filled with alien sounds.

Ghazala's instruments certainly look alien, with flashing lights, eyeballs, laser controls, and hand-inked patterns that resemble psychedelic extra-

"Their main target was my instrument. It symbolized a notion of music beyond verbal ballad, telling stories in a language they couldn't understand."

terrestrial animal prints. Ghazala considers his creations more organism than art object, with body modifications that allow them to "sing things they couldn't sing before."

Perhaps Ghazala's instruments have grown eyeballs and skins because they're evolving. Nearly 40 years after his first experiments, circuit bending has become a worldwide phenomenon. "As the art of circuit bending claims more territory, spreading through the music underground like alien bacteria," Ghazala says, "it's infecting more and more circuitry every day."

—Peter Kirn

➕ Ghazala's tutorial on how to build an Incantor is available at makezine.com/04/circuitbending.

Peter Kirn is a composer, musician, and media artist; his blog (createdigitalmusic.com) regularly covers circuit bending and other far-out music technologies.

SET UP.

MATERIALS

[A] Casio SK-series keyboard We used an SK-5, but they're all great for bending. Check local thrift stores or eBay.

[B] Stranded wire 22-gauge or thereabouts

[C] Small machine screws and nuts At least 77 of each

[D] Flat, non-conducting, easily drillable box For the patch bay, we used a translucent plastic storage box. The patch bay box should be big enough to hold and permit access to 77 screws without crowding.

[E] Flat, non-conducting box Like above, but smaller; for external key box controller (optional)

[F] Assorted switches, contacts, and other bend components For the control panel. We used 4 toggle switches, 1 potentiometer, 1 photometer, 2 doorknobs (for body contacts), and a push-button momentary switch to reset the device.

[G] Flat, non-conducting, easily drillable panel, about 5"x5" For the control panel, which fits over the keyboard's speaker area.

[H] Hose clamps (2)

[I] Alligator clips At least a few; the more, the merrier.

[J] Ethernet cable For external controllers (optional)

[K] Ethernet jack For external controllers (optional)

[L] Flexible plastic tubing

Momentary switches We used keyboard keys from an old Macintosh. In many older, heavier computer keyboards, each key is its own self-contained momentary switch (not shown).

TOOLS

[M] Soldering equipment
[N] Glue gun and hot glue
[O] Multimeter
[P] Drill and drill bits
[Q] Rotary tool and bits
[R] Electrical tape
Multiple voltage transformer (optional, not shown)

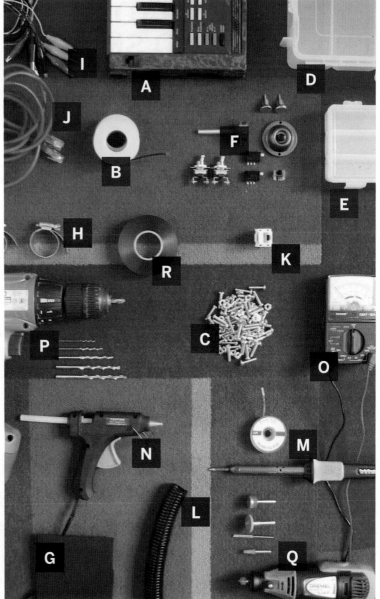

MAKE IT.

BEND YOUR CASIO SK KEYBOARD

START ⟫ **Time: A Weekend Complexity: Medium**

1. CONNECT THE PATCH WIRES

The chips on Casio SK keyboards contain vast potential for unruly sounds, and wiring up a patch bay unlocks all of it by making every possible pin-to-pin connection available.

1a. Find the main chips. Open the case, remove the motherboard, and identify the main sound processing chips; these are two large, rectangular chips next to each other in the middle of the board. Each has 14 pins per side, a total of 56 pins. Turn the board over and find the points of these pins on the underside. You will solder one wire to each of these.

1b. Solder and check wires. For each row of 14 main chip pins, solder 14 lengths of wire, each around 30 inches long, to the contact points underneath. Carefully check each pin for solder bridges. When you finish a row, replace the board and play the keyboard to see if it still works normally. If it makes any weird sounds, you probably bridged two pins together. You can also use your multimeter to test pairs of wires for continuity and resolder wherever you find one.

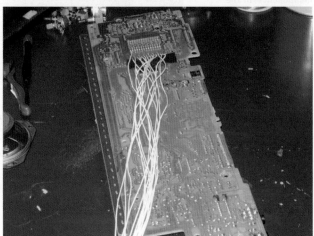

Photography by Cristiana Yambo

1c. Insulate. Once each row passes its test, cover its pins in electrical tape, and wrap the wires in electrical tape to create a nice cable. After you've soldered and checked all four rows, remove the tape covering the pins and test the unit again. When it passes the test, warm up your glue gun and cover each row in hot glue. This holds the wires in place and insulates the connections to prevent unintended short circuits.

2. BUILD AND CONNECT THE PATCH BAY

Some people construct patch bays using RCA jacks, but screws and alligator clips are cheaper, more compact, and they let you connect multiple clips to the same point, which vastly expands the sonic possibilities.

2a. Drill the holes. Measure, mark, and drill 60 holes in the patch bay box, four rows of 15, one for each of the 56 wires, plus four jumper points on the side to facilitate multiple connections. The holes should fit your screws snugly. Leave room for another, smaller set of holes for the control panel which you'll drill later.

2b. Insert the screws. Bolt screws into the holes, with the heads on the inside and the shafts pointing out.

2c. Make the patch cable. Count your control panel components, and then strip and cut enough wires to connect to each of these, plus two more for the tuning circuit; we used 17 total. You'll connect these wires later, so it helps to include some extra ones and label which ends connect to one another. Gather them together with four taped-together sets of patch wires, to make a super-cable that will run between the keyboard and patch bay. Secure each end with hose clamps, and push it through the plastic tube.

2d. Route the patch cable. Cut a large hole in the side of the patch bay box. Insert the patch cable and secure it by hot-gluing the hose clamp to the box.

2e. Connect the patch wires. Now it's time to wire the screws. Wrap the end of each patch wire clockwise underneath the head of each screw and tighten the bolt to secure the wire. Connect each group of 14 patch wires to a different row on the patch bay.

2f. Test the board. Using alligator clips, connect switches and pots between different patch points, play the keyboard, and see how it works. Operating the components and changing their configuration should yield different sounds.

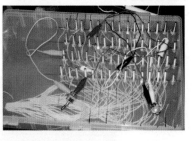

3. REMOVE THE TUNING POT

Many old electronic instruments easily lose their tuning and have a small potentiometer that adjusts them back into tune. We'll bring this component's connections out to our patch bay, so you can create bends that alter the pitch of the entire keyboard.

3a. Unsolder the pot. The SK's tuning pot is normally accessed through a hole in the underside of the keyboard, in the center, and you adjust it with a small screwdriver. Find the component on the board and use a solder sucker or desoldering braid to remove all of the solder.

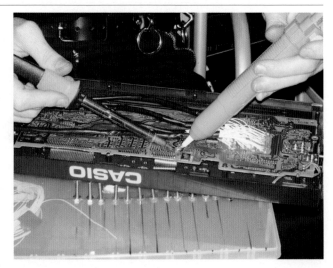

3b. Pull out the tuning pot. You will probably have to use some force. Then mark where it was located, to help you find it later.

4. BUILD THE CONTROL PANEL

The control panel houses interface components which you can hook into any circuit bends via the patch bay. It also has the reset button, a momentary switch that restarts the instrument after a crash.

4a. Cut out the speaker grill. Using a rotary tool, cut out the speaker grill. This hole is going to be used to house the interface elements.

4b. Build the control panel. Attach your components to the panel. For ours, we drilled holes in a piece of hard plastic and then hot-glued in our components: four toggle switches, a pair of doorknobs for body contacts, a potentiometer (separate from the tuning pot), a photometer for light-sensitive distortions, and a momentary switch, which will act as a reset button.

5. CONNECT THE CONTROL PANEL

5a. Finish the patch bay. Drill holes and install screws in the patch bay, as in Step 2. You'll need the appropriate number of contact points for each component on the control panel, plus two more to lead to the tuning pot's contacts on the circuit board. Arrange the holes in the same way that your components are arranged on the control panel. This makes it easy to remember which contact point on the bay corresponds to which control.

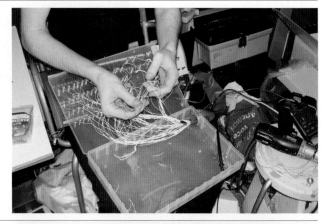

5b. Wire the control panel. Now you're going to use the extra wires in the cable. If they aren't labeled, use a multimeter to identify matching ends. Then connect the control panel components to the new patch bay array, following your logical mapping. Screw the wires to the patch bay points as before, and solder the other ends to the components.

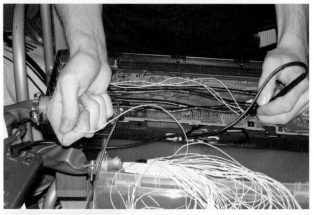

5c. Connect the tuning circuit. Screw two remaining extra wires to the tuning-circuit points on the patch bay and solder the other ends to the tuning pot contacts on the board.

6. CONNECT THE RESET, AND CLOSE IT UP

Bent instruments sometimes crash in a way that disables the power switch. One solution is to wire a momentary switch between the positive power line and the circuit board, but with the SK-5 it's easier to install a switch that connects the positive and negative power lines, momentarily shorting them out.

6a. Connect the reset button. Solder the reset button's contacts to the positive and negative terminals that lead from the battery compartment.

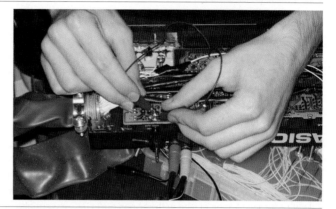

6b. Hot-glue the control panel to the case.

6c. Close up the case. Screw it together, switch it on, and then rock out with the ultimate SK!

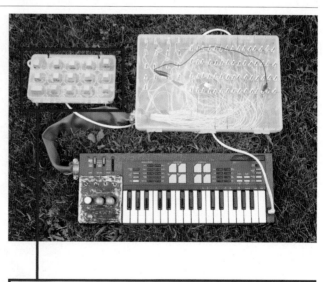

⊞ Extend the capabilities of your new keyboard by adding an external controller port and making two kinds of external controllers. Visit makezine.com/04/circuitbending for complete instructions.

FINISH ✕

NOW GO USE IT 》

USE IT.

JOIN THE CIRCUITS

MAKING TOYS PERFORMANCE-READY

One of the great features of this project is that the patch bay also allows you to connect other devices to the keyboard's circuitry. For example, we alligator-clipped the output of an iPod to our SK-5's tuning-pot contacts. This made the waveform of the music playing on the iPod modulate the pitch of the SK-5's output, an amazing effect.

Body contacts can be played with various parts of the body at different pressures for different effects. Touching a body contact with your tongue gives very low resistance, which can translate to an extremely high pitch — and electrocute your tongue a little bit.

SK-series keyboards, like the one used in this project, have wall warts and audio jacks. This means they won't run out of juice, and you can amplify them directly and record them by plugging them in. But many bend-worthy audio toys lack these two essentials. Here's how to remedy the situation.

Adding an Audio Out

With a toy that has only a speaker, you can make an audio out by connecting the two wires that feed the speaker to a standard mono audio jack or a cut audio cable. If the wires are colored, you can follow the convention of connecting red to the tip and black to the ring. But the order shouldn't matter; either way will work fine.

With some toys, output to the speaker might be high enough to cause distortion. To bring it down to regular analog audio signal levels, add a resistor in-line to the wire that connects to the tip. Experiment with different values to find one that works.

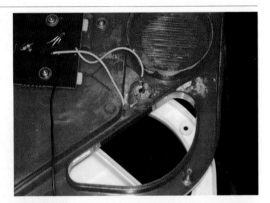

Adding a Power Source

Add the battery voltages and find an adapter with the same total voltage — you may already have one from some old appliance. Confirm its voltage with a multimeter (labels can be incorrect) and determine its polarity, which is usually but not always tip positive, ring negative. If it checks out, solder wires from the negative and positive ends of the toy's battery tray to the corresponding contacts of the DC in.

Toys sometimes list an amperage. You can usually ignore this, but if the circuitry is especially sensitive, test the power adapter's amperage along with the voltage. Again, trust the multimeter, not the label.

WELCOME TO THE WARPED WORLD OF CIRCUIT BENDING

If you've grown weary of computer software for look-alike, sound-alike musical instruments, you might have something to learn from Adrian Dimond. His latest project is making modified experimental sound creation devices out of Furby toys.

Dimond, aka Xdugef (xdugef.com), unwinds from his day job as a motion graphics artist by producing raunchy electrical noise. His tools of choice range from modified CD players and ancient reel-to-reel tapes to educational phonics toys and a flanger effects box. That might sound archaic, but Dimond is hardly alone. Fueled by online instructions for device modification, circuit bending has become a phenomenon. Nine Inch Nails uses bent instruments now, and you can hear experimental artist Oval's skipping CD players in a Calvin Klein fragrance ad. Bending's exposure may have gone mainstream, but the sound is far from conventional: bent creations are adored for their noisy, unpredictable nature.

Popular instruments for beginning benders include the plentiful Casio CZ-1 keyboard and the venerable Texas Instruments "Speak &" talking toy series (Speak & Spell, Speak & Read, Speak & Math). Whereas bending once involved countless hours of adjusting circuits with what-if pokes and prods, online treasure maps and tutorials make it easy for newcomers to the field. And once you've cut your teeth with one of the more well-documented devices, plenty of unexplored frontiers remain, particularly the more complex digital devices.

The gamut of bending projects is virtually limitless. Some works are art objects as much as instruments, sculptural creations with elaborate cases and what Dimond describes as "Duchampian collage." Others are more utilitarian, salvaging plastic cases or simply painting absurd illustrations on the bent object. Many are tongue-in-cheek, merciless modifications of toys.

Dr. Age of the U.K. band Cementimental has warped Nintendo games into video art, and built bizarre instruments out of a Ghostbuster voice modifier toy, a Pikachu keychain, and a plastic toilet, which he converted into an optical Theremin. Techdweeb (techdweeb.com) has taken on dancing daisy toys and the infamous talking fish Billy Bass. Carrion Sound (carrionsound.com) has done unspeakable things to Howdy the Talking Pony.

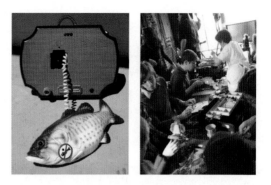

It's a little premature to talk about a circuit bending "scene" yet, particularly since it's hard to know where to draw the lines between "true" circuit bending, device modification, instrumental abuse, and more general noise art. But amidst the countless experimental performances around the world, there's clearly a growing interest in festivals and workshops focused on bending. New York City is a major epicenter, with numerous concerts and classes and the Bent Festival, held in 2004 and 2005 at The Tank (thetanknyc.com). Bending festivals are also in the works for California and Virginia, and Los Angeles' Il Corral (ilcorral.net) has established itself as a West Coast hub for noise and sound art.

In the meantime, benders share hundreds of websites featuring music, schematics, photo galleries of various creations, and (sometimes heated) message boards and mailing lists.

—*Peter Kirn*

RESOURCES

Circuit-Bending: Build Your Own Alien Instruments, by Reed Ghazala

Reed Ghazala's site:
anti-theory.com/bentsound

Other bending and noise art sites:
cementimental.com/links.html
harshnoise.com

Discussion groups:
groups.yahoo.com/group/benders
groups.yahoo.com/group/bendersanonymous
launch.groups.yahoo.com/group/Roil_Noise
p206.ezboard.com/biheartnoise

Composer, musician, and media artist Peter Kirn profiled Reed Ghazala on page 104.

Homebrew Game Design

Turning wacky ideas into fun board games.
By James Ernest

SO, YOU'D LIKE TO INVENT YOUR OWN board game? Great idea! Here's how to do it — or how I do it, anyway.

1. Determine Who Your Audience Is
Know your audience. If you're making a game to play with your family and friends, that's great! That's the best way to start. But if you aim to sell to a game publisher, it's never too early to think about how to position your game in the marketplace.

For example, when I started Cheapass Games in 1996, I knew I wanted to target "core hobby gamers" like my friends and me — college-age males who frequent game stores. Because gamers already have their own dice and game pieces, I knew we could have fun just with new sets of rules and cards or a board. This simplicity meant that games could be sold cheap, in plain envelopes, and displayed near the cash register of the game store to serve as impulse purchases. That's how Cheapass was born.

2. Outline the Story
My games sell on story, so the first thing I do is try to come up with a good one: zombie fast-food workers fight over a single brain (*Give Me the Brain*) or infected mad cows graze fields to discover unexploded landmines (*Unexploded Cow*). A game's theme is the first impression a new player will have of the game, so it's important to nail it.

Once you choose the theme, think about what game mechanics will fit it — things like turns, dice rolls, and auctions. Some designers base their games around the mechanics, but for me, "Do you want to play a game about pirates?" is more compelling than "Do you want to play a game with a unique movement point system?"

3. Imagine the Ideal Experience
What is the playing experience? What are players "doing," and how do they win? For example, in a space-mining game, do you want to spend more time flying spaceships or managing inventory?

How much luck, skill, or complexity do you want? "Skill factor" games have no luck at all, just pure intellectual competition, as with chess. At the opposite end, other games let players passively follow where the cards or dice take them, don't make you work too hard, and with a little luck, let anyone win.

4. Decide on the Format
The next choice is the overall format of the game. Do you want to design a card game? A board game? Something else? Here is a quick rundown of the most common formats. There are other formats, too, and you can always invent your own.

Board games: Games like Monopoly, Scrabble, or chess, that use a central board for static, common information, such as a map, or for tracking game progress, such as score, army size, and so on.

Card games: Cards handle randomizable, quantized, or secret information. Randomizable means that the cards can be shuffled; quantized means that the game content is broken down into pieces rather than being presented all at once; and secret means that players can hold information that isn't known by everyone. Card games tend to be simpler

Illustration by Melinda Beck

and quicker than board games, though not always.

Collectible card games like Magic: The Gathering and Pokemon can be complex because the card sets are huge. As randomizers, cards work well if you want hidden results, complex information, or control over how often a particular result shows up.

Dice games: Dice are another randomizer and are better than cards for outcomes that are openly vis-

> ❝ A good game is a road made of doors, with a clear goal and multiple ways to achieve it. ❞

ible and statistically independent from one another. Some games rely purely on dice (Yahtzee), while others use them within a larger game (Monopoly). Spinners and wheels can function in the same way.

Role-playing games (RPG): A paper-based RPG is an adventure-style game in which one player (the gamemaster) leads the others through a campaign setting that he controls. It's not a competition between the gamemaster and the other players; he just wants to give the players a fun experience.

Party games: Party or "parlor" games are simple games such as charades that can be played by a large group. Party games usually don't need much equipment, but mass-market party games (Pictionary, Balderdash, Time's Up) include cards and other components so that there's something in the box. With so few rules and minimal equipment, these may seem easy to design, but simple games are often the hardest to get right.

5. Sketch Out Some Rules

Write an outline of the game rules. Even if they all change, you need a place to start. Explain the number of players, the game components, setting up, and the order of play. Describe what you are doing, game length, and how you win.

Also, remember to make the game fair. This is one of the reasons we like games — they're a sanctuary, perhaps, from an unfair world.

I like to say that there are two kinds of novice game designers: those who make houses out of doors, and those who make roads out of walls. The "house made of doors" game has no clear goal but offers many ways to get there. The "road made of walls" has a clear objective, the end of the road, with obstacles along the way. But a good game combines these two — a road made of doors, with a clear goal and multiple ways to achieve it.

First-timers also get "first game syndrome," which is like making a salad with every ingredient in the kitchen. They include all the rules they can think of, and the game bogs down. So, after writing down all the rules you want to try, pare them down until there are just enough to make a playable game.

If you're stuck for game mechanics, consider your story and what it suggests. To help you along, here are three basic types of mechanics, based on the "road made of doors" metaphor. Robust games mix these together to produce complex choices.

Two doors (random): A player is offered two identical doors, side by side. One leads forward, the other leads back, and that player has no way of

BUILDING A PROTOTYPE

At some point, you will want to build a prototype. When you should start building it depends on the game. For example, real-time games with a speed component need art designed early and made very legible because readability is important. Here are a few pointers:

Artwork: There's a huge assortment of clip art online. For homemade games, or even prototypes for game companies, you don't need to buy your art. And for a game you want to sell, licensing clip art costs next to nothing.

Cards: These can be hand-cut from paper or made from index cards, but if they will be shuffled a lot, use sturdy, flexible paper and cut them to exact dimensions with a decent cutter. Rounded corners also help.

You can also make them out of existing playing cards, which are printed on laminated card stock — a plastic-coated paper that you unfortunately can't buy in an art store. Your local card room probably gives away old poker decks for free, and any game store probably also has a huge stack of common, low-value cards from the latest collectible card game on their way to the trash.

For cards with black and white artwork, I laser print on 110 lb. index, a smooth card stock that comes in lots of colors. For color cards, I print label sheets and stick them on old poker decks after cutting the labels down to fit. This gives you custom color artwork on a very sturdy deck. You can also print on normal paper, and insert it into plastic card-collector sleeves, available at game stores, along with a playing card as a backing sheet.

knowing or influencing which is which. Such luck-based mechanics can paradoxically make a game both less challenging and more entertaining.

Hurdle and mud pit (statistical): You have a choice in how to move ahead: you can try to jump the hurdle or walk through the mud pit. The hurdle requires a die roll — if you roll a six, you clear it and move on; otherwise, you make no progress. The mud pit, on the other hand, takes exactly four turns to cross.

This type of choice is interesting because the "right" answer varies with the situation. Statistically, both choices are about the same (you have about a 52% chance of rolling a six within four turns), so the optimal strategy changes based on the game state. For example, if everyone else is already in the mud and you absolutely have to be the first player across, you need to take your chances with the hurdle.

Taxicabs (political): Several taxicabs are available, but sharing a ride with another player is better than traveling alone. A game mechanic like this gets players to negotiate with each other to their mutual benefit, while still trying to win the game individually.

6. Play-Testing (Repeat)

Game design is an iterative process that needs testing with real people, who won't always act the way you expect! In the first tests, be prepared to make sweeping changes. If a game isn't working, abandon it mid-game, fix the problem, and start again.

In a game with secret information, it helps to start play-testing "open handed," where players show their secrets and talk about what they are doing, explaining their reasoning along the way.

When something doesn't work, resist the urge to add a new rule to fix it. Rules for special cases are hard to remember and tend to create new problems.

It's better to change the underlying cause. For inspiration, look back at your story and try to identify other aspects that you can bring out.

Here's an example: Suppose you're writing a dice-based horse-racing game, and you find that halfway through, the player in the front almost always wins. (In a random race, odds strongly favor the early leader.) You might be tempted to introduce a "headwind," the code name for a kludgy rule that arbitrarily slows down the front-runner.

But horse races aren't about the weather. The real problem is that random races are too predictable and have no strategy. What if instead of moving the horses, the dice gave them "energy" that players could spend early to sprint ahead or conserve for later? Using the dice in this way makes the game state more complex, and lets players come from behind, thanks to wise resource management instead of a patchy and artificial rule.

For play-test groups, it's good to mix casual gamers with experienced game mechanics people. Naive users are useful for judging overall fun, and the hardcore gamers are good at analyzing the rules and coming up with fixes. Sometimes an astute gamer will suggest debugging a rule in a way that no one else understands. In this case, try playing the suggested way and see if it works.

The last step in the play-testing process is blind-testing, in which you give the rule book and components to a group that's never seen the game before, and watch them play. If the group understands the rules and has fun, you've got a winner!

James Ernest is the president and lead designer for Cheapass Games (cheapass.com) in Seattle. His portfolio includes Kill Doctor Lucky, Give Me the Brain, and Unexploded Cow.

Cards should be easy to read. Don't handwrite them unless your handwriting is very good. If players will be holding a hand of these cards, "index" them by putting key information in the upper-left corners (held both ways) so that it's visible when the cards are fanned.

Boards: Mass-produced game boards are made of a printed label stuck to a sheet of cardboard. You can get the same effect by printing on full-sheet labels or using spray glue on matte board or other heavy cardboard. For larger boards, print your graphics as several sheets and tile them together, or find a large-format printer.

To waterproof your board (inkjet prints run if they get wet!), cover it with a sheet of clear contact paper, and then trim the contact paper, print paper, and backing all at once for a clean edge. Make the board foldable by cutting it in half and reattaching the back seam with packing tape or cloth tape. For a four-fold board, pre-cut the board along one dimension before you add the contact paper, which will form the forward-folding seam. Then finish the board, cut it in half the other way, and use backing tape for the backward-folding seams.

Other components: Game components can be salvaged from other games you own or from games at the local thrift shop. Things like dice, play money, counters, and chips are widely available, and game stores also sell other components. You may have to make some pieces by hand. I frequently use Lego to build prototypes, but that's just because I have a lot of it.

Pinball, Resurrected

Restoring a crusty, beat up *Cyclone*.
By Bill Bumgarner

THE FLIPPER'S SWEET SPOT STRIKES THE steel sphere, sending it through a gantlet of colorful obstacles. Electrical relays fire, bells ring, lights blink, sounds play, and the world vibrates with a dizzying barrage that only sharpens your focus. You parse the steel ball's trajectory, then adjust its course by nudging the 300-pound machine underneath; there's nothing "virtual" here. You are playing pinball, and it envelops you in a sense-filling physicality that video games can't begin to approach.

Even though the once-ubiquitous pinball machine is a rare sight in arcades and bars today, you can still fulfill your wildest pinball dreams by buying and restoring one yourself. The machines and the parts are out there, and many games in good or reparable condition are collecting dust, waiting for a loving home. Old pinball machines are tough, designed to survive with little maintenance in rowdy, smoky roadhouses. They're also built for repair; different models share many of the same mechanical and electrical subsystems, which you can easily swap around. I've enjoyed restoring several pinball machines, and my most recent adoption was a Williams *Cyclone*, a great solid-state pinball machine from 1988. Here's how I did it, along with more general advice on pinball restoration.

Find a Machine

Pinball machines can cost anywhere from $100 for an as-is electromechanical to nearly $10,000 for a particularly collectable game in good condition. Prices vary wildly, so unless you absolutely must have a particular, hard-to-find game, you should

be prepared to look at many before finding a good one to buy. For first-timers, even those with backgrounds in electronics or mechanics, I recommend spending the extra dollars to buy a machine that's at least in decent shape. There will still be plenty of things to tune.

You can buy pinball machines in better condition from restorers, who range from hobbyists to full-time professionals. The most pristine are "home use only" machines, which were never operated in public places. You can also pick up machines from arcade operators and "route operators" who place them in bars and restaurants. In my experience, you should be more careful with these. I have seen evidence of numerous dubious ad hoc "repairs" done to on-route machines in order to avoid the effort and expense of taking them off-location for proper maintenance. The worst offense is the use of spray lubricants such as WD-40. Pinball machines were designed as sealed boxes with mechanical systems that don't require oiling, so they could work maintenance-free in smoke-filled bars for months. Extraneous lubricant sucks up dust and smoke, and turns into a thick, damaging sludge.

To find machines for sale, I scanned craigslist.org, the rec.games.pinball forum, and *Mr. Pinball* classifieds (see makezine.com/08/pinball for more resources). I prefer late-80s and 90s Williams/Bally pinball machines, and my search led to Thursday Night Pinball, a group in San Jose that meets in a battery shop. The group's organizer had a Williams *Cyclone* for sale for $1,000. *Cyclone* is one of my favorites; it's a brilliant machine from

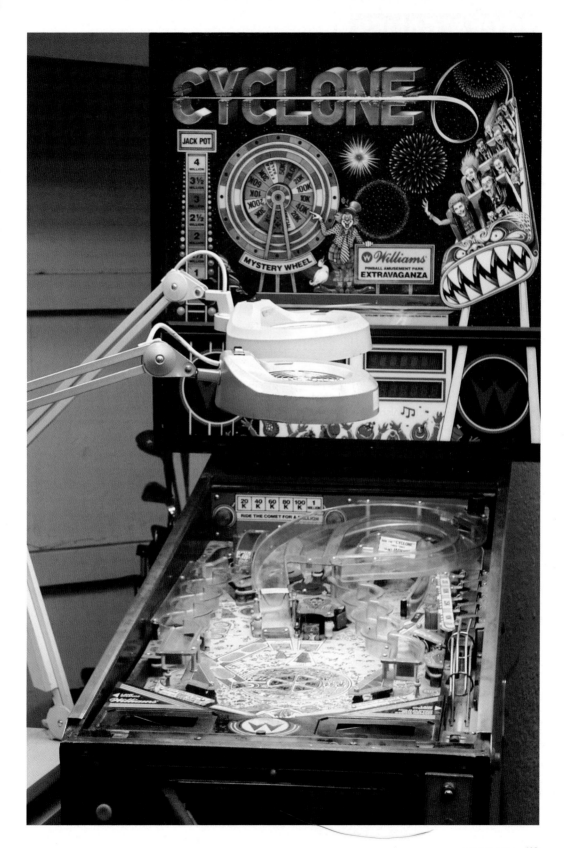

1988 with a fairgrounds roller-coaster theme, a kitschy midway soundtrack and voice-overs, ramps, a Ferris wheel, and a spinning wheel of chance on the backglass.

I played the machine and found no catastrophic problems. All parts were present, though extremely dirty, and the playfield had minimal paint cracking and no bare spots. But the machine would still need a lot of work. The Ferris wheel didn't spin, rubber bumper bits were broken or missing, one string of lights flickered intermittently, the flippers were weak, and the Player #3 scoring display didn't work. Pretty typical for a machine that had spent time on route.

I handed over the cash, and readied the machine for transport by folding the backbox down and unscrewing the legs. Like this, it measured 53" by 29" square, and just fit into my Subaru Forester.

Initial Play Test

After transferring the machine safely to my garage, the first step was to play it for a few days and note any problem areas — which is also a good way to meet the neighbors and their kids. When play-testing a neglected machine, you should watch for any relay coils that are energized all the time.

Pinball machines have dozens of these coils, which are what make playfield objects move. Any feature that's stuck in the wrong position — like a bumper always popped out, a flipper stuck up, or a target that won't drop — indicates an always-energized coil underneath, which will also typically buzz and be warm to the touch.

If you find one, it probably means either a short circuit or a blown transistor is leading to the coil. You should turn the machine off immediately and try to locate the problem using the schematic and a multimeter (see "Special Tools," page 125). Pinball machine coils are not designed to be on all the time; when they are, they overheat, strain the power supply, and can damage other components or scorch the playfield.

As you play-test, learn the rhythm of your machine. If you feel something change — like a flipper or bumper losing strength, or a ramp starting to wiggle — this is often the initial symptom of something that could grow to become a major problem later.

Get Documentation

Before going further, you need the manual and schematic for the game you're restoring. Fortunately, my *Cyclone* came with both. If it hadn't,

I would have looked for a printable copy online, at Marvin's Marvelous Mechanical Museum or the Internet Pinball Database, or else ordered a reproduction manual from Pinball Resource or Bay Area Amusements (see makezine.com/08/pinball). Marvin's site is a fantastic resource in general; it has saved me countless hours with its repair guides and other helpful info.

You will refer to the schematic often during restoration, but, unfortunately, it isn't always the gospel truth. A game's circuitry may be tweaked during its manufacture run, making it differ in small ways from the published schematic.

Run Diagnostics and Check Components

Check the manual or repair guide for how to run your machine's diagnostic routines. Every solid-state machine has these. On *Cyclone* and other solid-state Bally/Williams machines, a single dot appearing after the number of game credits indicates that the diagnostics software has found a problem. Switches inside the coin door let you enter diagnostic mode, and from there you can navigate menus to identify what's wrong.

Diagnostic mode also lets you test individual subsystems of the game, all the way down to individual lights and coils. With multimeter and schematic in hand, you can use the diagnostic mode to check everything out. Whether or not you are taking guidance from the diagnostic software, you'll want to check for proper voltages from the power supplies (there are several), check for bad bulbs, bad diodes on coils, or shorted/broken coils.

For the rectifiers, probe the output voltage of that rectifier's power supply section. With the multimeter in continuity testing mode, put one lead on ground (metal strap in the backbox is easiest) and touch the metal tab on each transistor. Any transistor that reads 0 ohms or causes the meter to buzz is bad, dead shorted, and will cause the corresponding coil to be energized all the time. Coils will typically have a diode across their two contacts and this diode will need to be unsoldered for full testing of the coil and diode, something I only do for coils that look particularly beat.

NOTE: Like any device that plugs into a wall, pinball machines can bite. Be careful when working inside the machine. Some electrical components may remain live while the machine is not in play, or even while it's turned off! Also, ground faults are common, so things that should not shock you

INSIDE THE CYCLONE

[A] Knocker that fires when you win a free game

[B] Sound board

[C] Fuse holder

[D] Mystery Wheel (spinning wheel of chance in the backglass)

[E] MPU/driver board. Contains CPU, software, switch matrix input, and driver output.

[F] Evil batteries that leak all over the main board. Moved to external battery holder in a Ziploc bag.

[G] Driver transistors that control the coils and flashlamps on the playfield. Flippers are not controlled by these in this model of machine.

[H] Auxiliary power board

[I] Wiring harness that connects the playfield to the MPU

[J] Bridge rectifier. One of many, but this one runs particularly hot.

[K] Big fat power supply capacitor

[L] Left speaker (midrange). Bass is in bottom cabinet.

[M] Display cable. Don't plug this one in backwards.

[N] Display board

[O] Right speaker (tweeter). Stereo effects are achieved by pitch, not true stereo.

sometimes will. Mechanically, pinball machines move a very heavy steel ball at high speed, so you can imagine what they will do to your finger.

Replace the Batteries

Like other solid-state pinball machines, *Cyclone* has batteries in the backglass that save settings and high scores when the machine is off. These batteries can leak acid and damage the CPU board. With the machine unplugged, our first task is to unscrew the backglass, open it up, and replace the batteries.

In my *Cyclone*, the battery holder was damaged from an acid leak, but the CPU board was still OK. I ripped the holder off the board and soldered in the wires of a new, off-board battery holder. I enclosed the new holder in a Ziploc bag and zip-tied it to a screw I drove into one interior side of the backbox. This arrangement will prevent any future acid-related damage, and I recommend it for any arcade machines that have batteries on their CPU boards.

Raise the Playfield

Remove the playfield glass by opening the coin door on the front of the machine and sliding the little lever inside that is just above and to the left. This unlatches the lock-down bar that spans the glass at the front of the machine. The glass will then slide out of the machine easily. Lean it out of the way, with the bottom edge resting on something soft like a towel.

It's important to remove the balls before flipping the playfield up all the way, lest they drop and break lamps and other components. To do this, lift the playfield a little, reach underneath, and flip the metal bit that kicks the balls into the launcher. With all the balls removed, you can swing the playfield up until it leans vertically against the backglass.

Initial Breakdown and Cleaning

I start cleaning at the bottom of the playfield, the flipper end, and work my way to the top. The bottom is simpler, and addressing it first gives me a feel for what kind of grime and other recurrent issues I'll be facing. With *Cyclone*, I quickly encountered a number of stuck screws, and it was better to practice handling these in a simple, open area.

With all the dirt, it soon became clear that I would need to remove every component from *Cyclone*'s playfield. I grabbed my digital camera and took over 100 pictures of the playfield and various assemblies in close-up, to help me reassemble everything later.

After documenting, I started by stripping the lower-left quarter of the playfield, the simplest quadrant. Using warm water and cotton rags, I gently took up a thick layer of dust and grime, only to reveal a more difficult cleaning problem underneath: the playfield was covered in thousands of tiny paint blobs, probable evidence of an old spray-painting mishap. No major discoloration, but definitely noticeable.

For the components themselves, I used warm water. I treated especially grungy plastic parts with Novus #1 or #2. If you don't completely replace the rubber rings, treat the grimy ones with rubber cleaner.

Playfield Detail

There are three kinds of pinball playfields: painted wood, Mylar, and Diamondplate. Each requires a slightly different cleaning technique. With a painted wood playfield, artwork is screened onto the wood with paint that's quite tough, but that steadily wears down under the abrasive powers of a rusty pinball or dragging flippers. To prevent this, some manufacturers began protecting the playfield with clear, adhesive-backed Mylar. And as of about 1990, Bally/Williams began covering the entire painted wood playfield with a thick coating of Varathane, branding it Diamondplate.

My *Cyclone* had plain painted wood, and happily the paint was still firmly bonded to the playfield. This let me remove most of the grime and spray paint with plain water. I used Wildcat Playfield Cleaner for my initial cleaning, followed by Mill Wax and Playfield Cleaner, a less harsh wax-based cleaner.

Note that Wildcat destroys plastics, so it should not be used to clean ramps or used near plastic parts. Mylar playfields become cloudy if you clean them with harsh cleaners like Wildcat. Diamondplate requires a good plastic polisher like Novus #1 or #2. All cleaners mentioned here are available from pinball parts suppliers.

New Lamps and Rings

While the playfield was stripped of parts, I replaced all the burned-out bulbs and changed all of the "general illumination" lamps (the ones that are on all the time) from #44 to #47. The latter use less current, which eases stress on the power supply. Cleaning the machine makes up for the slight decrease in light. Using brighter #44 lamps for the computer-controlled illumination makes special lighting effects stand out more.

After cleaning each section of the playfield surface and its resident components, I reassembled them, substituting new rubber rings from the replacement kit for the old ones.

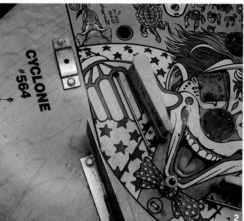

The art, playfield, and mechanics of *Cyclone*, mid-restoration. Every pinball machine is an antique model never to be made again. 1. Cameo appearance by Nancy and Ronald Reagan on backglass. 2. Left flipper exposed for cleaning. 3. Flipper mechanism with flipper shaft removed. 4. Newly cleaned playfield under removed left flipper.

Upper Playfield

Cleaning the upper playfield was just like with the lower playfield, but with more features to remove. *Cyclone* has a network of clear plastic ramps in its top half, many of which have integrated switches. Some of the switch leads disconnect under the playfield, which lets you remove sections of ramp with their switches attached. But they don't all work like that. Along with three separate ramp assemblies, I removed the ball launching assembly and several plastic-on-posts assemblies. Underneath

have stripped the playfield bare on both sides and coated it with an automotive clear coat. As it was, this wasn't worth doing, but I am on the lookout for an NOS (New Old Stock) bare playfield that I can clear-coat and swap in for that gleaming-perfection look.

Flipper Rebuild

Every time a flipper flips, a metal pawl is pulled through a relay coil and slams against a stop. After thousands of repetitions, the end of the pawl and the stop both change shape; and if the pawl isn't replaced, it eventually mushrooms and fragments around the edges. This can rip apart the coil, causing a short and possibly a power-supply failure. Also, *Cyclone* uses "Fliptronics I" flippers, in which the buttons directly switch on the 50V DC coil cur-

> ❝ A pinball machine is like a mid-70s British roadster: it may play just fine, but there is always something you can fix, clean, or tweak to make it work better. ❞

these toys I found a mess of rubber bits, lots of dirt, and even a few seed pods (no idea).

After cleaning and reassembling, I readjusted the ramps. With the playfield down, I took a ball, rolled it around the ramps, and fine-tuned their positions so that it wouldn't get stuck between the upper and lower ramp.

Next came the Ferris wheel. I found that it didn't spin because its drive belt was broken. Simple enough, but at $6, this special rubber belt turned out to be considerably more expensive than any other rubber pieces. Installation wasn't easy, either; I had to disassemble the wheel completely to put the belt around its hub, then pass the belt around a piece of metal that prevents the ball from falling out under the playfield, and finally attach it to the motor underneath.

Clean Underneath

The underside of the playfield is pretty straightforward to clean. On *Cyclone*, a ball guide connects the various under-the-playfield features together. It is made from the same plastic as the topside ramps, and it should be clear, though mine was close to opaque when I first opened the machine. A bit of Novus #2 cleared that up.

If my *Cyclone*'s paint were in better shape, I would

rent. This results in considerable arcing and wear on the switches, which need to be replaced as frequently as the mechanical components. It is possible to upgrade Flip I's to software-controlled Fliptronics II, which use low-voltage switches, but that's another project.

Rebuilding a flipper involves taking it apart and replacing all of the components that wear out with new ones, usually from a flipper rebuild kit. The tricky part is putting the flippers back together so that they're both at the same correct angle when in the "down" position and never grind against the playfield surface. My *Cyclone* machine showed some wear around the clown's mouth from improperly adjusted flippers, but it didn't go through the paint entirely, and I touched up the damage with some paint pens.

Finally, it was time to put the playfield back together, install the ball, and fire up the game. It's good to do this before replacing the glass because there's invariably something else that you need to tweak before you seal it all back up.

I have yet to figure out why *Cyclone*'s Player #3 display is dead, but the machine is otherwise in great condition: bright, responsive, and fully playable. Frankly, given that the machine works so well, fixing the display was bumped down the priority list.

Ongoing Maintenance

A pinball machine is like a mid-70s British roadster: it may play just fine, but there is always something you can fix, clean, or tweak to make it work better.

Maintaining a restored pinball machine is easy. Clean the playfield regularly with the proper cleaning compounds. Grime builds up in a pinball machine, and if a metal flake ever gets stuck in the grime, it can easily be driven into the playfield, creating a seed for more significant damage later. On a clean playfield, they just vibrate out of harm's way.

Conventions, Modifications, and Upgrades

There are pinball swap meets, conventions, and competitions held around the world. See the *Pinball News* site for upcoming events. There is also a huge pinball community online, including many of the original game designers. Pinball hackers have even produced an open source pinball emulator and control board, PinMame and PinMame-HW, that let a PC control a long list of pinballs — an easier way to restore if an original board is missing or trashed.

A lot of machines also have relatively easy hacks and upgrades. In some cases, the original machine design included extra switches and coils that offered more advanced play or special effects. These were sometimes removed during manufacturing to reduce costs, but the software was typically left untouched.

Custom versions of software also add features to some popular hobbyist machines. For example, a popular home version of *Twilight Zone*'s software adds a "pause" feature: if you catch the ball on a flipper and hit a certain button, the machine will go silent and hold the ball until you are ready to resume the game! With a bit of research, you can find what additional features your game might support.

Because pinball machines are so modular, many of their control circuits and mechanical systems can be broken out to do real-world tasks. You can also buy these at reasonable prices, and full schematics and documentation are online. Pinball machine subsystems can perform some powerful tasks in the real world, as I hope to describe in another MAKE article.

You can spice up a machine quite a bit even without making any functional or permanent modifications; for example, many an *Addams Family* owner puts a funny character in the game's electric chair.

Bill Bumgarner plays with high voltage, cooks with fire, hacks code, corrals bugs with his son, and tries to make stuff do things that were never intended. See friday.com/bbum.

SPECIAL TOOLS

A multi-bit magnetic screwdriver, preferably with a telescoping pick-up tool. This is your most important tool by far. Pinball machines have countless screws in hard-to-reach locations. Worse, if you drop one while the playfield is rotated up, it will invariably "pachinko" its way through the maze of wires and parts, only to fall in some tiny crevice.

A multimeter that can handle both low and high voltages. Modern pinball machines use 5V DC to handle the logic and sensors, 18-50V DC to drive lamps and coils, and upwards of 400 volts to drive the gas-plasma scoring displays.

A big box of slow- and quick-blow 5×20mm glass fuses, of many different amperages. Pinball machines run within tight power consumption parameters, so you should never substitute fuse types or deviate from the proper amperages.

A full rubber ring kit for the machine you are restoring (pictured above). Many rubber rings are buried under hardware, making them difficult to access. Replacing them during restoration will save frustration later.

Flipper rebuild kit for the machine you are restoring. As with the rubber ring kit, these are available from pinball parts retailers. See makezine.com/08/pinball.

GAME BOY HACKS

Games, music, movies, photos, and eBooks, all on this versatile little device.
By Phillip Torrone

The old-school Game Boy Advance is great for plane rides and DMV queues. But hauling a billion cartridges around wastes precious man-bag space — and I tend to lose them. I looked for a way to put all my games on one cartridge, and found a cool solution that also lets me play music, view photos and movies, and even run UNIX.

The key to it all is a piece of hardware called the Flash Linker & Card Set (*gameboy-advance.net/ flash_card/gba_X-ROM.htm*). It costs $89 but it's worth it. Compatible with the Game Boy Advance and GBA SP, it's based on a special cartridge that fits into your Game Boy, holds 512MB of flash memory, and connects to your PC via USB. The flash cartridge can carry up to 32 of your cartridge games, and lets you access thousands more as downloadable freeware. With free emulators, you can translate and play games from other systems, and the cartridge can also hold videos, music, eBooks, and photo albums.

Getting Started

Before you plug anything in, download and install the software that controls the communication between the Game Boy and the PC. You'll need the X-ROM driver and LittleWriter, both available at the website listed above.

Use the cable to connect your Game Boy to your PC. (The instructions say you shouldn't plug into a USB hub, but it worked for me.) Then, power up the Game Boy and hold down both Start and Select. The LittleWriter app on the PC should detect the Game Boy, and once it loads, you can back up your game cartridges to the PC, or transfer new games stored on the PC over to the flash cartridges. The games are all stored and accessed as GBA (a.k.a. ROM) files. It took me about an hour to back up all of my game cartridges.

Free Game Boy Games

A Google search of "Gameboy Advance ROMs"

[sic] will yield thousands of games. I like a lot of the freeware homebrew titles on *gameboy-advance-roms.tk* and *pdroms.de*.

To copy GBAs into the flash cartridge, drag them into LittleWriter's ROM window, and click Flash. Booting up with this cartridge will then offer a menu with all the games. Freeware games are typically smaller than licensed cartridge titles, so you can fit more of them into flash.

Game Emulators

You can play games written for other platforms by translating them from their original formats into compatible .GBA ROM files, using an emulator. I've tried free GBA emulators for games from Sega Master System (SMS), Nintendo Entertainment System (NES), Super Nintendo Entertainment System (SNES), the original Game Boy, and even the old Sinclair ZX computer from 1982. (I'm especially fond of Super Metroid, from the SNES.) You can find emulators for these and other machines linked from *gameboy-advance.net/gba_ roms/emulated_on_gba.htm*.

...And Other Delights

The emulator principle extends to other media besides games, and free converters have sprung up that let you convert a variety of files types into .GBA files, including video clips, music, slideshows, and text-file eBooks. The site above links to these as well. For music, you first need to convert the file(s) to .PCM format in your favorite sound editor. The sample rates can be 16,000, 11,025, or 8,000 Hz — 11,025 worked fine for me. Stereo doubles the file size, so use mono for spoken word.

Meanwhile, there's even a UNIX project for the Game Boy Advance. I didn't have the proper files to compile the ROM to get this to run, but here's a guide: *kernelthread.com/publications/gbaunix*.

Phillip Torrone is associate editor of MAKE.

Within this joystick lies the heart and soul of a favorite old Commodore (no, not Lionel Ritchie).

HACKING THE C64 DTV

Retro-gaming joystick easily converts into full Commodore computer emulator.
By Mark R. Brown

Retro gaming is the latest craze, and Mammoth Toys is letting people rediscover 20-year-old videogames by making a joystick that mimics 30 classic Commodore 64 games, including Paradroid, Uridium, Impossible Mission, and World Championship Karate. It plugs right into your television — no console required. QVC sells the C64 DTV (Direct-to-Television) joystick for $30. At a buck a game, it's a real bargain.

The ASIC chip at the heart of the C64 DTV simulates the original C64 while doubling the RAM and offering 16 times the color palette. (That's, um, 128K and 256 colors.) But there's far more to the DTV, thanks to a development team that clearly was not satisfied with just software bonuses. They loaded the DTV with undocumented hooks and features, and, not surprisingly, the toy has a cult following of Commodore-happy joystick modders.

Without even cracking the case you can boot the joystick up in C64 mode, run "secret" games, find numerous Easter eggs, and summon an onscreen keyboard that lets you write and run BASIC programs from the C64 command line.

And if you pull the DTV apart and look inside, you'll see that the circuit board is littered with helpful labels and unused, hack-inviting solder pads. On the C64 DTV hacking forums that have sprung up online, people exchange notes on all sorts of modifications, from enhancing the audio to kludging up a flash memory module. In short, the C64 DTV is a hacker's dream: a Commodore 64 for the 21st century. Is it any wonder that the C64 community is excited about this thing?

Photography by Mark R. Brown

Hacking the Hardware

The DTV hacking community has come up with some pretty fancy modifications for the joystick platform. We're going to turn ours into a full-fledged Commodore C64 clone by adding a power supply, a disk drive, and a keyboard (not that we don't enjoy entering text with the joystick). If you can handle a screwdriver and a soldering iron, doing this is easy, and, after cloning the C64, you can jump off into your own custom hacks.

Parts List

A "wall wart" power transformer that puts out 5-volt DC or less, at 800 milliamps or better. You've probably got at least one of these in your junk pile. But before you go soldering, test it with a multimeter to make sure it's not putting out more than 5V. I've found the DTV to be pretty robust, but you don't want to fry its power regulator.

A PS/2 keyboard. You can use an older one with a large connector, or a new one with a small connector. Just stay away from really old XT-style keyboards. They won't work.

A Commodore 1541, 1571, or 1581 disk drive with at least one compatible floppy disk. If you don't have a drive in the attic, you can pick one up on eBay. Check that it comes with a cable. If not, you can look for a super-cheap, used Commodore drive cable (see links at end of article), or buy a new one for about $10.

TOOLS:	
Medium Phillips and small flat-head screwdrivers	Hobby knife
	Multimeter
	Solid hookup wire (24-gauge)
Soldering iron (less than 30 watts, with a fine tip) and solder	Shrink-wrap tubing or electrical tape
	Frosty beverage

Build the Clone

Start by taking out the DTV's batteries. Then unscrew four Phillips screws on the underside of the joystick's base, and separate the base from the top with a small flat-blade screwdriver, being careful not to pull any wires that connect the two.

Cut the plug off your power supply, plug it in, and use a multimeter to identify the positive and ground/negative wires. Then unplug it. Inside the DTV, use the knife to scrape the silicone off the battery contact terminals on the underside of the top half. Solder your transformer's positive wire to the terminal that connects to the red wire, and the ground wire to the terminal with the black wire. Plug in the supply again, and make sure the DTV's power LED lights up when you turn it on.

Power supply done; now, the keyboard. Cut your keyboard's connector off, strip some casing off the plug side, and use your multimeter to associate which color wires run to each of the different pin positions in the keyboard's plug. There are two sizes of PS/2 plugs, with different pinouts. Write down the correspondence between the colored wires in the cable and their functions, as listed below for your connector.

Big XT/AT Connector (plug side)

Pin	Function
1	Clock
2	Data
3	Not connected
4	Ground (power -)
5	+5VDC (power +)

Small PS/2 Connector (plug side)

Pin	Function
1	Data
2	Not connected
3	Ground (power -)
4	+5VDC (power +)
5	Clock
6	Not connected

Now it's time to remove the DTV's two circuit boards. Four Phillips screws hold the main board, and two more secure the smaller joystick button board. Remove them all, and lift out the board assembly, being careful not to let the seven red plastic joystick buttons fall onto the floor and roll under your workbench.

Flip the board assembly over and find the four rubber pads that serve as joystick contacts. Pop off the rubber pads located at 12 and 3 o'clock when the joystick board is facing up.

Find the five solder pads marked by the three arrows in the photo to the right; they're clearly marked KEYBOARDDATA, KEYBOARDCLOCK, IECCLK, IECDATA, and IECTAN. These are where you'll be connecting the keyboard and disk drive.

Prepare four 3" to 4" pieces of hookup wire. Then strip some casing from the keyboard cable,

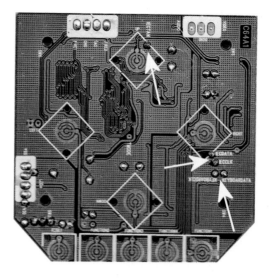

spread the wires, and refer to your color-function correspondence to find the Data, Clock, +5VDC, and Ground wires. Fire up the soldering iron and use the hookup wire to connect the first two to the KEYBOARDDATA and KEYBOARDCLOCK pads, respectively, and the +5VDC and Ground wires to the terminals that you already soldered your power supply to.

You can insert and solder the wires from either side of the board, but it will be easier to reassemble things if you insert them from the component side. Use heat-shrink tubing or electrical tape to wrap the joints. (Wiring the keyboard directly to the power terminals like this, bypassing the switch, means that it draws power whenever the DTV is plugged in — and that's OK.)

Plug in the power supply, turn on the DTV, and hold down the keyboard's K key as it powers up. The DTV should boot directly into BASIC mode. If you get the normal game menu, check your connections and try again. Some people report having to reboot several times before the DTV recognizes the attached keyboard. Then power down and unplug before continuing.

Now let's hook up the disk drive. As with the key-

board, you'll splice the cable and probe the wires to identify their different functions by their plug pin positions, using the chart below. We'll only use four: Ground, Attention, Clock, and Data.

Disk Drive Serial Connector (plug side)

Pin	Function
1	Service Request (SRQ)
2	Ground
3	Attention (ATN)
4	Clock
5	Data
6	Reset

You can probably guess what's next. Solder three more lengths of hook-up wire to the pads marked IECCLK, IECDATA, and IECTAN (which is a typo for IECATN), then connect the other ends to your drive cable's Clock, Data, and Attention wires. Finally, solder the Ground wire to the power supply ground terminal.

Now you just need to reassemble everything. Feed the tips of the connector pad plugs through the holes on the board, then pull them gently through from the other side. Replace all the red buttons, put the board in place, and secure it back with the screws. Drill, carve, or use a Dremel tool to cut a hole in the DTV's base that's big enough to feed the cables through.

Plug the other end of the drive cable into your Commodore drive and turn it on. Fire everything up, holding down the keyboard K to boot into C64 mode. Insert your floppy and enter LOAD "$",8. You should hear the drive spin, and the DTV should load the directory from the disk. Enter LIST to see the disk's contents.

Cable-tie the internal wires together, to provide enough strain relief that you don't accidentally pull them out of the board. Carefully route the wires out the hole, snap the base back on, then reinsert and tighten the four case screws.

You are now the proud owner of a brand-new, improved, 20-year-old computer. Ain't it cool?

Now What?

If you just want to download and run C64 programs, pick up an XE1541 cable, which hooks a Commodore drive to your PC so you can transfer downloaded disk images and programs to C64 floppies. Find out more at *sta.c64.org/xe1541.html*.

If you're an assembly language programmer, check out the free Turbo Macro Pro+DTV assembler at *style64.org*, which gives you access to the DTV's expanded color palette and extra memory. You might also want to hack an S-video output to replace the sometimes-funky DTV composite video and separate the mono audio output into three discrete synthesizer voices. Or you might hack in a second joystick. For these and more ideas, check out the sites listed below.

C64 DTV Links
C64 DTV Yahoo! group
games.groups.yahoo.com/group/DTVTalk

DTV hacking site and forum
galaxy22.dyndns.org/dtv

Project 64 — C64 doc archive
project64.c64.org

Funet C64 software archives (US mirror)
ibiblio.org/pub/micro/commodore

Mark R. Brown was Managing Editor of <i>.info</i>, the legendary Commodore computer magazine.

Secret Software Tricks

When you turn on the C64 DTV, you briefly see the famous Commodore "Chicken Head" logo and a credits screen. Then the C64's standard blue prompt screen appears, and the lines LOAD "*",1 and RUN are entered, as if by a ghostly typist. The screen clears after this, and you're presented with the regular game menu.

Interrupting the blue screen during startup will fork you over to lots of hidden extras. Hold down the left joystick button during startup, and you'll get to the regular games menu faster. Quickly wiggle the joystick left and right during startup, and you'll get to the C64 Mode menu, which offers six bonus games and "Basic Prompt."

Select Basic Prompt (the left joystick button selects), and you're in BASIC mode, a clone of the C64 blue screen and prompt, with the command-line interface. From here, you can summon a joystick-controlled onscreen keyboard by holding down the left joystick button. To type, navigate around the keyboard, then release the left joystick to enter the character you're on. Repeat as necessary. The SH (Shift) keys act like Shift Lock, remaining on (or off) until toggled back again. (The Commodore 64 keyboard is quite different from a PC keyboard. If you're unfamiliar with how it works, visit the online manual at *lemon64.com/manual*, and refer to section 2.1, Communicating with your 64: The Keyboard.)

From BASIC mode, you can list the programs in ROM by typing:
```
LOAD "$"
LIST
```
To run the "Entropy/Electron" demo, a favorite interactive art piece from the early "Demo Scene" era, go to BASIC mode and type:
```
0 POKE1,55:LOAD"ENTROPY",1
RUN
```
Then go through the screens by holding down the A button while pushing the joystick up for a few seconds each.

The C64 DTV also has numerous Easter eggs. Here are two, and others are documented at *xrl.us/fmzrn*. To see Commodore legend Jim Butterfield drinking beer with a friend, type:
```
LOAD"1337",1,1
RUN
```
Finally, speaking of elite, you can bring up a picture of some members of the C64 DTV development team from BASIC mode by typing:
```
LOAD"DTVTEAM",1,1
RUN
```
The full team includes Jeri Ellsworth (manager), Jason Compton, Adrian Gonzalez, Robin Harbron, Per Olofsson, and Mark Seelye.

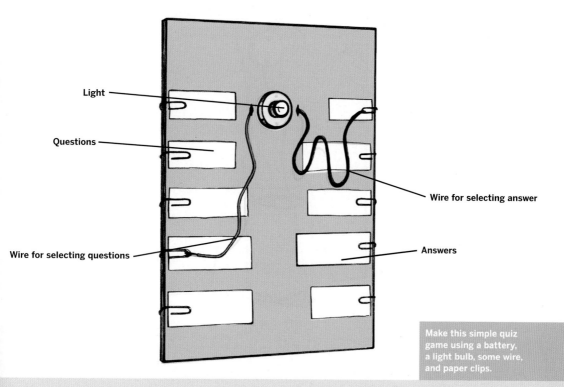

Light

Questions

Wire for selecting answer

Wire for selecting questions

Answers

Make this simple quiz game using a battery, a light bulb, some wire, and paper clips.

CIRCUIT QUIZ GAME

Teach kids about circuitry in 30 minutes.
By Ben Wheeler

Illustrations by Damien Scogin

Arthur C. Clarke's Third Law states that "any sufficiently advanced technology is indistinguishable from magic." Clarke's law could use a corollary: sometimes it doesn't take much for something to be advanced enough to seem mystical. Today's kids are used to every teddy bear and softball being Turing-complete, so hidden complexity doesn't impress them. To really dazzle them, you need older technologies, like circuitry, which feel magical by their very openness and simplicity.

Here's a classroom-tested example: a simple project that introduces kids to circuitry. It functions as a quiz game, where a light bulb signals that the correct answer has been selected. The circuit itself is very basic, but if you provide enough sections of quiz material, it's surprising

and satisfying to see it work. You can quickly rearrange the questions and swap them in and out, which makes for fast competition, impromptu topic-creation, and ample opportunities for ridiculing siblings.

Questions and answers are attached to a board by paper clips, with matching question-answer pairs connected in back. Two wires that dangle on the front of the board connect to the light bulb and the battery. When a player touches one of the wires to a question and the other to its correct answer, the circuit completes across the hidden wire behind the board, and the bulb lights.

Circuitry projects like this are also a great way to introduce kids to circuit diagrams, and show how they map to physical reality. Here's a simple circuit diagram for this project:

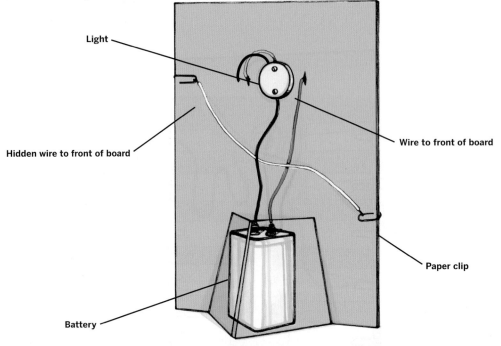

Light

Hidden wire to front of board

Wire to front of board

Paper clip

Battery

Behind the curtain: Compare the illustration on the previous page with the one shown here. The two wires on the front of the board (emanating from holes on either side of the light bulb) form a complete circuit and light the bulb only when they touch the paper clips that are connected by the wire behind the board. (For purposes of clarity, the paper clips for the wrong answers are not shown in this view.)

All of the circuit components in this project are available at RadioShack.

MATERIALS

6V light bulb (RadioShack #272-1128)

Light bulb base with screw terminals (RadioShack #272-357)

6V lantern battery (RadioShack #23-016)

Wire spool: 20-gauge, insulated (RadioShack #278-1225)

Two large pieces of cardboard

Paper clips

Duct tape

Paper and pens or printer, for labels

Total cost: $15

TOOLS

Hobby knife

Small screwdriver

Wire cutters/stripper

Make the Board

Cut two identical cardboard trapezoids, about 6" on the top and 9" on the bottom, with sloping sides. Tape them together along one side and to the bottom of the quiz board to form a stand.

Install the Light

Cut a small circle in the center of the quiz board for the light bulb base. Cut two small slits on either side of the circle. Attach two wires to the light base. Send one wire through a slit to the front of the quiz board. Screw in a bulb and fit the base into the circular hole.

Install the Battery

Put the battery inside the quiz stand. Attach the second wire from the bulb to a battery terminal. Connect a long wire from the other terminal through the other slit to the front of the board.

Wire the Answers

Make each question-answer wire by wrapping both ends of a wire around paper clips. Make sure the connections are snug. Attach the clips on alternate sides of the back of the quiz board.

Label the Quiz

Make paper labels for the quiz questions and answers and place them next to the corresponding paper clips. Place dummy paper clips next to all the non-matching questions and answers, so they all look identical from the front.

Ben Wheeler is a math teacher living in Brooklyn, N.Y.

Young kids have more
fun playing CD-ROMs
using a touchscreen.

STICKY FINGERS

Modifying tech for pre-mouse toddlers.
By Damien Stolarz

As anyone raising a toddler knows, baby minds are hungry sponges for information. As a parent, I try never to pass up a learning opportunity. I'm a big technology enthusiast, and so naturally I want my kids to know everything about the hardware and software that runs our world.

Many parents are (understandably) overprotective of their devices at the expense of their children's education. They worry that kids will put sandwiches in the VCR or even shock themselves. What they fail to realize is that only through working with objects and technology — by making mistakes and learning from them — will their children ever gain the necessary knowledge and competence. And in my own case, as a maker and tinkerer who has broken more than my fair share of gadgets in the process of learning how they tick, it would be

downright hypocritical of me to stop my children from breaking all my digital devices for the noble cause of their education.

Toddler Kiosk

As our daughter grew older, she wanted to imitate what her parents did. She liked to sit in front of our PCs, bang on the keyboard, and inform us that she was "working."

We bought her a brightly colored PS2 keyboard that has a picture next to every letter (apple for A, snake for S, dog for D, and so on). For a while, we would just pull out her keyboard and let her type on it when she wanted to work along with us.

After a while, we knew she needed her own computer — heck, she was already 2. So I dusted off an old G3 Macintosh and bought a handful of games designed for toddlers.

However, I saw how much she struggled with the so-called "intuitive" mouse user interface. She would move the mouse to the screen with no results; she would look at the mouse and move it, then look at the screen, lifting the mouse. With Mom and Dad's help, she got very interested in several of the interactive games, but we always had to be there and work the mouse for her.

I was familiar with a series of VGA touchscreen monitors, from a company called Newision.com, that were used for in-car computers. These 7" and 8" monitors are designed to stick to the dashboard of a car, but I found that they worked just as well stuck to the top of this G3 Macintosh. Since they provided a Macintosh touchscreen driver, I simply hooked up the display, and my daughter was immediately able to play the games all by herself.

I rigged up a launcher program that had big buttons, so she could even launch the games herself. And since one of the games asked her to type in her name (and I had the foresight as a parent to give her a name with only three letters), she quickly learned how to hunt and peck her name in and press Return.

Mobile Entertainment

One of my hobbies/professions is car hacking. Instead of purchasing the $1,800 factory DVD option with our new Dodge Caravan, I installed a Mac mini ($599) and a fold-down 15" touchscreen monitor ($199), so that my daughter could watch *Dora the Explorer* in hi-def resolution.

The essential feature of in-car entertainment systems these days is "multi-zone" entertainment, meaning the kids in the back have headphones so the driver and front passenger can listen to their own audio program.

I initially tried giving her headphones, but you will find that babies take a while to warm up to hats, headphones, sunglasses, or anything else you try to wrap around their heads. I wanted to find a way to create an isolated sound environment for her so we could listen to our own music.

Most toddler car seats consist of a washable, foam-filled padded cover wrapped around the plastic frame. Inspired by a noise-canceling headphones project (MAKE, Volume 05, page 56), I took the connecting loop out of a pair of inexpensive headphones, duct-taped them to the left and right

sides of the plastic car seat, and then put the foam back on. For $5, I had upgraded my daughter's car seat with a sophisticated "surround sound for babies."

I used a mini-jack extension cord to connect this to an in-car DVD player. Now, I can have my own Mac mini-based music in the front seat, and my daughter can watch her DVDs and hear them in the back seat. Because of the perfect positioning of the speakers on either side of her head, she has her own aural experience, apparently unaffected by the music in the front seat.

Toddler Shuffle

A toddler's rabid affinity for media makes her able to listen to the same song repeatedly, apparently indefinitely. On one of our daughter's CDs, she has found a two-minute song that she likes to hear over and over.

Unfortunately, the CD player in our car does not have a "repeat song" function, and despite the integrated steering wheel controls in the minivan, it is still unnerving to have to press "back" every two minutes during long drives.

I thought about getting my daughter an old iPod, but the interface would be too complicated. Fortunately, Apple came out with the iPod shuffle, which has no breakable screen, is small, and more importantly, has a very simple three-button, one-switch interface.

We had also by now acclimated her to headphones — not earbuds, but a small pair of over-the-top earphones.

We showed her that to get her music, she needed to move the switch on the back of the iPod (to turn it on) and then press the "play" triangle. And when her favorite song comes on, she can simply press the "back" arrow to repeat it.

She's still getting the hang of the iPod shuffle — she'll overshoot, rewinding too far — but we don't have to intervene nearly as much now. And when she needs help, we can usually talk her through how to get back to her song again ("Press the left arrow two times. Is that your song? Good."), and she can do it all by herself.

Damien Stolarz is an inventor with a decade of experience making different kinds of computers talk to each other. His book *Car PC Hacks* is published by O'Reilly Media, Inc.

Custom Travel-Game Mod
By Harry Miller

Make a travel edition of your favorite, and otherwise housebound, board game.

You will need: Magnetic metal boards (available at office supply stores), clear Contact paper, duct tape, adhesive magnetic paper or magnets, plastic clips (to hold cards or paper money together)

Can you take your favorite board game with you on a trip? If not, follow my instructions to make your own travel edition. I made a travel version of Rivers, Roads and Rails (by Ravensburger). The original version has to be played on the floor or on a table. You can take mine in a plane, a car, or a boat. When choosing your game, keep in mind that the fewer the pieces the better.

1. Make a magnetic board.
Rivers, Roads and Rails is composed of 140 2"×2" cardboard tiles, on which are drawn a river, a road, a railroad, or a combination of any of the three. The pieces for a travel edition need to be able to move without disrupting the game. Also, the playing space needs to be more confined. I found that 2 magnetic metal boards (about 9"×14") held together by a piece of duct tape were a small-enough space to play on if I shrunk the tiles down by half.

2. Shrink game pieces and playing field.
I used my printer/copier to shrink the tiles down by half to 1"×1" squares. If your game has an illustrated playing field, you'll have to do the same thing for the game board, and then stick it to the magnetic boards.

3. Make the game pieces magnetic.
Take a sheet of flexible adhesive magnet paper and stick the tiles onto them. Then cover them with clear Contact paper and cut them apart using a cutting board. (My mom helped a bit.) I sent a letter to Ravensburger with my design plans, hoping they will produce it. (Maybe I can even get some credit.)

For extra credit
+ A zip-lock plastic bag might work just fine as the travel bag, and some plastic potato-chip clips can keep your game cards together. On the other hand, a hand-sewn bag of vintage fabric and an old silver cigarette case may be a more elegant card holder.

You're done! Play the game to find any flaws. If there are problems, go back and make necessary adjustments until it's perfect!

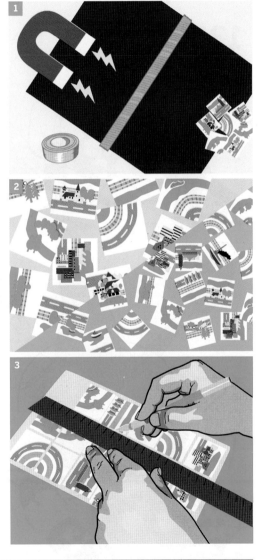

Illustrations by Dustin Hostetler

Harry Miller is a 12-year-old home-schooler who loves to invent, build, and destroy things. On sunny days he can be found terrorizing plastic soldiers with a magnifying glass.

<blockquote>
» **"Fast-reactor design: BEAMbots use simple circuits that interact with the world directly. Unlike control-freak robots, their brainless reflexive reactivity is the whole point."** Gareth Branwyn, *Two BEAMbots: Trimet and Solarroller*
</blockquote>

138

150

164

171

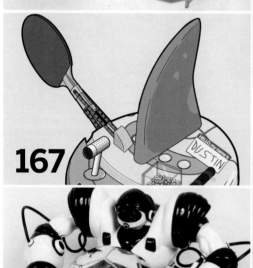

167

168

THE BEST OF ROBOTS PROJECTS

from the pages of MAKE

>>

A HUMAN IS HOW A ROBOT BUILDS A BETTER ROBOT

We think of robots as being staggeringly complex machines, but they don't have to be. Any machine that senses its environment in some way, organizes a reaction, and then executes it, based on the sensor input, can be considered a robot. In this section, we include a number of simple robots built with very common analog components and recycled techno-junk, an approach known as BEAM (Biology Electronics Aesthetics and Mechanics). The Robosapien and Roomba, two of the most successful robots to date, rely on a similarly simple, bare-bones approach. We'll show that you can upgrade these commercial robots to add some cool new capabilities.

TWO BEAMbots: TRIMET AND SOLARROLLER

By Gareth Branwyn

Solder together one simple circuit and use it to control two very different solar-powered robo-critters: a little satellite that scoots and bumps around, and a mini cart that just keeps a-rolling until the sun goes down. »

Set up: p.142 Make it: p.143 Use it: p.149

Photography by Douglas Adesko

GO SOLARENGINE!

The low-tech, analog, dumpster-diving, and hack-friendly world of BEAM robotics has produced a bestiary of bot types, including Symets, Rollers, Walkers, Jumpers, Climbers, Swimmers, Flyers, and Crawlers. Many of these creatures can be powered and controlled by a Solarengine, a simple and popular BEAM circuit that draws energy from a solar cell and temporarily stores and dispenses it using one or more capacitors.

We'll make a couple of voltage-triggered Solarengine circuits, and then build them into two little bots: a Trimet, which looks like a satellite in orbit as it's moved around by a spinning, top-like base, and a Solarroller, which drives straight ahead in fits and starts. These light-sensitive critters will look cool and très geeky on your desk, as long as you can keep them from wandering off the edge (they're both active diurnally, and they don't have an off switch).

Gareth Branwyn wrote "A Beginner's Guide to BEAM" (MAKE, Volume 06, page 54).

FAST REACTOR DESIGN

BEAMbots use simple circuits that interact with the world directly. Unlike control-freak robots, their brainless, reflexive reactivity is the whole point.

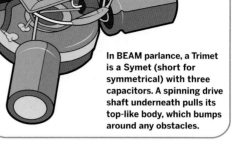

In BEAM parlance, a Trimet is a Symet (short for symmetrical) with three capacitors. A spinning drive shaft underneath pulls its top-like body, which bumps around any obstacles.

Solarrollers are little solar-powered race cars. At BEAM and other robot competitions, builders pit Solarrollers against each other in a kind of robo-mechanical Pinewood Derby.

Our Trimet and Solarroller bots are based on a voltage-triggered (Type 1) Solarengine. These circuits collect energy from a small solar cell, and periodically release it when there's enough stored up to actually do something, like run a motor.

The solar cell converts light into electrical energy, slowly juicing up the capacitor (or multiple capacitors).

The capacitor collects and stores a voltage, which discharges whenever the circuit is completed between its two terminals.

The 1381 voltage trigger measures the voltage across the capacitor, and sends a trigger signal once it's high enough (2.4 volts with a 1381-G trigger).

When the base pin of the 3904 transistor receives the trigger signal, it completes a connection that allows the capacitor's power to discharge through the motor.

The motor runs intermittently, whenever it receives a power dump from the capacitor.

During discharge, current flows to the base of the 3906 transistor. This takes the 1381 trigger offline, allowing it to reset, and routes current to the 3904 base, which keeps the motor circuit flowing until the cap is fully discharged.

The 2.2kΩ resistor reduces the voltage to the 3906 base pin, so it diverts less power away from the motor during discharge. This makes the circuit more efficient.
➕ Solarengine schematic: makezine.com/06/beambots

Illustration by Timmy Kucynda

SET UP.

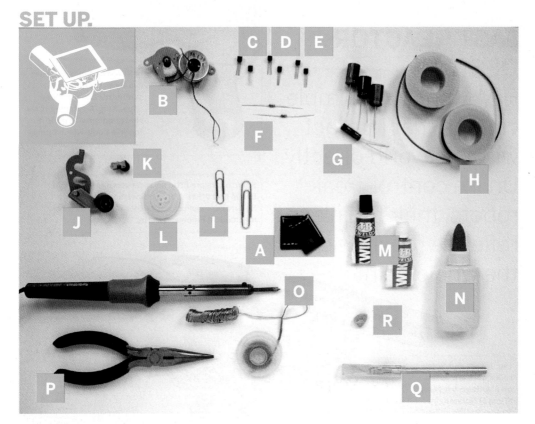

MATERIALS

The following parts will build two Solarengines. Just get one of each if you're only building the Trimet or the Solarroller, but not both. All part numbers refer to Solarbotics (solarbotics.com):

[A] 37×33mm polycrystalline solar cell, part #SCC3733 (2)

[B] Cassette motor (2) From an old Walkman or other player, part #MCM2

[C] 1381-G voltage trigger IC, part #1381-G (2)

[D] 2N3904 ("3904") NPN transistor, part #TR3904 (2)

[E] 2N3906 ("3906") PNP transistor, part #TR3906 (2)

[F] 2.2kΩ resistor (2)

[G] 4700µF capacitors (4) Or use three 4700µF capacitors (for the Trimet), and 1 "supercap" such as a 0.33F Gold Capacitor, part #CP.33F (for a higher-performance Solarroller)

[H] Hook-up wire, red and black 24-gauge stranded

[I] Paper clips (2) one small, one large

PARTS YOU'LL ONLY NEED FOR THE SOLARROLLER:

[J] Pinch roller and arm from a VCR, or similar Smooth rubber roller, about ⅝" in diameter and ⅝" wide

[K] Pinch roller and arm from a cassette player, or similar Smooth rubber roller, about ½" in diameter and ¼" wide

[L] Drive wheel of any lightweight material, with a diameter slightly greater than the motor casing Between 1½" and 1⅝" is good. An old VCR might have a suitable pulley, or try the disc that holds the control rods in a servomotor. You can also use a wheel from a toy, or any other right-sized plastic disc.

Rubber band

[M] Epoxy

[N] White glue

TOOLS

[O] Soldering equipment Iron, stand, solder, and solder-sucker, desoldering bulb, or braid

Dremel tool with grinding wheel, cut-off wheel, and router bits

"Third hand" tool with alligator clips Two are ideal

[P] Needlenose or long-nose pliers

Wire cutters

[Q] Hobby knife

Medium-grade sandpaper or metal file

Ruler

[R] Poster putty or tape

Safety glasses

BUILD YOUR BEAMBOTS

1. BUILD THE SOLARENGINE CONTROL CIRCUITS

We'll be freeforming these circuits, which means connecting components together directly, without a board. Normally I would breadboard and test my circuits before soldering, but this one is so simple and has so few parts that we can live dangerously. Parts are easily desoldered and resoldered if there's a problem.

1a. Face the two transistors up with their pins toward each other. Solder the base pin (middle) of the 3904 transistor to the collector pin of the 3906 (the right pin, as you read the printing).

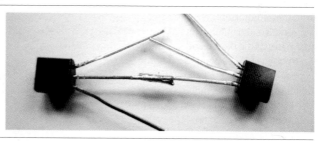

1b. Use needlenose pliers to gently bend the 3904 emitter pin (left) 90 degrees to the side and its collector (right) 90 degrees up. Bend the 3906 base pin (middle) 90 degrees up and its emitter (left) 90 degrees to the side. Solder the 2.2kΩ resistor from the 3904 collector to the 3906 base.

1c. Trim excess lead length from previous step. Place the 1381 voltage trigger to the right of the 3906, facing the same way. Solder its Pin 3 (right) to the 3904 emitter and its Pin 1 (left) to the 3906 collector. Finally, arc its Pin 2 (middle) around and solder it to the 3906 emitter (left). There's your basic circuit, ready for motor and power!

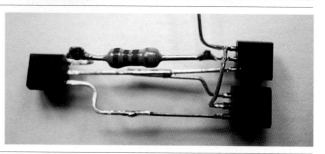

1d. If you're making both BEAMbots, build a second Solarengine circuit by repeating steps 1a–1c above. From here, you can continue on to step 2 to build a Trimet, or jump ahead to step 3 and build a Solarroller.

2. MAKE A TRIMET

2a. Prepare the motor by removing any mounting tabs with a Dremel grinding wheel. Then use sandpaper or a metal file to scuff the drive-shaft side of the case until you're down to the shiny metal underneath. Really scuff it up good; you'll be soldering capacitors directly to the case, and they'll need to hold as the Trimet drags and bumps around.

2b. Clip the negative/cathode leads on the three 4700µF capacitors so there's just enough wire to solder them to the motor casing. Bend the positive/anode leads up, making sure they comfortably clear the casing. Find 3 equidistant points at the perimeter of the motor, and solder the 3 cathodes to these points so that the capacitors form an equilateral triangle radiating out from the motor's center. Use generous gobs of solder, and use poster putty or tape to hold the caps in place while you solder.

2c. Center the circuit assembly over the motor, and solder a scrap lead from the 3904 emitter to the motor casing. This grounds the circuit, while also attaching it to the motor. For optimal balance, bend this connecting wire at 90 degrees, and try to position the circuit in the middle of the motor.

2d. The motor case is our circuit's ground (-); now let's work on the power (+) side. Take a small paper clip and bend it into a ring with the same diameter as the motor. (Conveniently, Walkman motors are the size of a quarter, so you can use one as a form to bend the clip around.) When you have a decent circle, solder it together.

2e. Bend and trim the capacitor anode leads evenly, so that they extend just above the control circuit. Solder the "power ring" to the ends of the 3 leads, preserving the equilateral symmetry.

2f. If you can bend the 3906 emitter lead to reach the paper-clip ring, do so, and solder it on. Otherwise, connect it with a short piece of wire or scrap component lead.

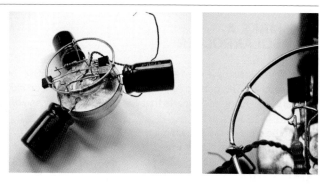

2g. Now, the solar cell. If yours has pre-tinned pads but no wires (most small cells come this way), start by soldering the 2 wires onto it — but be careful, because solar cells are fragile. Then solder the positive/red wire to the ring and the negative/black wire to the motor casing. Make the wires long enough so you can still work on the circuit, but short enough so they'll stow neatly underneath when you finally glue the solar cell down onto the ring.

2h. Connect the motor. Solder the negative/black motor wire to the point where the 2.2kΩ resistor meets the 3904 collector. Solder the positive/red motor wire to the paper-clip ring.

Now, place the solar cell on top of the Symet and shine a light on it, or put it in the sun. After 10 seconds or so, it should fire and scoot along, or spin around if you're holding it by the driveshaft underneath. If so, congratulations — you're the proud parent of a BEAMbot! You can go ahead and glue the solar cell onto the paper-clip ring. Or, if the cell stays in place without glue, leave it that way so that people can peek under the hood.

3. MAKE A SOLARROLLER

Solarroller builders have used all sorts of materials, from LEGO® bricks to soldered paper clips to computer mouse cases. This popular approach relies on parts from an old cassette player and VCR. Your mileage may vary, depending on the parts that you use for the body and drivetrain.

3a. Cut the arms on the 2 pinch rollers with a Dremel and cut-off wheel, so that they make full, flat contact against the motor casing. The Solarroller will stand on the triangular base that's formed by these 2 idler wheels and the larger drive wheel that will go onto the motor's drive shaft.

3b. Prepare your drive wheel. First, check that it will fit on the motor's drive shaft. (The hole in the hub of the disc I used was too small, so I reamed it out using a Dremel router bit.) Then glue a rubber band around the outside of the wheel, to improve traction. Cut the band, smear a thin layer of glue onto one side, and when it gets tacky, carefully roll the wheel over this "tire" until it comes full circle. Let the join overlap, then use a hobby knife to cut away excess rubber and make sure the ends are perfectly joined.

3c. Epoxy the 2 idler wheel arms into position on the motor casing, then fit the drive wheel onto the motor shaft without gluing it (use poster putty to hold it on, if needed). It is critical that all 3 wheels run parallel to each other and make full contact with flat ground when the Solarroller is standing. If you're using the Solarbotics motor, you can affix the larger roller arm to the motor's large mounting tab, pointing toward what will be the front, and leave the two other mounting tabs and holes pointing up on top, for attaching the circuit and solar panel.

3d. Cut about 4" of wire from a large paper clip and fashion it into a U shape. For the Solarbotics motor, it can be just wide enough to run between the two upper mounting holes. Trim the remaining piece of paper-clip wire and solder it across the U as a cross-brace, about ¾" from the open end.

3e. Epoxy a capacitor directly to the motor casing, running horizontally, on the side opposite the drive wheel. The leads should point backward, with the cathode (-) closer to the motor.

3f. Solder (or epoxy) the paper-clip frame atop the motor casing, using the two mounting holes if present. Since we didn't glue the drive wheel on yet, you can remove it to access the top of the motor. For extra sturdiness, you can position the frame so the cross-brace rests on the capacitor, and epoxy the brace onto the cap. Glue on the drive wheel.

3g. Position the Solarengine circuit underneath the paper-clip frame, next to the cap, on the side opposite the motor. Solder the 3906 emitter (left) pin to the positive/anode lead of the capacitor. The connection should be short enough so that the cap holds that end of the circuit up in the air.

3h. Turn the Solarengine upside down and solder a scrap component lead to the 3904 emitter pin at the point where it attaches to the 1381 trigger's Pin 3. Bend the capacitor's negative/cathode lead around the undercarriage side of the cap's barrel, and solder it to the lead you just connected to the 3904. This will anchor the other end of the circuit.

TRANSISTOR

Collector

Base

Emitter

Bipolar transistors can act as switches, connecting parts of a circuit just like a mechanical switch would. In an NPN transistor, applying a voltage with the positive side to the base and the negative side to the emitter allows current to flow from emitter to collector. A PNP transistor goes the opposite way; running a negative voltage across from the base to the emitter allows current to flow from emitter to collector.

3i. If your solar cell doesn't have wires, attach some to the pads marked (+) and (-). The wires only need to be long enough to reach the pins on the capacitor. Thread the solar cell's wires through the frame and epoxy the cell to the top. When the epoxy is set, solder the solar cell's positive to the cap's positive/anode and the cell's negative to the cap's negative/cathode.

3j. Finally, connect the motor. Solder the positive/red motor wire onto the 3906 emitter (left) pin and the negative/black wire to the 3904 collector.

Now, put the Solarroller on a flat surface in the sun, or shine a flashlight on the cell. After a little while, the circuit will trigger, the capacitor will dump, and your Solarroller will take off for a short run. Shine, wait, and repeat.

FINISH X

NOW GO USE IT »

BEAM ME UP, SCOTTY

TROUBLESHOOTING

If your BEAMbot doesn't make you beam, carefully examine all connections, resolder anything that looks weak, and separate any components that might be touching (shorting). It's a simple circuit, so not much can go wrong besides incorrect connections or bad joins.

FURTHER HACKING IDEAS

On the Trimet, add an outer paper-clip ring. This creates a bumper that will help prevent the robot from getting stuck.

On the Solarroller, replace the regular 4700µF capacitor with a "supercap" like a 0.33F Gold Capacitor, as shown in the project photos. These capacitors can take several minutes to juice up, but they'll make your Solarroller take off like a bat outta hell.

You can easily convert an old Sony Walkman into a great Solarroller. Leave the motor, roller wheels, and pulleys in the original frame's base piece, and use it as the vehicle's chassis.

Try Andrew Miller's more efficient variant of the basic Solarengine, which is almost as easy to build. You need a different resistor, an additional capacitor, and a diode, but you can lose the 3906 transistor. Varying the value of the small cap, between 0.47µF and 47µF, lets you "program" different discharge times. (See schematic at: makezine.com/06/beambots.)

Once you have the basic ideas down, you can go crazy, improvising BEAMbots with greater storage capacity, better obstacle-avoidance strategies, or swankier, more attention-getting designs. Here are some Symet and Solarroller variations (pictured at right).

RESOURCES

There are many more hacks and variations on these two project types, as well as other applications for the Solarengine. For more information, see "Getting Started in BEAM" (MAKE, Volume 06, page 57).

⊞ Schematic for Miller variant of Solarengine circuit: makezine.com/06/beambots.

MOUSEY THE JUNKBOT

By Gareth Branwyn

With a few spare parts, you can turn an old computer mouse into an amusing little robot. >>

Set up: p.153 Make it: p.154 Use it: p.163

THE FINE ART OF MAKING "FRANKENMICE"

This project turns an analog computer mouse into a robot that'll delight friends and wow workmates down on the cube farm. Mousey's behavior is fittingly mouse-like. It scoots quickly across the floor, thanks to lively little motors. And when the critter crashes into anything, it speeds off in the opposite direction.

The robot's "brains" are an ingenious hack based on an audio operational amplifier (op-amp), an 8-pin chip that's normally used to drive answering machine speakers and other lo-fi equipment. Following Randy Sargent's pioneering design, Mousey repurposes this chip to boost light-sensor input to motor-powerable levels. The result is simple, fast-reacting analog circuitry that fits inside a mouse case.

Gareth Branwyn writes about the intersection of technology and culture for *Wired* and other publications, and is a member of MAKE's Advisory Board. He is also "Cyborg-in-Chief" of *Streettech.com*.

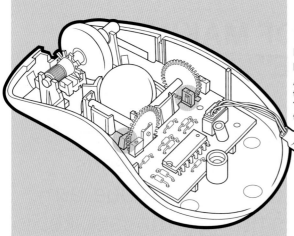

INSIDE MOUSEY

How a mild-mannered computer mouse becomes a fast, freewheeling photon-hog.

Analog (non-optical) mice pick up movements of the ball with two axles that turn gear-like wheels. The teeth rotate between IR emitters and receptors that capture the flickering shadows to read horizontal and vertical directions and speeds. Reverse-biasing the diode emitters turns them into Mousey's "eyes."

Mousey's bumper (from one of its buttons) empties a capacitor-full of current across a relay, temporarily crossing the motors' voltages and throwing Mousey into reverse.

Randy Sargent's Herbie was the first LM386-based bot. It finished last in the 1996 Robothon's line-following race, but went on to spawn many descendents, like this one by Grant McKee.

The eyes' light difference is amplified and tapped into the circuit between the two motors, wired in series. As one motor draws less power, the other uses more, steering the bot.

Illustration by Timmy Kucynda

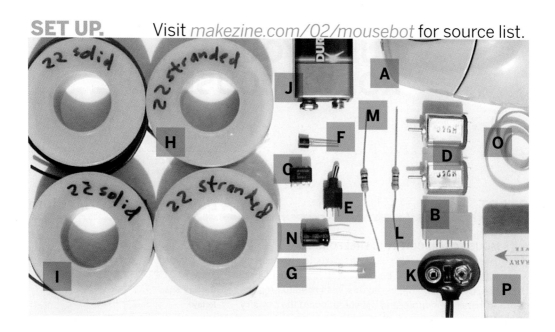

First, you'll need an analog (non-optical) mouse to cannibalize for its case and several parts inside. If you don't have an old mouse or two gathering dust, ask friends and colleagues. Otherwise, you can buy a new, super-cheap model such as the Kensington ValueMouse, which costs $10 and has enough space to fit all of your components inside. The bigger and more symmetrical the mouse, the easier the build will be. "Handed" mice with asymmetrical,

curved bodies present problems.

The other components can be scavenged, or purchased from an electronics retailer. For the motors and other specialty parts, we recommend Dave Hrynkiw's Solarbotics (solarbotics.com) as an excellent source. Where available, we've listed Solarbotics parts numbers for components, and they now offer a complete mousey kit for about $20 (without the mouse).

MATERIALS:

[A] Mouse case

2 Light sensors
From mouse

SPST touch switch
From mouse

[B] Double-pole, double-throw (DPDT) 5-volt relay
From analog modem, or Solarbotics #RE1

[C] LM386 audio operational amplifier (op-amp)
From answering machine, speakerphone, intercom, etc., or Solarbotics #LM386

[D] 2 Small 4.5 VDC motors From motorized toys, or Solarbotics #RM1A / Mabuchi FF-030-PN

[E] SPST toggle switch
Solarbotics #SWT2

[F] 2N3904 or PN2222 NPN-type transistor
Solarbotics #TR3904/ TR2222

[G] Light-emitting diode (LED)

[H] 2 Spools of 22 to 24-gauge stranded hook-up wire Ideally, 1 black and 1 red

[I] 4 Pieces of 22-gauge, solid-core hook-up wire Ideally, 2 red and 2 black, 6½" long

[J] 9V battery

[K] 9V battery snap

[L] 1kΩ to 20kΩ resistor

[M] 1kΩ resistor

[N] 10μF to 100μF electrolytic capacitor

[O] Rubber band or other tire-making material

[P] Small piece of plastic At least ¼" × 2½" of hard, springy, thin plastic, like .030" Plasticard stock, or an old credit card

Piece of Velcro or two-way tape (optional)

TOOLS:

Phillips screwdriver
For disassembling mouse

Dremel tool
With bits and cutting discs

Needlenose pliers

Digital multimeter (DMM)

X-ACTO/hobby knife

Soldering iron

Solder sucker or desoldering bulb

Wire cutters/wire snips

Breadboard, hook-up wire

Superglue, epoxy, or other contact cement

Poster putty, electrical tape, cellophane tape

Ruler

Protective goggles, mask

BUILD YOUR ROBOT MOUSE

START ›·›   **Time: A Day Complexity: Medium**

1. MOUSEY'S CIRCUITRY IS FREEFORMED

This means that we'll solder the parts to each other without a circuit board, building everything up right inside the mouse case. But before we do this, we'll need to prep the case and install the motors, and then breadboard the circuitry separately to make sure everything works.

Before unholstering your Dremel tool, you'll need to determine if the mouse has enough space inside. Unscrew the mouse case and eyeball it to make sure that it will hold the two DC motors and a 9-volt battery. Screws may be hiding under little nylon feet or tape strips on the bottom of the mouse. Save these bits so you can put them back at the end of the build; they'll help reduce friction.

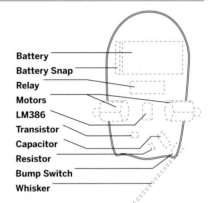

Battery
Battery Snap
Relay
Motors
LM386
Transistor
Capacitor
Resistor
Bump Switch
Whisker

Mouse case and parts that need to fit inside.

Cleared-out mouse case.

2. PERFORM AN ALIEN MOUSE AUTOPSY

Once you have a suitable candidate, remove all of the mechanics and electronics. Unhook the mouse cable from its plug-type connector, pop out the scroll wheel (if it has one), and then pry out the PCB (printed circuit board). Set these parts aside. Then use your Dremel and cut-off wheel to hollow out the case, removing all of the plastic mounts and partitions inside, except for any screw post(s) that hold the case together. Do the same for the top half, although you may want to leave the mounts that hold the buttons in place.

Note: Plastic dust is nasty stuff, so work on newspaper and wear goggles and a mask.

3. ADD THE POWER SWITCH

The last piece of preparatory bodywork is adding the power switch, a large toggle placed rear topside so it looks like a tail. Find an appropriate mouse-tail location, then drill a hole in the case big enough for the switch. If the switch has a threaded bushing and two nuts, take one nut off, insert the bushing up through the hole, and then tighten the nut back down onto the outside of the case. In some cases, a plastic screw post interferes with the tail area. If so, you can cut out the post and reconnect the top and bottom halves with tape or glue.

The top of our mouse case with its toggle "tail" installed.

Illustrations by Mark Frauenfelder

4. MOTOR AND BATTERY PLACEMENT

Now we're ready to figure out the arrangement of the bigger components and cut openings for the motors. Mouse shapes vary, so you'll use some judgment here, but the two motors should be oriented perpendicular to the centerline of the body, so the bot travels in a straight line. Also be sure to leave enough space behind the motors for the battery.

Once you've placed the motors and battery, you're ready to cut openings for the axles and wheels, which are simply the drive shafts and gears of the motors.

You'll want to angle the shafts coming out of the mouse body so they support the bot and set a proper speed. The steeper the angle, the less rubber will meet the road, which slows the bot down — but this is good, since many builders have complained that Mousey moves too fast. If you're using the lively Solarbotics RM1 motors, 60 degrees is about right, as shown.

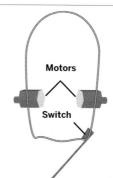

Motors

Switch

Eventual whisker

Motor placement, angle, and switch placement are very important for making Mousey work properly.

Use poster putty to hold the motors in place temporarily. Then get down at eye level and make sure the gear "wheels" are making good, level contact with the table. Once the motors are positioned properly, glue them in place.

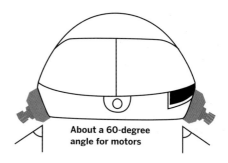

About a 60-degree angle for motors

5. MAKE THE BUMP SWITCH AND TIRES

Your mousebot will have a giant "whisker" — a bump switch (courtesy of one of the mouse's button switches) that triggers Mousey's scuttle-away behavior. Look on the mouse PCB (see photo in Step 7) for a tiny plastic box that clicks when

Finished motor and bump switch installation. Shown with battery test fit.

you press it down; then desolder it. Once you have the switch removed, attach the base with putty to one side of your mouse's front end. Tape the strip of hard plastic in place, so that it covers the tiny switch button and runs along the front of the mouse like a wide bumper. The idea is to have the switch triggered by a bump anywhere along the length of the "whisker," so when you press in the plastic, you should hear an itty-bitty click. Tweak this arrangement until it looks good. Once you have your placement, drill a small opening in the mouse case bottom for the switch to stick out. Also cut the plastic strip down to size, about ¼" x 2½".

The last mechanical modification needed for the bottom half is adding tires. Find a rubber band with the same width as the sprockets on the drive shafts, and then cut it to length, wrap it around, and glue it on. You can make the wheels thicker by continuing to wrap the band around itself. Rubber or plastic tubing also makes good tires, as does corrugated tubing from a Lego Mindstorms robot kit or the rubber cylinders from Dremel drum sander bits.

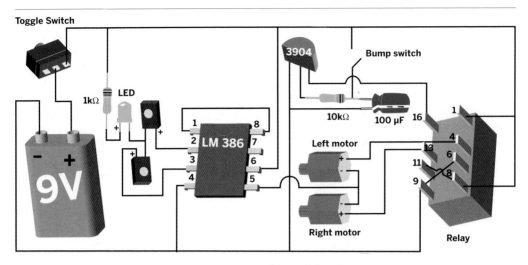

Toggle Switch · **3904** · **Bump switch** · **LED** · **1kΩ** · **9V** · **LM 386** · 1 8 2 7 3 6 4 5 · **10kΩ** · **100 μF** · 16 · **Left motor** · + - · **Right motor** · + - · 1 4 6 8 9 11 13 · **Relay**

6. UNDERSTAND MOUSEY'S BRAIN

The LM386 op-amp, the main component of Mousey's control circuit, "listens" to two input signals. If one signal is lower than the other, the chip boosts that signal to equalize the one output. In our case, the inputs are light values rather than audio. If we hook this output to two DC motors, we have a little brain that reads input from two light sensors, compares them, and boosts the power on the dimmer side. This creates a robot that follows a light source, auto-correcting itself as it moves.

The bump switch triggers a relay that reverses the two motors' inputs for a few seconds. This makes Mousey scuttle away from light after any collision, adding to its lifelike behavior. The diagram above shows the circuit diagram for Mousey's brain.

Use this diagram as a reference as you build your mousebot.

A larger version of this image can be found at http://xrl.us/fkxi.

For up-to-date information on Mousey, including new hacks, visit: streettech.com/robotbook

BEAM ROBOTICS: SURVIVAL OF THE FUNNEST

Mousey comes out of the BEAM design tradition, a biology-inspired doctrine which frowns on microprocessors in favor of simple analog control, in order to create robots that act and react with the physical world directly, perhaps instinctively.

BEAM's natural selection process occurs at conventions and gatherings like Robothon, where bots compete against one another in races, "sumo" matches, high jumps, rope climbs, and other Olympics-style events.

Through BEAM's 16 years of evolution, BEAMers worldwide have designed and refined numerous species of inexpensive and easy-to-build robo-critters, including photovores such as Mousey, four- and six-legged walkers, sun-powered solarollers, and swimming aquavores.

Mousey's circuitry is based on Randy Sargent's line-follower bot Herbie, which competed in the Seattle Robothon in 1996. Many variations of the design followed,

including Dave Hrynkiw's Herbie Photovore. Following Dave's example, we built ours with as much techno-junk as possible, including an old computer mouse and a 5-volt double-pole, double-throw (DPDT) relay – a component found inside most analog modems.

BEAM Resources
The acronym BEAM stands for "Biology, Electronics, Aesthetics, and Mechanics" and was coined by Mark Tilden.

Solarbotics: The main BEAM portal Solarbotics.net

Yahoo! Groups: BEAM Robotics
groups.yahoo.com/group/beam

Annual Robothon Event
robothon.org

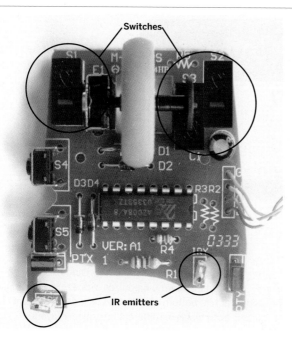

Switches

IR emitters

A pair of IR emitters will serve as your robot's eyes. Note their likely location on the mouse's PCB.

7. CREATE MOUSEY'S EYES

For Mousey's eyes, we can use the mouse's own two IR emitters, a.k.a. phototransistors. During normal computer mousing, these shine infrared through the mouse's perforated encoder wheels, which is then received by photodetectors on the other side.

Like many fundamental devices, these emitters can work as both transmitters and receivers. As receivers, they're more robust and less specialized than the mouse's dedicated internal photoreceivers, and this makes them a better choice for Mousey's eyes to the outside world. On most mice, the emitters are clear plastic boxes with a tiny dome protruding from one face, while the detectors are solid black.

Find the clear emitters and desolder them from the PCB. You are now the proud owner of a pair of robot eyeballs.

8. GIVE MOUSEY EYESTALKS

Our IR emitters only have two stubby little pins coming out. We need to give Mousey some optic nerves — eyestalks that jut from the front of its body. These not only look cool, but also allow you to adjust Mousey's sensitivity to light by bending the stalks around.

First we need to determine which pin on each emitter is positive and which is negative. Set your digital multimeter to Diode Check mode, and touch the probes to each pin. If the read-out is "OL" (no connection), reverse the probes. When connected correctly, you should get a reading of about 1V, with the red probe indicating the anode (or positive) pin. If your DMM doesn't have Diode Check, look for a positive voltage of about 0.6V when the red probe is on the anode.

To create the stalks, cut four 6½" pieces of 22-gauge, solid-core hook-up wire. If you have red and black, cut two of each color. Solid core is better than stranded in this case, because it makes stiffer stalks that hold their shape when you mold them.

Our finished eyestalks, ready to shed some light on our control circuit.

Solder the red wire to the cathode (-) pins on the emitters and the black wires to the anode (+) pins. The colors are switched because we're reverse-biasing the diodes; with current flowing in the normal direction, additional electrons excited by light in the diode's junction get lost in the flow, but with current trickling the opposite way, the difference is more noticeable, making the circuit more sensitive. When the wires are soldered in place, twist them together and strip some of the jacket off of the other ends.

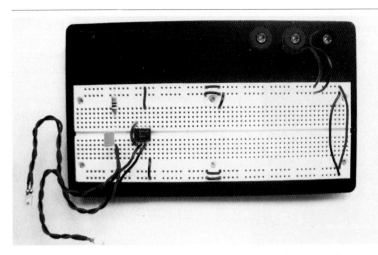

The first part of Mousey's brain: sensors and main control circuit.

9. HOOK UP THE OP-AMP

With all of your electronic components in hand, we're ready to breadboard. Here are the steps to install the op-amp chip and main control circuit:

9a. Install the LM386 chip across the trench on your breadboard. With all ICs, pins are numbered counter-clockwise around, starting at the little dimple.

9b. Connect tie-points for Pins 1 and 8 together with a piece of hook-up wire. These two pins control the op-amp's gain; by connecting them with a jumper, we're increasing the circuit's sensitivity to the input.

9c. Connect the eyestalks by taking the black wires from each and connecting them to tie points for Pins 2 and 3 (the op-amp's inputs). Connect the red wires together by plugging them into a node about

five or six rows left of the chip. Our horizontally oriented board is organized with +/- power supply at top/bottom and all chips facing left. Translate accordingly for different breadboard layouts.

9d. Plug the negative lead of an LED (the shorter end) into the node with the two red eyestalk wires, and the positive lead into a new node on the opposite side of the trench. Then take a 1k-ohm resistor and plug one end into the LED's positive node, and the other end into the positive/upper power bus. These components constitute a sensitivity-boosting subcircuit originally developed by Wilf Rigter.

9e. Finish this part of your circuit by connecting the power pin of the LM386 (Pin 6) to the positive power bus, and the ground (Pin 4) to the lower/negative bus. We'll connect the battery later.

10. CREATE THE RUNAWAY CIRCUIT

If we hooked up Mousey's motors and battery at this point, it would simply chase a light source. Now we'll make it more interesting by adding Mousey's whisker-triggered "fear" reflex. To create the runaway circuit, we need the bump switch you already pulled, a 5V DPDT relay, a transistor, and a simple timer consisting of a capacitor and a resistor. When the switch is triggered, the transistor enables the runaway circuit, where the capacitor powers Mousey's motors in reverse. When the capacitor has fully discharged a few seconds later, the transistor switches motor control back to the regular, light-following circuit.

The resistance and capacitance determine the rate and amount of current discharged, and you can play with different resistor and cap values until you find the runaway behavior you want. Try resistors in the 1k- to 20k-ohm range, and capacitors in the 10- to 100-microfarad range. With both, the higher the value, the longer the discharge time. We used a 10k-ohm resistor and a 100-microfarad capacitor, which gave about 8 seconds of fast backing up. Here are the steps for breadboarding the runaway circuit:

10a. The relay's pins are spaced apart widely, so we'll refer to pins by their breadboard locations. Plug in the relay about six nodes to the right of the LM386, or 1–16 (although the relay actually has only eight pins).

10b. Cross a wire from Pin 8 to Pin 11 and another from Pin 6 to Pin 9. These two wires will reverse the motor connections when the relay is engaged.

10c. Plug the capacitor's positive lead into an unused row just left of the relay, and the cathode to the negative power bus. On electrolytic caps, the cathode is usually marked with a stripe or (-) symbol.

10d. Plug in one end of the higher-ohm resistor to connect with the capacitor anode, and jump the other end over the trench to a new node on the other side.

10e. Spread the transistor's pins and plug it in with the flat side facing the trench, above the relay, such that the center pin (base) connects to the resistor lead, the left pin (emitter) is in an unused node, and the right pin (collector) connects to Pin 16 of the relay.

10f. Plug one hook-up wire into the bottom resistor and capacitor node, somewhere between the two, and a second wire up to the positive power bus. Bend the tips of the wires so they can touch, but keep them separated. These wires will act as the

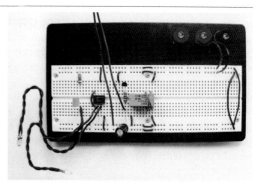

Our breadboard with control chip, timer, and relay circuits installed.

bump switch when you touch them together. We're being lazy and assuming that the switch works, but you can hook the wires up to it to make sure.

10g. Run two wires to connect Pin 1 and Pin 8 on the relay with the top/positive power bus. Connect Pin 9 to the negative bus. Finally, connect the transistor's left pin (emitter) to the bottom/negative bus. This connects the relay and transistor to power. That's it — look over your cool robot brain!

11. CONNECT THE MOTORS AND POWER

Now we're ready to connect the motors and power and see if it all works. Take the right motor and connect its negative terminal to Pin 5 of the LM386 chip and its positive terminal to Pin 13 on the relay. Take the left motor and connect its negative to Pin 5 of the chip, and positive to Pin 4 on the relay. On many motors, the positive terminal is marked with a dimple or a (+) symbol.

Finally, connect the 9V battery to the board via a battery snap or clips, recalling that the battery's "outie" snap is its negative pole. Your breadboard should look like the image at right, and the motors should run. If so, congratulations! Get yourself a flashlight and start having fun moving the beam around Mousey's light sensors, noticing the speed changes. Then touch the switch wires together, hear the relay click, and see the motors reverse their direction.

If all did not go well, check that everything's where it should be, with the capacitor, resistors, and transistor in the proper holes and power running in the right direction. Some breadboards split their power

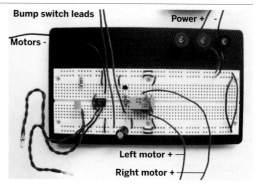

Bump switch leads · Power +/ -

Motors -

Left motor + —
Right motor +

Finished breadboard circuit with motors and power attached.

busses into multiple segments; in this case, you need to connect the battery to each occupied segment of the power bus, or else wire them together. Use a fresh battery, and probe around with the multimeter to make sure that the right amount of power is getting where it should. If the eyes don't work, check the eyestalk solder joins, and if necessary, swap the eyeballs out for another set from another old mouse. Some definitely work better than others.

12. FREEFORM MOUSEY'S CONTROL CIRCUIT

Now that we have a light-hungry robot brain, we need to install it in our mouse body so that it can feed (cue *Night of the Living Dead* sound effects here). In general, we'll want to use a lighter wire, such as stranded 22-gauge, to tuck into the case and put less stress on the solder joints.

Before soldering, test fit all the parts inside your case, starting with the battery, motors, and bump switch. Then position the other components around these. The resistor/LED sensitivity-booster circuit will fit against the top half. As you arrange, check that the case still closes, and leave some headroom for the wires. When you're happy with your arrangement, empty the case and install the battery using two-way tape, Velcro tape, or poster putty. That way, you can replace it when Mousey gets that run-down feeling.

13. INSTALL THE RELAY

To prepare the relay for installation, put it in "dead bug mode" (on its back), and solder short lengths of solid-core wire to the bottom four pins (the switch pins) in an X configuration, as shown.

13a. Solder the transistor's collector (the right pin when you're looking at the flat side with the pins pointing down) to the top-left coil pin on the relay, Pin 16 on the breadboard. Solder a 4" piece of black wire (denoting negative) to the transistor's emitter. This will connect to Pin 4 of the IC and negative power.

13b. Solder a short red wire connecting the top and bottom pins on the relay's right side, Pins 1 and 8. Solder a 2" black (negative) wire onto the bottom-left pin, Pin 9, and then a 3" red wire onto the bottom-right, Pin 8.

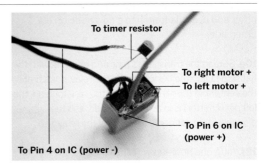

To timer resistor
To right motor +
To left motor +
To Pin 6 on IC (power +)
To Pin 4 on IC (power -)

13c. Glue the relay into the case, in dead bug mode, and allow it to dry before soldering anything else to it. We glued ours between the motors.

13d. Using red wire, solder the left motor's positive terminal to the second pin down on the right side (Pin 4 on the breadboard), and solder the right motor's positive to the opposite pin on the relay, Pin 13.

14. CONNECT THE SWITCH COMPONENTS

With the relay close to the front, we can chain together the timer resistor, capacitor, and bump switch without needing additional wires. As with the relay, we'll attach components "out of body" first, for easier soldering.

14a. Solder a 4" black wire to the capacitor's negative lead (which should be marked).

14b. Using a multimeter on your 3-pin bump switch, determine which side pin connects with the middle pin when you click, and clip off the other side pin.

14c. Solder the cap's positive lead to the remaining side pin of the bump switch, and solder one end of the timer resistor to the same pole.

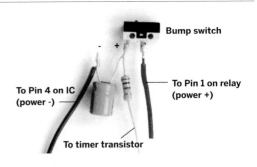

Bump switch
To Pin 4 on IC (power -)
To Pin 1 on relay (power +)
To timer transistor

14d. Solder a 2" red lead to the middle bump switch pin, and then glue the switch into the body, through the hole you cut earlier.

14e. Solder a lead between the transistor's middle pin and the free end of the timer resistor.

15. POWER TO THE MOTORS

15a. Solder two 2" black wires to the motors' negative terminals, then solder the stripped ends of these two wires together side-by-side.

15b. Solder a third, 3" black wire to these joined ends, then solder it to the control chip's output pin (Pin 5).

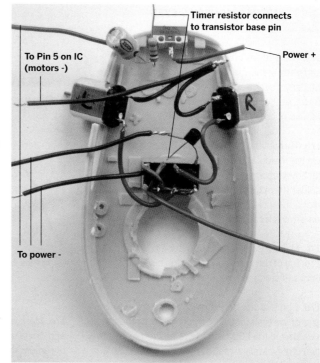

Timer resistor connects to transistor base pin

Power +

To Pin 5 on IC (motors -)

To power -

16. INSTALL THE LM386 CONTROL CHIP

16a. Bend Pins 1 and 8 of the op-amp chip down and solder them together.

16b. Find the black wires from the transistor, the relay, and the capacitor, strip the ends, and solder them all together side-by-side.

16c. Solder the battery snap's negative wire to this same junction.

16d. Solder a 1" black wire to Pin 4 of the op-amp, and the other end to the negative wire junction.

16e. Solder the red wire from the relay to Pin 6 of the chip. Then glue the chip into the mouse case in dead bug mode.

That's it for Mousey's bottom half!

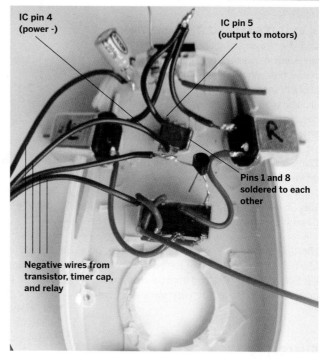

IC pin 4 (power -)

IC pin 5 (output to motors)

Pins 1 and 8 soldered to each other

Negative wires from transistor, timer cap, and relay

The LM386 control chip wired and ready for action.

17. INSTALL MOUSEY'S EYES

17a. The buttons on most computer mice are separate, semi-attached pieces of plastic. To give Mousey's eyes a solid foundation, glue the buttons down, wait until dry, and then drill small holes in Mousey's lid to thread the eyestalks through.

17b. Thread about 1¾" of stalk through each hole. On the inside, trim the two red wires so that they just overlap against the underside of the lid, then solder them together. Run the black wires back along the inside and bend them down where the op-amp is located (but don't solder them yet).

17c. Make the sensitivity booster circuit by cutting a 1" piece of red wire, and soldering one end to the 1k-ohm resistor and the other end to the LED's anode.

17d. Connect the booster by soldering the free end of the resistor to the middle pole of the toggle switch and the LED cathode to the junction of the two red eyestalk wires.

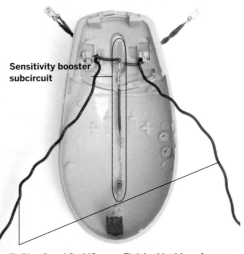

Sensitivity booster subcircuit

To Pins 2 and 3 of IC   Finished insides of mouse top with eyestalk placement, sensitivity booster, and power switch.

17e. Mark where the LED sits, gently bend it aside, and drill a hole in the case for the LED to poke out of (unless it can already come up through the scroll wheel slot). Push the LED through and hold it in place with electrical tape.

18. IT'S ALL ABOUT CONNECTIONS

We almost got bot! Now install the front whisker and make the final connections between power, the switch, and the control chip. There's no photo of these final steps, because they happen inside a semi-closed mouse. But you're such a circuit-hackin' fool by now that you don't need us anymore.

18a. Solder the black eyestalk wires to Pins 2 and 3 on the LM386.

18b. Solder the red battery wire to either of the side poles of the toggle switch.

18c. Solder a red wire from the toggle's center pole to Pin 6 of the IC, or to either Pin 1 or Pin 8 of the relay. Solder another red lead from the unconnected bump switch pin to one of these same locations.

Congratulations! It's a slightly anxious, light-seeking robot.

18d. Cover all exposed leads and junctions with electrical tape to prevent shorts. Then glue or loosely tape your plastic "whisker" to the bumper switch, so that it clicks on impact.

18e. Finally, snap in the battery, and screw or tape the two mouse halves back together. Then put Mousey on the floor, switch it on, and watch it go.

FINISH ☒

NOW GO USE IT ››

ENJOY YOUR ROBOT MOUSE

MOUSEY GAMES

If all went well, Mousey the Junkbot's behavior will be apparent once you flip its tail. The robot should zoom away and eventually hone in on the brightest area in the room. It works best if you limit Mousey's surroundings to just one source of illumination — one light or sun-soaked window. Here are some other fun experiments:

Put Mousey in the hallway and close all doors except one. Make the open room as bright as possible, and see if Mousey eventually scuttles in there. Try orienting Mousey in different starting positions.

Tune Mousey's light sensitivity by bending the eyestalks. Move the stalks farther apart, closer together, and bent in different directions until you get the steering you're looking for.

Use a flashlight to lure Mousey around. This will drive pets insane! But be careful; agitated pets will attack your robot and try to rip its components out.

TROUBLESHOOTING A WAYWARD MOUSEY

If you turn on Mousey and nothing happens (cue laughing clarinet, "Wha-wha-WHAAAA"), or if it acts strangely, turn it off immediately. Something went wrong with the build. Here are a few things to check:

First, ask yourself the tech-support alpha question: is it plugged in? Make sure that the battery is new, the battery snap is well-seated, and its positive and negative wires are properly connected. Then make sure that bare wires, pins, and solder joints are not making unauthorized contact with one another. One sign that you may have such a short circuit is if the battery gets warm.

Next, double-check all solder connections against the instructions. Besides being in the right places, they should all be fat, shiny, healthy-looking joins. Use the multimeter to check resistances, and resolder anything suspicious.

If Mousey frantically spins in a tight circle, you've probably hooked the motors up incorrectly. Reverse the wires that connect to the motor on the side that's going backwards.

If it's a broader circle, the motors might be wired correctly, but just not level with each other. If so, reglue the motors so they're symmetrical and make sure the tires are the same size.

If Mousey's always heading backwards, swap the wiring on both motors.

RESOURCES

This project is adapted from my book *Absolute Beginner's Guide to Building Robots*. You can find schematics and installation instructions for additional Mousey hacks on my robot page at Street Tech, streettech.com/robotbook. More cool hardware hacks live in Dave Hrynkiw's *Junkbots, Bugbots & Bots on Wheels*.

To find other ideas for hacking your Mousey, and other LM386-based bots, Google "robot +LM386," "herbie +LM386," and "Randy Sargent +robot."

To learn more about DC motors, and see a dissected version of the motor used in this project, see xrl.us/fkxh.

Pummer, Dude!

Part robotic plant life, part techno-sculpture, these desktop toys are easy and fun to make.
By Gareth Branwyn

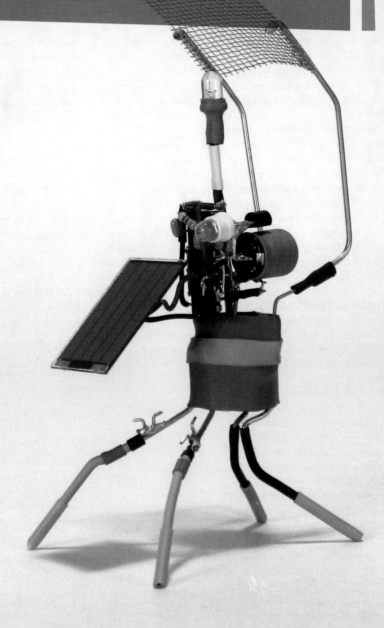

I N MAKE VOLUME 06, I WROTE ABOUT BEAM, a branch of robotics built on low-end, mainly analog electronics that is inspired by biology. I described how to build two types of bots in the BEAM taxonomy: Solarrollers and Symets. One of the more obscure members of the BEAM family tree is the Sitter, an immobile robot with few or no moving parts.

One of my favorite types of Sitters is the Pummer, a nocturnal, robotic plant that soaks up the sun during the day; stores that energy in batteries or capacitors; and then, when it senses darkness, feeds power to a light which pulses, or "pumms," away in the dark. Since the electronics are simple and minimal, you can have fun with the design of your Pummer, creating a swanky piece of high-tech art that will intrigue everyone who sees it adorning your geekosphere.

How a Pummer Works

In "A Beginner's Guide to BEAM" (MAKE, Volume 06, page 54), we talked about different types of Solarengines (SE), which are simple power circuits for actuating miniature robots. We mentioned the nocturnal type of Solarengine. This is the variety of SE used in many Pummers. All SE circuits work in much the same way: the solar cell captures light energy, converts it to electrical energy, and sends it to storage, either in capacitors or rechargeable batteries. When a trigger value is reached, the stored energy gets sent off to do some sort of work. In a voltage-triggered SE, the trigger is a set voltage ceiling. In a nocturnal SE, the trigger is a threshold value of light.

Looking at the circuit diagram on the following page, you might be asking yourself: where is the sensor that tells the Pummer that it's lights out and time to get with the pummin'? Ingeniously, the solar cell and the circuit itself serve this purpose.

During the day, when light hits the cell and the cell is sending juice to storage, the diode in the circuit keeps the enable line set to high. When the level of light/current reaching the cell/circuit falls below a certain value (as set by the value of the parallel resistor), the enable goes low, triggering the discharge cycle and the pumming of the LED(s). The diode, being a sort of one-way valve in a circuit, prevents the current from flowing back into the charging part of the circuit; it has no place else to go but along the discharge path.

Pummer Circuits

There are a number of different Pummer circuits you can use, from simple ones that power a single

LED, to more sophisticated ones designed to maximize power collecting and discharging, and ones that can power multiple LEDs. The one shown here, used in the Solarbotics Bicore Experimenters BCP Applications Project (see makezine.com/08/pummer), balances simplicity with circuit efficiency and bang-for-buck; i.e., it makes a pretty damn cool Pummer without too many building headaches.

This nocturnal SE circuit makes use of another hallmark BEAM circuit, the bicore, which is the basic "neuron" of BEAM "intelligence" (see MAKE, Volume 06, page 54 and page 58). Here, the two-state oscillator is used to create the flashing/pumming behavior. The C1 and C2 caps are used to set the blink/pause rates, and C3 handles the "decay" rate of the pumms. You can play around with these rates by trying different capacitor values on a breadboard.

Other Pummer circuits, including those that can handle multiple LEDs, can be found on Solarbotics.net, in /library/circuits. Costa Rica BEAM (costaricabeam.solarbotics.net) has a fairly thorough library of schematics for Pummers, including a circuit for making a Type 1 Solarengine (which uses a 1381 voltage trigger) into a darkness-activated power circuit.

Pummer Designs

One of the cooler aspects of a Pummer is that, because it's a Sitter and has no moving parts and no concerns over weight, etc., the design and aesthetics of the robot can take center stage. You can build Pummers to look any way you want. A lot of builders, inspired by the idea of Pummers being a sort of robotic plant life, put the LED(s) on a long stalk or on multiple stalks. But Pummers have also been built in the shape of modern sculptures, hexagons, triangles, cubes, even a dragon with solar

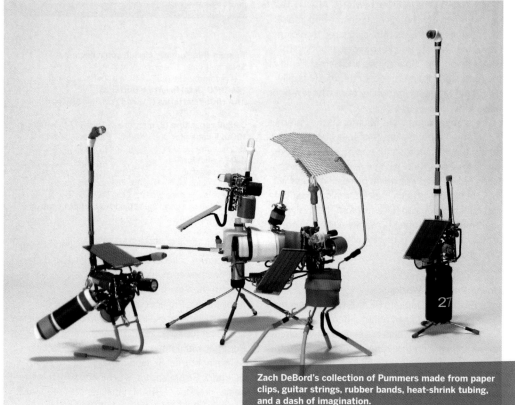

Zach DeBord's collection of Pummers made from paper clips, guitar strings, rubber bands, heat-shrink tubing, and a dash of imagination.

Single-LED High-Efficiency Pummer

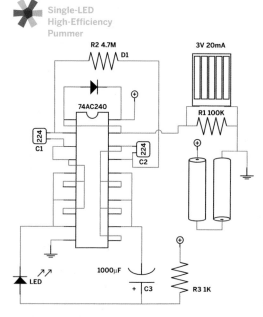

cells on the wings and glowing LED eyes. Really, your imagination and building skills are the only limitations.

A large majority of Pummers are built using paperclips as the main building component. Zach DeBord, a BEAM builder from Chicago (whose Pummers are pictured here) writes: "Buy a pack of jumbo and regular sized paper clips. For the $2 you spend, you'll be able to build a whole fleet of robots. I almost exclusively use paper clips and guitar strings for my creations."

Other common structural components are rubber bands and heat-shrink tubing. "An assortment pack of heat shrink (available at RadioShack and other places) goes a long way," says DeBord. "Not only are your bots more interesting looking, but you can use tubing in key places to reinforce weak joints."

➕ **For more Pummer resources visit makezine.com/08/pummer.**

Gareth Branwyn writes about the intersection of technology and culture for *Wired* and other publications, and is a member of MAKE's Advisory Board. He is also "Cyborg-in-Chief" of Streettech.com.

ROOMBA HACKS

DON'T LET YOUR ROOMBA JUST COLLECT DUST WHEN YOU CAN HACK, MOD, AND TAKE OVER THE WORLD WITH YOUR OWN (CLEANING) ROBOT ARMY.

By Phillip Torrone & Tod E. Kurt

In May of 2006, iRobot, makers of the Roomba robotic vacuum, announced they had shipped more than 2 million cleaning robots, making Roomba one of the (if not *the*) most successful domestic robots in history. With 2 million of anything that can be taken apart, it was only a matter of time before dozens of Roomba hacks hit the net.

Courting this audience, iRobot opened up the interface to all current Roomba models, and released an educational version called the Create. With so many ways to hack these suckers, makers responded by building more projects and developing software. Here's a roundup of some of the interesting projects.

Roomba Costumes
myroombud.com
MyRoomBud was started by a couple of kids who wanted to earn money to buy cowboy boots. They make and design handmade Roomba costumes, including frogs, pigs, tigers, cows, ladybugs, and rabbits. Their motto is, "If you don't dress up your Roomba, it's just a naked vacuum."

As Seen Through the Eyes of the Roomba
roombacam.com
Roomba-Cam is a new site that catalogs videos shot from the point of view of the Roomba. The Roomba kitchen tour puts you in the (autonomous) seat of your vacuuming robot, but nothing is more thrilling than the "nighttime infrared cat hunt."

"The Robot that Vacuums and Serves Web Pages"
makezine.com/go/roombanet
Despite many warnings from science fiction movies and books, Bryan Adams, an MIT Ph.D. student, decided it would be a good idea to use a neural network on a Gumstix Linux board to control a Roomba. Enter RoombaNet. Currently the bot is cleaning Sarah Connor's apartment.

Wii-mote Control
spazout.com/roomba
While Sony and Microsoft have been duking it out over polygons and megahertz, Nintendo brought to market one of the best gaming systems in recent history, the Wii. The motion-sensitive controller makes everything more fun; with Roomba Wii you simply flick your wrist and command your Roomba to do anything. Source code included.

If you have a new Apple MacBook, you can use the tilt sensor in a similar way.

Roomba Music
makezine.com/go/roomidi
The Roomba has a piezo beeper that can play tunes. You've heard it. And its motors make noise. Why not put them under MIDI control? RoombaMidi is Java-based; RoombaMidi2 is written in Objective C and C. Both create a virtual MIDI instrument for use by any Mac OS X MIDI sequencer, like Ableton Live, Logic, and so on.

Cellphone-Controlled Roomba
makezine.com/go/rcontrol
RoombaCtrl is a small Java program for your Bluetooth- and J2ME-compatible phone that works with the build-your-own Bluetooth adapter, as shown in *Hacking Roomba*, or the pre-built RooTooth available at Spark Fun (sparkfun.com).

Cylon Roomba
makezine.com/go/cylon
How does the Cylon base ship keep itself so tidy? With a Cylon Roomba, of course. This tutorial has all you need to make a pulsing LED Cylon Roomba, perfect for cleaning your very own Gaius Baltar or Number Six flat.

Roomba, Get Me a Beer
makezine.com/go/roomba
One of the first robotics projects we all seem to gravitate toward is building a line-following bot. This how-to for modding the Roomba IR sensors shows you how to make your Roomba follow lines on the floor ... maybe even all the way to the fridge.

Bionic Hamster
makezine.com/go/irobot
Using the new iRobot Create programmable robot, a hamster can run around in a clear plastic sphere controlling where the robot goes and how fast it can get there.

Robot Roomba Chimp
makezine.com/go/chimp
We're not living on the moon, or driving flying cars, but we do have an animatronic chimp head that screams atop a Roomba as it cleans.

Tod E. Kurt is the author of *Hacking Roomba*. Phil Torrone is a pioneer of Roomba Frogger and Roomba Cockfighting.

Illustration by Tim Lillis

HACK A ROBOSAPIEN

This maker-friendly bot begs to be opened up. By Dave Prochnow

One-and-a-half million Robosapiens were sold last year. Appealing to both adults and kids, Robosapien has probably gone further toward inspiring future roboticists than the most ambitious educational product.

But that's only half the story. Robosapien is one of those few toys that you absolutely must take apart, for three reasons. First, as you disassemble Robosapien, you will be truly amazed at the simple beauty of its design. Counterbalancing springs, integrated plastic strain reliefs, and intricately geared servo motors will delight even the most jaded toy buyer. This robot ain't no bucket of bolts.

The second reason for opening up Robosapien is learning the basics of robot and toy design. Yes, the insides of this robot are well documented, ensuring a good, competent education in robotics — that is,

if you're willing to pick up a screwdriver and open it up.

Finally, you've got to open up Robosapien if you want to become one with this robot's greatest inner strength — it wants to be hacked. I should know: with the blessings of WowWee Ltd. and Robosapien's inventor, Mark W. Tilden, I wrote *The Official Robosapien Hacker's Guide* (TAB Electronics, 2005), which details over a dozen

> **WARNING: Before you begin either of these hacks, beware that, if you don't know exactly what you're doing, you could damage your Robosapien. While these instructions make every effort at holding your hand through the process, one errant soldering mistake could render your robot a gigantic paperweight.**

modifications, construction projects, and hacks that can be performed on this remarkable robot. Regrettably, there were two hacks I couldn't thoroughly discuss in this book. Here they are.

What's the Frequency, Robosapien?

By far, the easier of these two Robosapien hacks is the replacement of the main processor crystal. Labeled "Y1" on the Robosapien main circuit board, this crystal is actually a monolithic capacitor. As such, it doesn't require a big leap of imagination to think that switching this capacitor crystal's value could result in a different Robosapien "personality."

And that's exactly what will happen: the frequency of the robot's operation can be slowed down or sped up by almost 50% just by using different-sized capacitors. Plus, the Robosapien IR remote control will still work. This hack allows you to vary the speed of motor actions for either fast and lightweight or slow and powerful designs.

Open, Sesame

The first step (and one of the hardest in this hack) is to get inside Robosapien to the main circuit board. All you will need for this portion of the hack is a No. 0 Phillips screwdriver.

Before you begin any hacking surgery, make sure you remove the four D-cell batteries from the Robosapien's feet. With your patient now suitably anesthetized, there are four screws that hold the back plate to the Robosapien body — one in each shoulder and two in the waist.

Once you remove these screws, the back plate will lift off. Be careful, however; the power switch wiring harness (this also holds the speaker wiring) is attached to the main circuit board. Just pull the main circuit board plug for the power switch harness, and the back plate can be removed. Set both the front and back plates aside.

Gain Some Capacitance

The main circuit board is located on the back of the Robosapien. Take a moment to study all of the lovingly applied labeling that WowWee Ltd. added to the main circuit board — all of this done to help you, the hacker.

Locate the crystal capacitor. It is to the left of the IC U3 and labeled "Y1." Although this crystal looks like a capacitor, it is actually a ceramic reso-

nator. Take a pair of diagonal cutters, snip the Y1 capacitor off, and remove it from the circuit board. You will now solder either a new ceramic resonator or a capacitor in its place. I began with a .22µF monolithic capacitor. Alternatively, you can add an inexpensive ceramic resonator (digikey.com). Your beginning frequency for the resonator would be around 4MHz. Just solder this replacement capacitor or resonator crystal to the decapitated leads from the old crystal.

Plug-n-Go

If you really want to experiment with a wide variety of capacitor crystals, you might wish to solder two header pins to the Y1 crystal passthrough pads. Then you can just temporarily attach your capacitors or resonators to these header pins, until you find the perfect hack.

The Robosapien main circuit board is located on the robot's back. The crystal capacitor is labeled Y1 on the circuit board.

Typically, resonators with higher frequencies (6MHz to 12MHz, for example) will result in a "faster" Robosapien. (NOTE: The IR remote control might not function properly at frequencies higher than 6MHz.) Lower frequency resonators (2MHz to 3.58MHz) will make the robot behave more slowly.

This same principle holds true for the replacement capacitors, as well. For example, try a 2.2µF capacitor to increase the speed of Robosapien. Just remember to hold onto the original Y1 crystal, so that you can return your robot to its factory state of mind.

The Soccersapien

How about hacking Robosapien into a powerful, fast-moving, soccer-playing robot? Yes, it can be done and it's easier than you think. Be forewarned, however, that this is a much more elaborate hack. If the complexity of the previous hack left your head reeling, then you might want to hold off on this one until you get some more circuit-building experience.

This hack will make it possible to triple the walking speed of Robosapien by using NiCad rechargeable batteries, two H-bridge post buffers, and two Radio-Control (RC) grade, high-torque, high-RPM motors in the hip gearboxes. One of the most common places to find these motors is inside an RC car. With motors in hand, you will need to build two H-bridge circuits and attach these directly to the motors.

In order to gain access to those hip gearbox motors inside Robosapien, you will have to remove the front and back plates as described earlier, as well as dropping the robot's trousers. Removing these "trousers" can be a tricky proposition because two screws are hidden under black plastic plugs on the backside bottom plate.

After disassembling countless Robosapiens, I have seen these plugs both slipped in and glued in place. While the slipped-in plugs can be easily pulled out with a small knife blade, the glued-in plugs must be drilled out. I use a hand drill equipped with a small bit. Don't opt for a power drill for this step — too much speed will melt the plastic. Just a simple hand twist or two and you'll have an easy-access hole through each plug. Now increase the diameter of your drill bit to accommodate the size of your screwdriver shaft, and slowly ream out the hole to its final dimension.

If you happen to scratch or mar the black plastic, don't fret. This plastic is actually painted black and can be quickly re-covered with some gentle sanding and a little dab of black gloss paint.

I've included a sample H-bridge schematic diagram to help you build this circuit. Wiring each H-bridge to the Robosapien main circuit board is deceptively easy. Each drive input for the H-bridge circuit is connected to the motor lines that are currently in Robosapien (i.e., "LEG-L" and "LEG-R"). Therefore, just snip the leads from the two motors in each Robosapien hip gearbox. Remove the old motors and insert your new high-torque motors. Just make sure that these motors have a static resistance greater than 4

ohms, are approximately ¾-⅞ inch in diameter, no greater than 2 inches in length, and have a pinion gear attached to the shaft (e.g., ideally any Mabuchi FA-130 series motor). Alternatively, you can use the existing motors and couple the H-bridge circuits to them, but the performance improvement isn't nearly as dramatic as with the higher-torque motors.

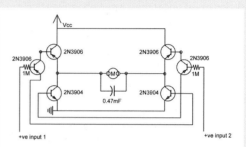

Schematic diagram of an H-bridge circuit. (Schematic redrawn from original design by Mark W. Tilden ©1997)

Greasy Bot Stuff

Now that your new motors are ready to go, you must find some power to drive the H-bridge circuits. Both Vcc and ground can be found on the two outside connectors located along the lower edge of the Robosapien main circuit board ("L-SW-GND-C" and "SPK-VDD-Fr-VCC"). Just remember that a big bypass capacitor (e.g., 330µF 16V electrolytic capacitor) must be attached to Vcc to avoid frying the Robosapien brain.

Now Fire 'em Up

It's off to the races. Before you send Robosapien forth fleet footed, make sure you test this new high-speed "mover" on a slick surface. Alternatively, you can coat the footpads on the bottom of each battery compartment with something slippery. Insert your NiCad batteries and get ready to be impressed — all over again. The second "running" forward mode will now give the robot more than enough speed for robot soccer applications.

Dave Prochnow is author of *The Official Robosapien Hacker's Guide* (TAB Electronics, 2005). You can learn more about this book and other projects at pco2go.com.

Vibrobot
By Mark Frauenfelder

1+2+3

Make a twitchy, bug-like robot with a toy motor and a mint tin.

You will need: Metal candy mint tin, wire coat hanger, 1.5V motor from a battery-powered toy, small metal washers (4), small bolts and nuts (2), about 1' of insulated wire, paper clip, ¼M flat plastic faucet washers ³⁷⁄₆₄" OD (3), AA battery, hot glue gun, hot glue, cable tie

When my 3-year-old daughter dropped the $1 battery-powered fan I bought her, the plastic case cracked, ruining it. I promised her I'd make something even better using the fan's motor. I'm a fan of Chico Bicalho's wonderful windup toys, so I made a robot inspired by his designs. I call mine the Vibrobot, and you can make one in a couple of hours or less.

1. Prepare the candy tin.

Sand the paint off the tin, if you wish. Punch 2 holes through the bottom of the tin on either end, using a hammer and a Phillips screwdriver. You'll use these holes to attach the legs. Punch a hole through the lid near one end. This hole is for routing the wires.

2. Make the legs.

Snip off 2 long pieces of wire from a coat hanger and bend each into a V-shape. Bend the tip of the V into a right angle, and then bend a little "foot" at each end (Figure A). Attach the legs to the holes in the tin using bolts, nuts, and metal washers (Figure B). Add a dollop of hot glue to each foot to give them rubber tips.

3. Install the motor.

Push a paper clip through one of the plastic flat washers, and attach the washer to the spindle of the motor. Solder 2 wires to the 1.5V battery, insert the battery in the candy tin, and thread both wires through the hole in the lid. Solder one wire to a lead on the motor, and solder a third loose wire to the other motor lead. Put 2 plastic flat washers between the motor and the candy tin, and secure the motor to the tin using a cable tie.

To operate the Vibrobot, twist the loose battery wire and the loose motor wire together (you can also solder an alligator clip to one of the wires for a switch). Experiment with the critter by gently bending the paper clip and legs into different shapes and observing the effects. Watch a video at makezine.com/10/123_vibrobot.

Mark Frauenfelder is editor-in-chief of MAKE.

Photography by Carla Sinclair

> "The cracker box amp I built cost $5 ... The amp circuit unleashes the full potential of the beast and creates ¼ watt of arena-shaking power. Think of it as a silicon shrunken head of the Marshall stack that Jimi Hendrix played at Monterey." Blind Lightnin' Pete, *The $5 Cracker Box Amplifier*

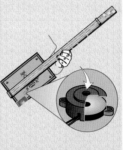

174

186

194

198

206

THE BEST OF MUSIC PROJECTS

from the pages of MAKE

MAKING THE MACHINES THAT MAKE THE MUSIC

Music is clearly one of life's greatest pleasures, either listening to it or creating it yourself. Now imagine not only making music but having made the instruments that carry the melodious sound waves you're cranking out. Or the pride of having modded or built your own home music system or portable amp. And then there's makin' MIDI monkeys. Strangling a sock monkey to make music? We're betting that's a musical pleasure you haven't enjoyed... at least not until you've applied our how-to.

CIGAR BOX GUITAR

By Ed Vogel

Sweet-sounding, three-stringed mini guitar revives an American musical tradition. >>

Set up: p.177 Make it: p.178   Use it: p.184

ROCK 'N' ROLL-YOUR-OWN

As a volunteer music teacher, I sometimes meet kids who can't afford instruments. So I decided to design one that they could inexpensively build themselves, based on the traditional cigar box guitar.

Before the 1950s, when factory guitars became less expensive, many folk musicians built their own stringed instruments. Wooden cigar boxes, which were solidly constructed connoisseur objects, became a popular choice for the instruments' bodies. Thus, an American tradition was born, and today, the cigar box guitar is enjoying a folk revival.

My guitar is a simple, three-stringed design that uses only one power tool and common hardware. Despite its low cost, this guitar plays real music and will hold its tuning for a couple of days. A kid can build it (and play it), and so can you.

Hear the cigar box guitar at makezine.com /04/cigarbox.

Ed Vogel lives in Minneapolis and believes that nothing may just be the next big thing.

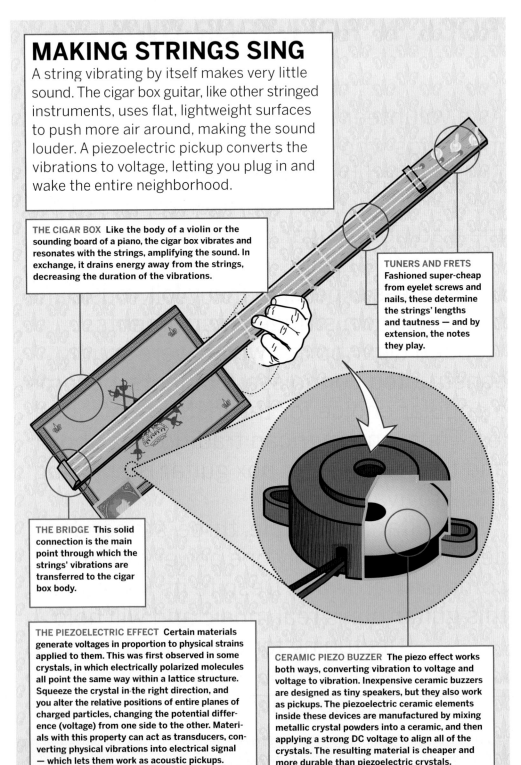

MAKING STRINGS SING

A string vibrating by itself makes very little sound. The cigar box guitar, like other stringed instruments, uses flat, lightweight surfaces to push more air around, making the sound louder. A piezoelectric pickup converts the vibrations to voltage, letting you plug in and wake the entire neighborhood.

THE CIGAR BOX Like the body of a violin or the sounding board of a piano, the cigar box vibrates and resonates with the strings, amplifying the sound. In exchange, it drains energy away from the strings, decreasing the duration of the vibrations.

TUNERS AND FRETS Fashioned super-cheap from eyelet screws and nails, these determine the strings' lengths and tautness — and by extension, the notes they play.

THE BRIDGE This solid connection is the main point through which the strings' vibrations are transferred to the cigar box body.

THE PIEZOELECTRIC EFFECT Certain materials generate voltages in proportion to physical strains applied to them. This was first observed in some crystals, in which electrically polarized molecules all point the same way within a lattice structure. Squeeze the crystal in the right direction, and you alter the relative positions of entire planes of charged particles, changing the potential difference (voltage) from one side to the other. Materials with this property can act as transducers, converting physical vibrations into electrical signal — which lets them work as acoustic pickups.

CERAMIC PIEZO BUZZER The piezo effect works both ways, converting vibration to voltage and voltage to vibration. Inexpensive ceramic buzzers are designed as tiny speakers, but they also work as pickups. The piezoelectric ceramic elements inside these devices are manufactured by mixing metallic crystal powders into a ceramic, and then applying a strong DC voltage to align all of the crystals. The resulting material is cheaper and more durable than piezoelectric crystals.

Illustration by Dustin Amery Hostetler / UPSO.org

SET UP.

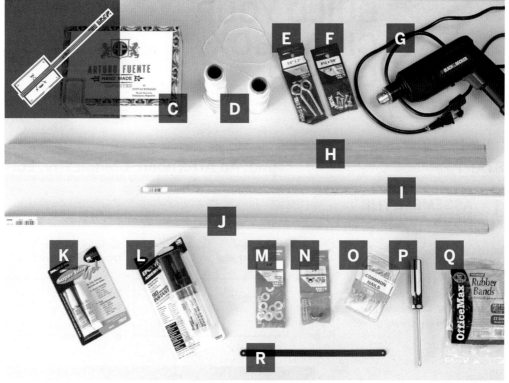

MATERIALS

[A] Pen, pencil, or markers **(not shown)**

[B] Scissors **(not shown)**

[C] Cigar box
Many tobacconists will give cigar boxes away free. I usually end up paying a dollar for them at a place near my house. If you can't find a cigar box, an intact pizza box will also work.

[D] Mason twine, #15 and #18
Available at hardware stores, this is used by bricklayers and cement workers to mark lines.

[E] ¼"x3" eyebolts and nuts **(3)**

[F] #12x⅝" wood screws **(3)**

[G] Drill and drill bits

[H] 3' length of 1x2 red oak
This will be the guitar's neck. The lumber's actual measurements are ¾" by 1½", but 1x2 is how it's named.

[I] ¼" square hardwood stock, at least 1½" long
This will be the nut, at the top of the neck.

[J] ½" square hardwood stock, at least 1½" long
This will be the bridge.

[K] Super glue

[L] 90-second epoxy **(or 5-minute epoxy, for a little more positioning time)**

[M] ¼" washers **(6)**

[N] ¼" wing nuts **(3)**

[O] 2" common nails **(at least 3)**

[P] Phillips screwdriver

[Q] 3½" x 1" (size 33) rubber bands

[R] Hacksaw blade or hacksaw

OPTIONAL

1500-3000Hz piezoelectric element, a.k.a. piezo buzzer **For amplifier pickup.**

¼" phono jack

Soldering iron, solder, and wire

MAKE IT.

BUILD YOUR CIGAR BOX GUITAR

START ⠶⠶   **Time: An Afternoon Complexity: Low**

1. ASSEMBLE THE NECK AND BODY

1a. Using the diagram at right, drill holes at each end of the 1x2. You'll drill six holes in two rows at the tail, to anchor the strings below the bridge, and six more in two diagonal rows where the tuning pegs will be.

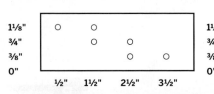

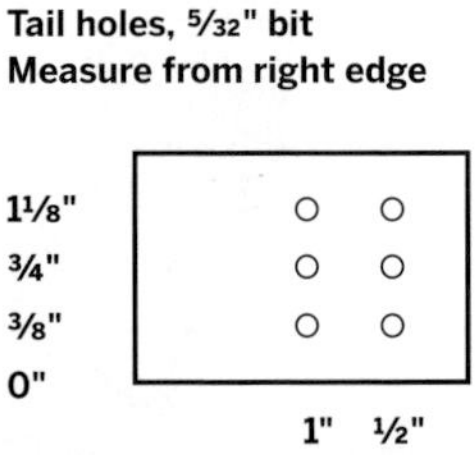

Tuner holes, ¼" bit Measure from left edge				Tail holes, ⁵/₃₂" bit Measure from right edge		
1⅛"	○	○		1⅛"	○	○
¾"		○	○	¾"	○	○
⅜"			○ ○	⅜"	○	○
0"				0"		
	½" 1½" 2½" 3½"				1" ½"	

1b. If you want to add an electric pickup to your guitar, skip ahead to **step 4**. Otherwise, super-glue the cigar box shut.

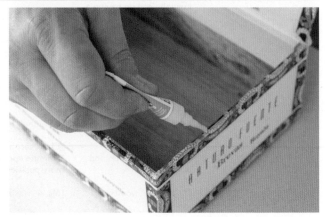

1c. Set the neck squarely on the box so that its six holes are just clear of one of the box's ends. Mark the box along both sides of the neck, so you know where to put the glue.

1d. Mix up some epoxy. I recommend using half a tube, since there may be some gaps to fill.

1e. Apply a generous amount of epoxy to the cigar box, position the neck on top, and weigh it down with a phone book or other weight. With 90-second epoxy, I wait 5 minutes to get a decent cure.

1f. Use the pen to mark the width of the neck (1½") on the ½" square stock. This will be the bridge.

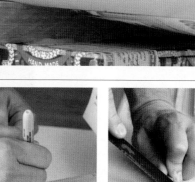

1g. Use the hacksaw blade to cut the bridge.

1h. Put down three dots of super glue for the bridge about ½" up from the six holes at the tail end of the neck.

1i. Set the bridge down on the glue and hold it long enough to sing "Twinkle, Twinkle, Little Star" twice. This song will help you tune later on.

1j. Repeat steps **1f** through **1i**, using the ¼" stock. Glue the nut six inches from the opposite end of the neck.

2. STRING THE GUITAR

2a. Take an eyebolt, spin a nut down the threads, add a washer, and then insert it up into a tuner hole at the end of the neck. Put another washer on top and spin on a wing nut.

2b. Repeat with the other two bolts, and tighten all three to light-finger tight. Are you still humming "Twinkle, Twinkle"? It can become a real earworm. This will work in your favor later, for tuning and playing.

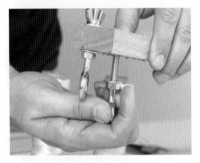

2c. Cut a piece of #18 mason twine about 5 feet long. You won't need all of it, but ends get frayed and we need some slack for pulling on. Thread the string through the empty tuner hole closest to the nut, and tie the end to the adjacent eyebolt in back. Make sure the knot is on the side where the "eye" starts its bend, so that it won't slip out when we tighten the string.

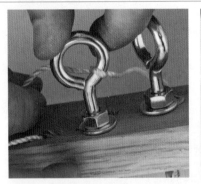

2d. Spin the eyebolt clockwise three times to get some string wrapped on. Tighten the wing nut to firm-finger tightness. Pull the string over the nut.

2e. Pull the string over the bridge and thread the other end down the corresponding hole just below.

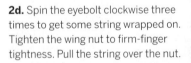

2f. Get a screw started in the hole on the other side, but leave some of the threaded part showing so it's easy to wrap the string around it.

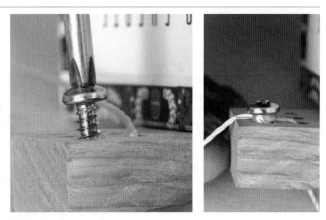

2g. Here is where some slack is handy. Make a loose loop of string around the screw, and then wrap the slack around your hand so you can pull the string tight while you tighten the screw to secure the string.

2h. Congratulations! You've just installed your bass string. Repeat steps **2c** through **2g**, using the #15 mason twine, for the tenor and alto strings.

LEGO INSTRUMENTS

People have been building playable musical instruments from LEGO for some time. Henry Lim's LEGO Harpsichord (MAKE, Volume 04, page 23) and Brad's LEGO Guitar have been written about many times, but they're only the start.

Telerobotic Glockenspiel Player: **XILO is a slightly frightening-looking device that uses the LEGO camera to watch the player and translate his or her movements into real-time Glockenspiel playing. Eight metal tubes lifted from a toy xylophone surround a two-motor whacker in the middle. One motor rotates a mallet into position, and the other brings it down onto one of the tubes to play the corresponding note.**

Robotic Ukulele Players: **Brown University students Bryant Choung and Amelia Wong built a ukulele-playing robot out of LEGO in 2003, and more recently, Middlebury College students Mike Rimoin and Jarvis Lagmans built a smaller one that plays three-chord, such as "Stir It Up" and "Rivers of Babylon."**

Singing LEGO Blocks: **The LEGO Mindstorms RCX block (the brain of the system) has a built-in speaker, and Ralph Hempel has written a webpage that explains in eye-watering detail how to make music with it. Remarkably, there doesn't seem to be a MIDI interface for the RCX; at least not yet...**

♫ **Get links to all these LEGO musical instrument websites at** makezine.com/04/lego.

Tom Whitwell is the founder of musicthing.co.uk. Reprinted with permission from Whitwell's weekly column on engadget.com.

LEGO Dulcimer: **What is it with medieval instruments and LEGO? Mountain dulcimer enthusiast Peter Always built himself a bright yellow dulcimer, with properly spaced frets made of little grey tiles.**

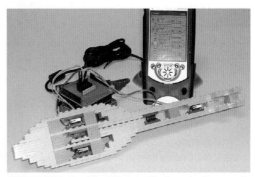

LEGO MIDI Guitar: **The "Lifelong Kindergarten" group at MIT developed this LEGO-based system. The idea was that children could use special LEGO components to build their own experimental instruments, which would work as MIDI controllers.**

—Tom Whitwell

3. ADD THE FRETS

You can usually just find the proper fret locations by singing the major scale. This picture shows approximately where they will go, but it'll take some tweaking to get it right. On my guitar, the five frets went from Re at 2½ inches down from the nut, to La at 11¾ inches down. Try plucking the string and listening for a tone, then marking a dot at the spot with a pencil.

3a. To attach a fret, first take a rubber band and double-loop it.

3b. Fit the looped rubber band to a nail near the head.

3c. Place the nail at a fret position, and pull the rubber band up from underneath the neck. Stretch and loop it around the pointed end of the nail. Notice how the nail point is pointing up. This is a good thing.

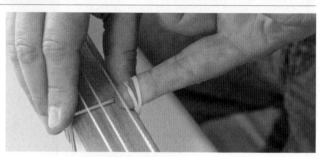

3d. Repeat steps **3a** through **3c** above to attach frets for Re, Mi, Fa, Sol, and La. Sing those notes a few times to get that going in your head because it is going to help you tune your guitar. You can add frets farther up, if you want, and the reason my guitar only has frets up to La, instead of up the full scale to Do, is almost maniacal: those are the notes you need to play "Twinkle, Twinkle, Little Star." And nothing else matters.

4. ELECTRIFY!
(OPTIONAL)

You can testify; now you must electrify! Here's how to add a quick-and-dirty (and I do mean dirty) pickup, so you can play your cigar box guitar through an amp. If you already know you'll want to play electric, it's easier to perform these steps first, before you build the rest of the guitar.

4a. Drill a ⅜" hole in the tail end of the box and mount the ¼" phono jack. (If you've already glued the box shut, you'll have to jimmy it back open with a hobby knife.)

4b. Glue the piezo element inside the box as shown, and solder two wires to connect the piezo pickup to the phono jack terminals. Then glue the box shut. If you haven't already built the rest of your guitar, jump back to step **1c**, and continue from there.

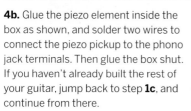

4c. Plug in and rock out! If you find that the pickup is picking up sounds other than the guitar, try covering the sound hole of the piezo element with a couple of pieces of duct tape.

BEYOND THE BOX

Other Easy Materials for Sound Boards and Sound Chambers

The cigar box is admirably suited to the job of radiating string sounds into the surrounding air. You can get the most sound out of one by using a short neck that stops mid-box, attaching the bridge directly to the box (rather than to the neck), and adding a sound hole in front. But that calls for a more complex design than what appears in this article. After you've successfully made one or two of the standard models, maybe you'll find yourself getting more ambitious!

If you can't readily lay your hands on a cigar box, or you just want to explore some other possibilities, here are a few materials and approaches to consider:

An inexpensive styrofoam picnic cooler makes a surprisingly effective sound radiator. Simply glue or screw the cooler to one or both ends of a stick-guitar, for respectable volume and a pleasant tone.

Lightweight metal pots, pans, or bowls are also effective. Attach them concave side out, facing the strings. With aluminum mixing bowls (and especially if you're using steel strings), you can get an exotic sound by placing a little water in the bottom and playing lap-style, tipping the instrument this way and that.

Kinda crazy, but an inflated balloon pressed against the strings where they cross the bridge will greatly increase the volume and add an interesting, percussive tone. With a simple stick guitar, secure a sausage-shaped balloon in place with a long, thin rubber band running around the back of the fretboard. Take a moment to adjust the location for optimal results.

—*Bart Hopkin*

Bart Hopkin runs Experimental Musical Instruments, an organization that produces books, CDs, and other materials relating to new and unusual musical instruments of all sorts. For information, write emi@windworld.com or visit windworld.com.

USE IT.

TUNE, AND PLAY A TUNE

TUNING

Tuning any musical instrument can be tricky, so I am going to offer two methods: you can tune by ear, using "Twinkle, Twinkle," or you can use a guitar tuner. If both methods still leave you unsatisfied, take your guitar to a music store and have them do it. I have never done this myself, but it might freak them out, which would be well worth the trip.

If trying to tune is making you crazy, put the guitar down for an hour, or even a day. Practice singing the scale and "Twinkle, Twinkle, Little Star." You are not tone deaf. If you were, you would not be able to tell the difference between someone asking you a question and giving you a command. Give in to the earworm and be patient.

After coming back to your guitar, just monkey around with it. You built it; it's yours, and you can do whatever you want! Try over-tightening the strings, and see what the notes sound like until they break. How loose can a string be and still play a recognizable musical sound?

Plan A: To tune by ear

a. Look at the picture you used to position the frets for **step 3**. To tweak in the frets, you'll play the lowest string, the bass string, which is the one nearest the heads of the nails.

b. Turn the bass eyebolt and play the string until you start to hear a relatively clear tone.

c. Tighten the bass wing nut.

d. Use a screwdriver to get some leverage and tighten it a little more. Notice how the tone gets a little higher in pitch as you tighten?

e. Pluck the bass string a few times and sing the first two notes of "Twinkle, Twinkle, Little Star" (which is, of course, the two-syllable word "twinkle").

f. Now tighten the middle string with the screwdriver the same way you tightened the bass string, until it sounds like the third and fourth notes of "Twinkle, Twinkle, Little Star," the second "twinkle." You may find yourself singing "Twinkle, Twinkle" along with the bass and middle strings 10 or 20 times. If you are truly going nuts at this point, proceed to **Plan B: Using a guitar tuner**.

g. The last string is a little trickier, because it requires that the first fret ("Re") is properly set. Additionally, you are going to sing in a different key. Use the middle string to play the first two notes of "Twinkle, Twinkle." Then, while fretting the third string up one fret, tighten it to play the second "twinkle." Your left hand holds the third string down behind the first fret, as your right hand alternates between plucking the string and turning the peg. Again, repetition will make it work.

h. If you feel like you've got it (and even if you don't), strum all three strings. It should sound nice. Another thing that should sound nice is fretting the third string on any fret while strumming all three strings. This little bit of magic lets you play melody and harmony easily, while only fretting one string. This is how the dulcimer works.

Plan B: Using a guitar tuner

a. Get a guitar tuner at a music store, or borrow one.

b. Tighten the bass string until it starts to sound clear.

c. Place the tuner on the cigar box and strum the bass. You should still be below A.

d. Tighten the bass string until you get an A.

e. Repeat **steps b** through **d** for the middle string until you get an E.

f. Repeat **steps b** through **d** for the top string until you get an A again. It will be an octave higher.

Now that the strings are tuned, you can use the tuner to set the frets:

g. Fret the third string to the first fret.

h. Strum and look at the tuner. If it shows a tone lower than B, move the fret down the neck toward the cigar box. If it shows a tone higher than B, move it up the neck toward the tuners.

i. Repeat **steps g** and **h**, moving down the frets, tuning to C#, D, E, F, and so on.

CRAFT BOX GUITAR

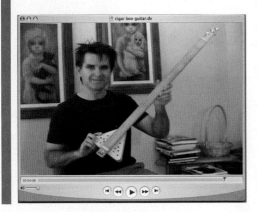 Watch MAKE Editor-In-Chief Mark Frauenfelder demonstrate his variant on the cigar box guitar at **makezine.com/04/cigarbox**.

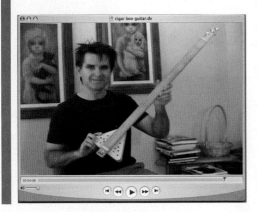

PLAY A TUNE

Can you guess which one? I'll bet you can! Yes, it is "Twinkle, Twinkle, Little Star." You can do it all on your guitar's bottom string. All you have to do is play the notes that appear above the words notated below:

Do   Do   Sol   Sol La   La Sol
Twin - kle   twin - kle   lit - tle   star,

Fa   Fa Mi   Mi   Re   Re Do
How I   won - der what you are.

La La Fa   Fa   Mi Mi Re
Up a - bove the world, so high,

La La Fa   Fa   Mi Mi Re
Like a dia - mond in   the sky.

(And so on.)

If you have a hard time remembering which fret is Do, Re, Mi, and so on, then just pencil them in on the neck. It's your guitar, and you can do what you want! To get a more bluegrass sound, you can strum all three strings down and up, double time, for each note.

TAKE IT ON THE ROAD

All-Cigar Box Guitar music festivals have been held recently in Carrolton, Ky.; Huntsville, Ala.; and Red Lion, Penn., and more are being planned for 2006. Check the cigarboxguitars Yahoo! Group for details and updates.

RESOURCES

CBG sites: cigarboxguitars.com, geocities.com/cigarboxguitar

CBG Yahoo! Group: groups.yahoo.com/groups/cigarboxguitars

CBG-building tutorial: cigarboxguitars.com

Musical Instrument Design by Bart Hopkin

Cardboard Folk Instruments to Make and Play by Dennis Waring

Sound clip of author Ed Vogel playing "Twinkle, Twinkle": www.geocities.com/ed_vogel/cbg1.WMA

THE $5 CRACKER BOX AMPLIFIER

By Ed Vogel & Blind Lightnin' Pete

BIG SOUNDS FROM A SMALL PACKAGE

In MAKE, Volume 04, I presented my version of the venerable cigar box guitar. The instructions for the project included adding an electric pickup so you could play the guitar through an amplifier.

People from around the world emailed me to tell me they'd built cigar box guitars based on my instructions. I struck up a conversation with one gentleman from Europe who goes by the moniker Blind Lightnin' Pete. He made a couple of beautiful cigar box guitars, including one he calls the Vintage Blues Texas Rattlesnake Special model. He then went one step further, and built a cracker box guitar amplifier. This outstanding little amp cost all of $5 to build (depending on where you get the parts). Pete kindly allowed me to modify his design and present it as a project for you to build. (See page 193 for a word from Pete about the origins of the cracker box amp.)

My amp differs a little from Pete's because I wanted to make a workable little practice amp with parts and tools that could be purchased "one-stop shop" at RadioShack and built in an hour.

Set up: p.189   Make it: p.190   Use it: p.193

Ed Vogel lives in Minneapolis and believes that nothing may just be the next big thing.
Blind Lightnin' Pete is the online pseudonym of Howlin' Mississippi Slim.

Photograph by Sam Murphy

THIS BOX ROCKS

Hi-Carb Sound

The heart of this surprisingly loud, clear-sounding, battery-powered guitar amp is National Semiconductor's LM386 series low-voltage op-amp IC. Two potentiometers in the circuit control the gain and the volume. For the cleanest sound, turn down the gain knob all the way and turn up the volume knob to the maximum. Then slowly turn up the gain. For a raunchier, distorted sound, start with the volume knob all the way down, and the gain knob at maximum. Then crank up the volume. You can achieve lots of different sounds by playing with the knobs. Experiment!

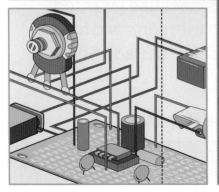

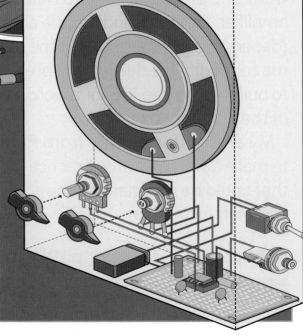

Illustration by Timmy Kucynda

SET UP.

MATERIALS

[A] A box of some sort or another (cracker box shown)

[B] Toggle switch, single pole single throw

[C] 9V battery

[D] Battery connector

[E] 0.047µF capacitor

[F] 220µF capacitor (biggest)

[G] 0.01µf capacitor

[H] 100µf capacitor

[I] Hookup wire, 20 or 22 gauge AWG solid core is best.

[J] 5KΩ potentiometer (audio or log taper)

[K] 25-ohm (25Ω) rheostat

[L] LM386N audio amplifier

[M] 8-pin DIP IC socket

[N] Chicken head knobs (2)

[O] Prototyping PC board

[P] Soldering iron

[Q] Solder

[R] Speaker, 8Ω impedance

[S] 10Ω resistor

[T] ¼" mono phone jack

[NOT SHOWN]
Speaker grill (optional)
Glue gun

MAKE IT.

BUILDING THE CRACKER BOX AMP

START ⋙
Time: An Afternoon   Complexity: Medium

1. MAKE THE CIRCUIT

1a. Make a copy of this schematic, or download the PDF at makezine. com/09/crackerboxamp and print it out.

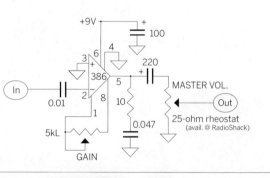

1b. Install the socket in the printed circuit board.

1c. Solder it down.

1d. Install the chip. I like having the chip in the printed circuit board while I build because there can be no doubt as to where pin 1 is. This is also why I install parts and make wire connections on the top of the printed circuit board.

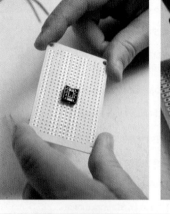

1e. Install the 0.01µF capacitor so one leg connects to pin 2 of the chip and one leg is in a "proto row." Flip it over and solder it.

Photography by Ed Vogel

1f. Install the 9V battery clip and mark a plus sign for the red wire and a minus sign for the black wire.

1g. Install the 10Ω resistor and the 0.047µF capacitor. Take advantage of the "proto rows" to make the connections:

- Chip pin 5 to one leg of the 10Ω resistor.
- The other leg of the 10Ω resistor to one leg of the 0.047µF capacitor.
- The other leg of the 0.047µF capacitor to "ground."

For our purposes "ground," which is shown on the schematic as a triangle with the point down, is the long "proto row" we marked with a minus sign.

1h. Use this same technique to install and make connections.

✳ **TIP: Every time you install a part or make a connection, mark it off on the schematic ("Little Gem" schematic courtesy of runoffgroove.com/ littlegem.html).**

1i. Solder the wires to the phone jack. Use green for signal and black for ground.

1j. Install the wired phone jack to the circuit. You should end up with something that looks like the photo to the far right here.

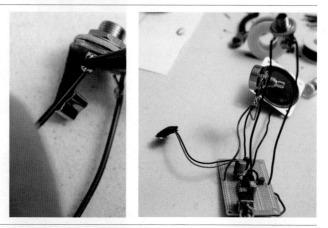

1k. Cut the red lead and install the switch.

2. BUILD THE ENCLOSURE

2a. Make holes in the side of your box to fit the potentiometer, rheostat, and phone jack.

✱ **TIP: Why bother with drills and X-Acto knives when you can use your soldering iron to make holes?**

2b. Make holes for the "speaker grill." You are going to find some hanging chads on the inside of the box. Reach in there with the soldering iron and burn them off.

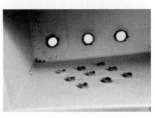

2c. Make a hole for the switch.

2d. Pop your circuit into the box.

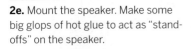

2e. Mount the speaker. Make some big glops of hot glue to act as "stand-offs" on the speaker.

2f. Mount the switch.

2g. Install the chicken head knobs.

NOTE: It's a proven fact that chicken head knobs greatly enhance the vintage sound of an amplifier. Use them liberally.

FINISH ☒

NOW INSTALL A BATTERY AND GO USE IT! »

THE ORIGINS OF THE CRACKER BOX AMP

Blind Lightnin' Pete's cracker box amp.

A FEW WORDS FROM BLIND LIGHTNIN' PETE

The cracker box amp I built cost $5. It uses an 8-pin National Semiconductor LM386 series low-voltage op-amp IC. The amp circuit unleashes the full potential of this beast and creates ¼ watt of arena-shaking power. Think of it as sort of a silicon shrunken head of the Marshall stack that Jimi Hendrix played at Monterey.

This integrated circuit has provided the basis for low-power solid-state amplifiers in recent years, including the famous Smokey Amp and a few of the designs at runoffgroove.com.

You can buy an LM386 for under a buck; it's a standard RadioShack item, the same one that was used in the MAKE project for turning your old computer mouse into a robot (see "Mousey the Junkbot", page 150). Our favorite hobbyist robot supply source, Solarbotics (solarbotics.com), sells them for 75¢ a piece.

I added a couple of capacitors, a couple of resistors, an LED, a ¼" jack, a potentiometer, and a $2 speaker, wrapped it all in a big blob of solder, crammed it in whatever empty box was laying around, and voilà!

The pot controls the gain, and it goes from California clean vintage Fender to Santana Mesa Boogie crunch to Hendrix Marshall. It runs off any combination of batteries — I usually use a 9V, but it's possible to get a cleaner tone with 12V (8 AA batteries in series). I have used it to drive a 4×12 Marshall cabinet, and it gets pretty loud. Not loud enough to compete with a rock drummer, but loud enough for me not to hear my wife screaming "turn it down," which I guess is enough for household use. If we had any neighbors, I could raise some complaints from it. Let's just say that even with a 2" speaker it's plenty loud enough for most apartment dwellers.

Interested in learning a bit more about the LM386? A great place to start is National Semiconductor's website (national.com) where you can download the data sheet. Even better, if you take the time to register on the site, they will generously send you a few samples for free!

Ask for the LM386N-4 series, as these are rated to handle up to 18V. Although any of the LM386 chips will work wonderfully for our hi-gain design, several experimenters and makers have found that cleaner tones with more headroom are achievable by running the circuit with a few extra volts.

OTHER OP-AMP 386 PROJECTS

Home-built bat detector:
bertrik.sikken.nl/bat/my_div.htm
Mini bench amp to test audio circuits:
makezine.com/go/minibench
Headphone amplifier:
radiowrench.com/sonic/so02144.html

See videos of Blind Lightnin' Pete playing his cigar box guitar through his cracker box amp at makezine.com/09/crackerboxamp.

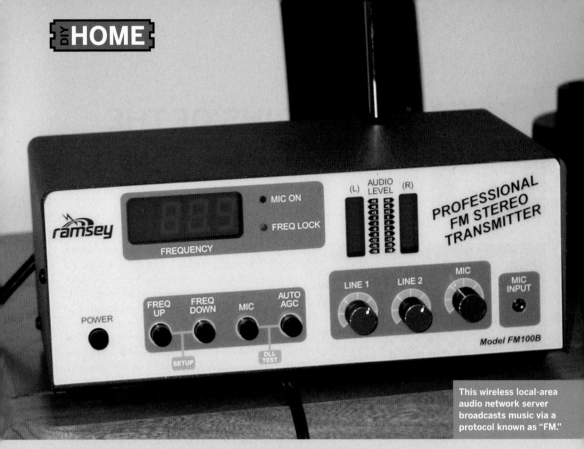

This wireless local-area audio network server broadcasts music via a protocol known as "FM."

HOMECASTING DIGITAL MUSIC

Good old FM beats wi-fi for sending streamed music around your house.
By William Gurstelle

I subscribe to Real Networks' internet-based music service Rhapsody, which lets me stream a wide variety of music through to my computer's speakers. Digital music services like Rhapsody access millions of songs at the touch of a computer mouse. At ten bucks a month for unlimited use, this seems like a deal. The problem is, I don't listen to music only at my computer — I want to listen on my living room stereo, in the driveway as I work on my car, in the kitchen, in the basement, and so on. You need to untether the music from the computer's often-mediocre sound system in order to get the full benefit of your customized, on-demand music programming.

Rhapsody recommends using a multi-room digital music system call Sonos (sonos.com), but it can cost thousands of dollars. Products like Sonos distribute the music stream via Ethernet, either cabled orwireless. This makes listening outside problematic, because even if I have wi-fi set up to transmit, I still need a digital device and electrical power out where I'm listening. Using wi-fi also makes it impossible to listen using just portable headphones.

Legal FM Transmitters
A more cost-effective and flexible solution is a small FM radio transmitter. This trick is familiar to iPod

Photography by William Gurstelle

The completed power supply assembly from the Ramsey FM100B Super Pro FM Radio Station Kit.

users who play songs through their car sound systems, but broadcasting throughout your house and yard requires a bit more power.

Legal FM transmitters are a nearly perfect solution. They are relatively inexpensive, and they use the most ubiquitous of music players — FM radios.

I tried a few of these transmitters, starting with the C. Crane FM transmitter ($70, hobbytron.com or amazon.com). The Crane was plug-and-play simple, but its tiny size was matched by tiny power. Its out-of-the-box range was just a few feet. To increase this, I peeled off the metallic FCC label on the bottom, and used a screwdriver to turn the small potentiometer hidden underneath. But even at its max, the Crane's range is limited and will reach only a few adjacent rooms.

I also tried the much larger Ramsey FM100B ($250, hobbytron.com), which puts out a quarter watt of power, the maximum the FCC allows for unlicensed radio transmitters.

Some Assembly Required

The FM100B comes unassembled, as a large kit with over 800 solder points. According to the instruction sheet, it takes ten hours for an advanced hobbyist to assemble the kit, and 24 hours for a beginner. Given the number of components and the amount of soldering, these estimates seemed optimistic.

Unfortunately, the kit I received contained a bad volume control and a bad fuse holder. This added some troubleshooting time to the build, but the company was responsive and replaced the defective parts quickly.

I enlisted the aid of my friends Al and Rick, who had previous experience building Ramsey FM transmitters. With a kit this complex, they knew that we needed to set up an efficient assembly line. They prepared a work area with adequate lighting and work surfaces, and then began work in earnest, systematically identifying components (resistors, capacitors, switches, integrated circuits, etc.), soldering them into place, checking items off the list, and testing connections.

The kit directions divide the assembly process into discrete sections organized by subsystem, including the display board, mixer, filters, and transmitter. Each section clearly documents what needs to be accomplished, and builders with reasonable soldering skills should have no problem following along. As a bonus, the manual provides

My friends Al and Rick, who had built FM transmitters before, set up an efficient assembly line; Rick wrangled the components, and then Al soldered them into place (Figure A). The Ramsey FM100B's main board, in the early stages of assembly (Figure B). The completed transmitter, ready to fill the airwaves (Figure C).

technical background as to what the subsystems do and how they work, turning assembly into a learning experience.

Many hours later, the kit was assembled and debugged. It was time to try it out. The FM100B has several input options. I hooked my computer into its stereo RCA jacks with a mini-to-RCA adapter cable ($8 at RadioShack).

The next step is to select a broadcast frequency, using the digital frequency selector on the front panel of the transmitter. I live in a big city, where the FM spectrum is crowded, but I found a fairly empty space toward the lower end of the FM band. I dialed it into the FM100B, launched Rhapsody on my PC, queued up several songs, and started broadcasting. Then I tuned the FM radios in my house to the broadcast frequency. The sound on several radios was pure and clean, as clear as any other FM station. A few radios did not work as well at first, but after fiddling with the placement of the FM100B's antenna, I decreased or eliminated the static and fade.

There is definitely a feng shui aspect to antenna placement, as the proximity of walls, windows, metal, and concrete have a big effect on the trans-

mitter's ability to "hit" individual receivers.

The FM100B has a microphone input and mixer knobs, so you can do sophisticated radio station tricks. Assembling and troubleshooting the FM100B was a great learning experience, and it gave the desired result: freedom of movement indoors and out, without interrupting the music.

DON'T PUMP UP THE VOLUME

Home FM transmitters have low power because Federal Communications Commission regulations restrict the ability of any device to cause interference with legal radio signals, especially those put out by licensed stations. The maximum power an unlicensed transmitter can legally develop is 0.08 watts of signal. That's not much to work with, but it keeps the radio spectrum free from interference. Compare this with the 110,000-watt FM signal radiating from a big station such as Los Angeles' KPFK or San Francisco's KQED.

William Gurstelle's latest book is *Adventures from the Technology Underground: Catapults, Pulsejets, Rail Guns, Flamethrowers, Tesla Coils, Air Cannons, and the Garage Warriors Who Love Them.*

For more info, corrections, and discussion on this piece, please visit makezine.com/06/diyhome_casting

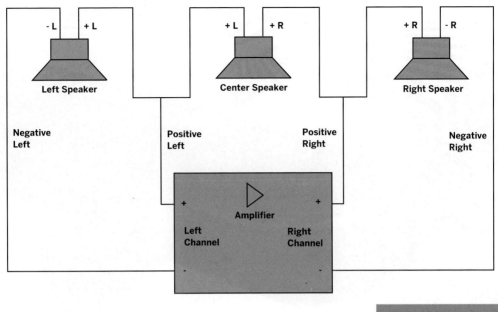

- L **+ L**

Left Speaker

+ L **+ R**

Center Speaker

+ R **- R**

Right Speaker

Negative
Left

Positive
Left

Positive
Right

Negative
Right

▷
Amplifier

+ +

Left
Channel

Right
Channel

− −

SURROUND SOUND, QUICK N' DIRTY

A simple wiring trick derives center channels. By Michael McDonald

Back in the late 1970s, I worked as a movie theater projectionist. That's when I learned this dead-simple trick for wiring a center-channel speaker to augment the right and left channels.

To do it, you wire the left and right speakers to the amplifier as usual, red to red and black to black. Then wire a middle speaker by connecting one pole to the amp's right red output and the other pole to the left red output.

This will take the signal that's shared by the left and right speakers, more or less, and put it in the middle. Then left and right speakers will produce only what is discretely left and right. Needless to say, the center speaker must be capable of reproducing a full range of sound. Add a fifth and sixth

speaker to a quad system, and you'll get true surround sound. The separation is great, even with low-dollar stereos (don't tell the Dolby and THX people). A good six-speaker arrangement would be to use two excellent speakers for front and rear channels with a four-channel stereo, and four less expensive speakers in the corners.

Audiophile purists may complain that adding speakers decreases the overall impedance and throws off the load balance between amp and speakers. But I've been using this hack in cars and homes for decades, always to great effect.

Michael McDonald is an inveterate geek who works for the federal government in Washington, D.C., as an IT specialist.

Illustration by Michael McDonald

Your MP3 player's music will sound minty fresh when played through this li'l headphone amp.

MINT-TIN AMP

Pocket amplifier punches up headphones.
By Warren Young

Headphone amps make portable listening good and loud. Commercial audiophile models can cost $200+, or you can build a great-sounding amp inside a mint tin for around $30, following Chu Moy's popular design. Powered by a 9-volt battery, this amp drives high-impedance headphones to thunderous volumes from even weak sources.

To make one, you need an op-amp chip (like the TI/Burr-Brown OPA132), capacitors, resistors, an LED, and a small prototyping board, plus optional knobs, switches, and other bits, all easily obtained. And, of course, you also need a pocket-sized box, like a Penguin or Altoids tin. See the website listed at the end for a full parts list, along with layout diagrams and more detailed instructions.

Prepare the Protoboard

Start with a small prototyping board such as

RadioShack's model #276-150 — anything that has at least 12 rows of holes and fits into your case. Larger protoboards can be cut down to size with an X-Acto knife. Then solder nine jumper wires as shown in Fig. 1 on page 200.

The jumpers lower down along the edges are what I call "M-jumpers." They tie three two-hole pads together to form a single pad with three free holes. You can make these by taking a one-inch piece of stiff wire, folding it in half, pinching the kink tight with pliers, and then bowing the two ends over. You may also need pliers to stuff the thick middle bit into the hole. Some people make the same connection by threading "S-jumpers" through to the other side, but this makes the jump less visible.

Build the Power Supply

Solder in the capacitors, and use them to orient

Photography and illustrations by Warren Young

everything else. Make sure you put the legs in the right holes — the jumpers you placed previously dictate the path of current flow, so there's no room for "creativity" now. All the electrolytic capacitors are polarized, so you must orient them properly. If one leg is shorter than the other, the short one is negative, and there should also be a stripe or other mark on the negative side.

Now solder the LED and the current-limiting resistor. As with the capacitors, the short/negative leg of the LED should lead to the negative side of the power supply, with resistors in series in between. If you're mounting the LED on the lid of the mint tin, solder the resistor close to the board and cover the joint with heat-shrink tubing. That way, its legs won't flex and snap from repeated openings of the case (see Fig. 1).

On the underside of the board, solder two 2-inch hookup wires to each side of the third row, in the holes marked V+ and V- (for voltage) in Fig. 1. Leave the other ends dangling loose, but don't let them touch while power is applied.

Now figure out a temporary way to add power to the board. For example, you can solder a 9V battery clip's leads into the holes marked "Batt +" and "Batt -." This only needs to work until you start working on the case.

To test the power, set your meter to DC volts, apply power to the board, and measure from the signal ground ("Ground" in the diagram) to the hookup wires going to the V+ and V- holes. With a fresh 9V battery, you should see about +4.5V DC and -4.5V DC at each capacitor, respectively, and their magnitudes should be nearly identical. If you're off by more than a tenth of a volt, check the wiring and look for solder bridges, wayward drips of solder shorting out to neighboring nodes. Fix any problems and unplug the power supply before continuing.

Add the Amplifier

Solder the IC socket to the board with the notch away from the power supply. Leave one or two rows of holes between it and the power supply caps; use two rows if your caps are exceptionally large. Insert the op-amp into the socket. While placing the other components, refer to the pinout diagram on the op-amp's datasheet.

Next, add all the resistors (see Fig. 2 on page 200). Notice in the photo on page 202 that I used jumpers

in place of R5. I almost never install these resistors; they're there to quiet the low-level hiss that you hear with some low-impedance headphones. If you hear a low hiss at normal volume with the audio source disconnected, you can try adding two 47 to 100Ω resistors in the R5 positions. But don't do this otherwise because it will raise the amp's output impedance, inhibiting control over the headphones.

For the remaining resistors, I add them in matched pairs: for each resistor in the left channel, I use my ohmmeter to find another that measures as close to identical as possible for the right channel. I haven't scientifically studied whether this really helps, but it's easy and quick, so why not? It's one less thing to blame if the final product's sound has flaws.

Now add the input capacitors (C2). These aren't polar like the electrolytic power supply caps, so you can orient them any way you like. Axial capacitors are easiest to fit into tight spaces, but most caps can be made to work with a bit of creative lead bending.

Next, take the hookup wires coming from the V+ and V- points in the power supply area and run them to pins 4 and 8 of the op-amp, the chip's V+ and V- points. I recommend you do this on the bottom side of the board, since you'll be adding more wires to the top later. The more wires you can put on the bottom, the cleaner the top side of the finished amp will be, facilitating any repairs and tweaks later.

Finally, add test points at R.out and L.out, and also at R.in and L.in if an alligator clip can't attach to the input capacitor leads directly (see Fig. 2). I use half-inch pieces of stiff wire, usually clippings from resistor and capacitor legs, bent in an upside-down "U." These are temporary, used only for the signal test, the next step.

Test the Amp

Now you're ready to see if you have an amplifier yet or not! Apply power to the board, and let it sit a bit. Then, carefully touch the op-amp and all the resistors' bodies to make sure nothing is overheating. You shouldn't feel any heat at all; if you do, unplug power and find out what's going wrong.

I use a portable CD player for testing since they have volume controls, useful when you're testing a circuit without a volume control. Turn the player's volume all the way down and start it playing.

MINI-CMOY POCKET AMP WIRING DIAGRAM

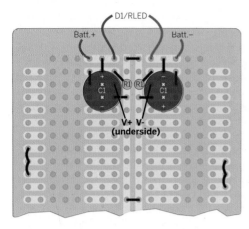

Fig. 1 Add power supply.

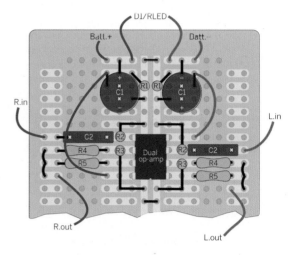

Fig. 2 Add amplifier section.

Fig. 3 Add panel components, enclose, and enjoy!

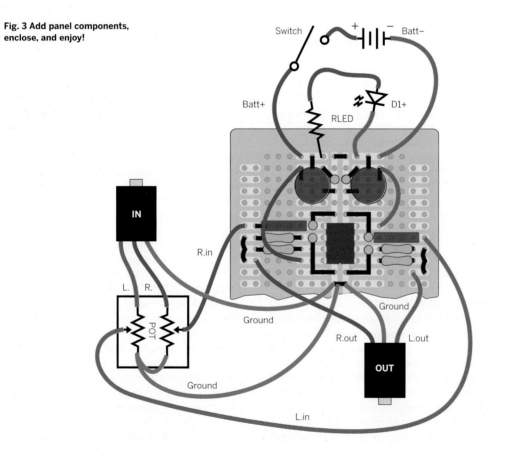

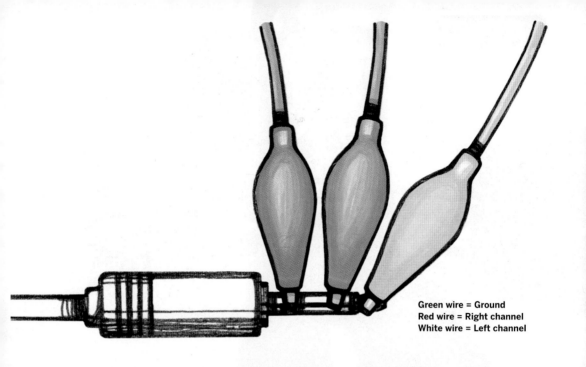

Green wire = Ground
Red wire = Right channel
White wire = Left channel

Hook the source into the circuit. There is no one right way to do this, but I use six alligator jumpers and a mini-to-mini audio jack cable. I plug the cable into the player's headphone out and connect three jumpers to the other end. For ⅛" and ¼" stereo mini plugs, the tip carries the left channel, the "ring" further down is the right, and the long remainder, the "sleeve," is the ground. I clip one jumper to each of these, and then clip the other ends to the corresponding test points and across the ground jumper on the board, as shown above.

Then I use the other three alligator jumpers to tie my amp's output points to a pair of cheap headphones. Don't use your $200 cans for first tests — if something's hooked up wrong, you can blow your cans' drivers out.

With the headphones sitting on the table, slowly turn up the volume on the player until you can hear some sound out of the headphones. Put them on now and adjust the volume. If you get good sound, you're done! You might listen for a while longer and try to stress the amp a bit, but basically, you hooked the amp up right the first try. Once you're satisfied that the amp circuit is behaving properly, try your good headphones with the amp. Sometimes efficient, cheapie 'phones will work fine, but the higher load of big, audiophile headphones causes problems.

Amps that don't work right exhibit different symptoms. With some, you'll hear nothing at all, or the sound will be faint or scratchy, even with the player volume turned up. Others sound fine at very low volumes, but distort when louder. And in some cases, the amp plays for a short while, but then stops. This last case comes from the op-amp shutting down; op-amps often have power-protection circuitry, which is triggered by various wiring faults.

Troubleshooting is difficult, and I can't cover it deeply here. But the main things to check are that all the connections to ground are solid, and that you don't have signals or power going to ground when they shouldn't. Check all connections with your meter; sometimes a connection will look right but have high resistance, in which case it needs to be re-soldered.

If your wiring fully checks out, try adding a second battery temporarily, in series with the first. If that cleans the sound up, your circuit or op-amp are marginal. You can either keep on using extra

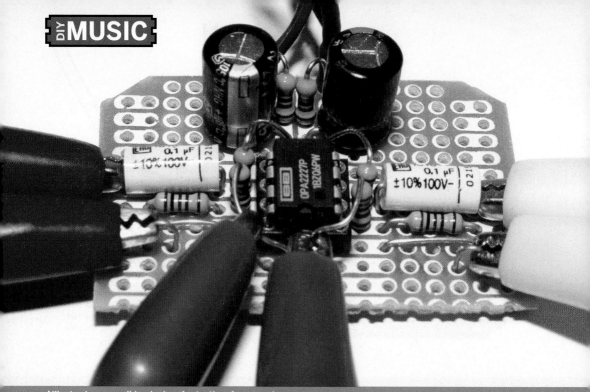

Alligator jumpers all hooked up for testing, from music source (I use a CD player) and headphones. Be sure to use a cheap pair of headphones, because an improperly wired circuit can destroy the headphone's drivers.

voltage and build for two batteries or try to improve your implementation. See the article, "Basic Troubleshooting for Headphone Amplifiers," at tangentsoft.net/audio/trouble.html for more advice.

Set Up the Enclosure

With the amplifier board built and tested, you can decide where on the case you want to place the panel components — the volume control, power switch, LED, and I/O jacks. To minimize the tangle of hookup wire, the exterior ports for these components should be positioned close to where they hook up on the board. Beyond that, it's your aesthetic judgment. Once you have your layout, drill the necessary holes and set it all up with the board in the case and the panel components fastened into their holes, but not connected to the board. This ensures that everything's going to fit before you start soldering again.

Add the Panel Components

When wiring the panel components into the circuit, add them one at a time, and retest the circuit after each. If you connect all the components at once and then have a problem, troubleshooting

becomes very difficult. Also, be sure to test each component on its own, out in the air, and then again once it's fastened in place. This identifies weak wires, bad chassis grounding, and other issues. I've built several amps that worked fine when the panel components were still flopping around on their hookup wires, but failed once the amp was battened down within its case. Fig. 3 on page 200 shows where all the panel components (the switch, LED, in and out jacks, and volume control potentiometer) hook up to the PC board.

First, I hook up the LED and the power switch to the power supply. Once these are working, I add the input and output jacks, remembering the standard stereo plug tip/ring/sleeve ordering described above. I connect and test each one at a time, and use extra-long wires for the inputs. That way, it's easy to splice in the potentiometer, which is the next and final step.

The potentiometer has six pins, three in a row for each channel. The middle pin, the "wiper," connects to the circuit board's input capacitor, at L.in or R.in on the diagram. The pins on either side of the wiper connect to the input jack and to ground.

The completed amp. Unless you want to build it in a day, including the time to get the parts, there's little reason to get RadioShack parts; the big mail-order houses have everything RadioShack has and more, with cheaper prices and better quality. I highly suggest you mail order everything you can, if you can stand waiting a week for the parts to arrive.

If you want volume to increase as you turn the knob clockwise, the ground usually needs to connect to the pin that's on the left, as you face the knob with the pins pointing downward. But it's best to determine the pin arrangement by checking the component's datasheet or testing with an ohmmeter; lowering the resistance between the side pin and the wiper means turning the volume up. If you reverse these connections, the amp will work fine, but the volume knob will operate in the wrong direction.

That's it — now you've got an amp! You can enhance it with various tweaks described on the website, including tuning the gain, adding a DC power jack, using different caps and resistors, and improving the virtual ground circuit.

For the full project tutorial site, including a parts list, background articles and other references, and more detailed instructions, visit

➕ makezine.com/04/headphoneamp.

Silly Putty's been around for 50 years, and in that time it's become an international toy classic. Now you can purchase a 5-pound block of the original Silly Putty at crayolastore.com/category. asp?NAV=PUTTY. The block comes in a box; no plastic eggs included. There's also a Bouncing Putty Mailing List at bulkputty.org/mail/, with instructions on how to get 100 pounds of Dow Corning's Coral Putty at bulkputty.org/ordering/dow.html.

Warren Young is a software developer who used to think the maxim "beware of programmers who carry soldering irons" was funny. He lives in Aztec, NM.

WORLD'S LOUDEST IPOD

iBump crossover lets you crank it up without distortion. By Tom Anderson and Wendell Anderson

Getting a big sound from portable music players should be easier. Fancy new plastic mini-speakers are everywhere, but what if you want to plug into a big old stereo and truly rock the house?

An inexpensive Y adapter cable can split a player's headphone mini-jack into two RCA plugs for left and right channels, but you'll get better sound if you put the low frequencies into a subwoofer and send just the mids and highs to the stereo speakers. That's because the large movements that a speaker cone must make to produce low frequencies will distort the mids and highs — especially at high volume. Our iBump improves this situation with an active crossover circuit, which separates audio into high- and low-frequency channels. Connect the low end to a powered subwoofer, and you'll have proper thump!

The iBump's crossover circuit is based on the classic Linkwitz-Riley audiophile design, which you can read about at Siegfried Linkwitz's site (linkwitzlab.com/filters.htm).

Some newer audio amplifiers include separate outputs for powered subwoofers and filters to keep the lows from the main speakers. We tested these and found that our iBump provides much better sound. Try it and see for yourself. This crossover sounds natural without being boomy or leaving a gap between the highs and lows.

If you don't have a powered subwoofer, you can make one using our article "Resurrecting This Old Amp,"(MAKE, Volume 02, page 110). Start with an eBay bass amp, add large speakers, and presto: you'll have yourself a 'quake. Or buy powered subwoofers from Klipsch, JBL, and others.

Photograph by Wendell Anderson

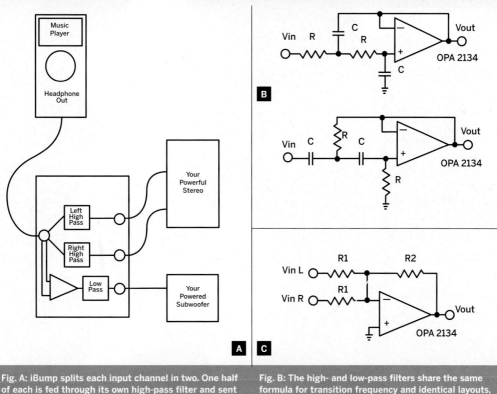

Fig. A: iBump splits each input channel in two. One half of each is fed through its own high-pass filter and sent to the stereo speakers. The other halves are summed, routed through a low-pass, and sent to the subwoofer.

Fig. B: The high- and low-pass filters share the same formula for transition frequency and identical layouts, but with resistors and capacitors swapped. **Fig. C:** The summer circuit combines left and right channels.

Build It

The iBump is an assemble-and-solder project on a single printed circuit board (PCB), with no moving parts. You can build the circuit from our schematic at makezine.com/08/ibump. You can also order the iBump bare PCB, or buy the board pre-assembled and packaged at the same URL. The iBump kit is designed for learning audio electronics, with clear labels and test points that make it easy to probe the circuit with a multimeter or oscilloscope. Once you've assembled the circuitry, enclose the board in a small case.

How It Works

The iBump's all-analog circuitry uses matching high- and low-pass filters to block low and high frequencies. Given resistor value R and capacitor value C, the formula for where they cross over is: $1/(2\pi RC)$. Our iBump uses 23.7kΩ resistors and 0.068μF caps, which gives a crossover frequency of 99Hz, which is close to a bass guitar's G string.

The summer circuit combines left and right inputs before sending signal through the low-pass and to the subwoofer. Human ears are no good at locating low frequencies, so mono is fine here.

Our circuits are all based on the OPA2134

op-amp from Texas Instruments. This is a great amplifier chip with low noise and 0.00008% distortion. The 5532 op-amp from Philips is a less expensive choice, and the high-end OPA2134 may be overkill, but that's the whole point, right?

High Volume Fun

CAUTION: Do not deafen yourself or anyone else with the magnificent power of your iBumped stereo. Wear ear protection.

To test and adjust the iBump, download and play the test tones from makezine.com/08/ibump and use a sound-pressure level (SPL) meter. You can be an audiophile and tweak the levels to achieve a flat frequency response. Or you can just crank up the subwoofer until things start falling off your shelves.

Challenge your neighbors to see whose stereo is loudest when you listen from the street. When the police come, prove that you're not exceeding local ordinances by showing measured sound-pressure levels from your SPL meter!

Tom Anderson and Wendell Anderson are engineers who like to develop audio hardware and software projects.

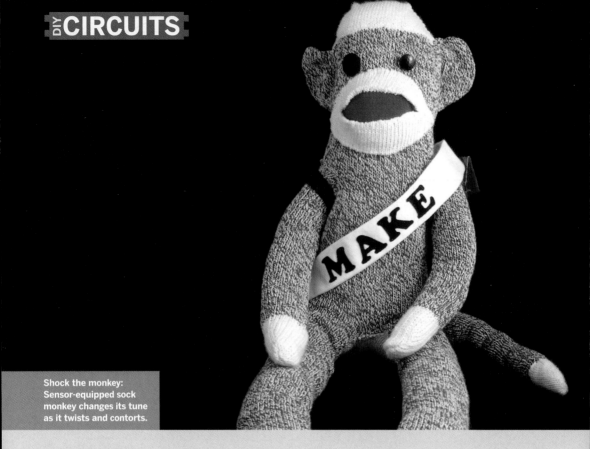

Shock the monkey: Sensor-equipped sock monkey changes its tune as it twists and contorts.

MIDI CONTROLLER MONKEY

A/V monkeyshines with flex sensors and a MIDIsense board. By Peter Kirn

Who said input devices have to be hard and mechanical? Here's one that's as soft, bendable, and easy to play with as a plush toy. In my MIDI primer (MAKE, Volume 07, page 158), we saw how to use the MIDI data specification, originally designed for music, to interconnect both musical and non-musical hardware and software. Now, we'll use this approach to construct a sock monkey instrument you can use to control visuals and sound.

The monkey has flex sensors sewn into its limbs, and it wears a sensor interface board as a backpack. The sensors detect the monkey's movements, and the board converts the readings into MIDI data. This lets the monkey conduct audio-visual symphonies and perform other MIDI magic.

Circuit Bending

Designers have long used off-the-shelf sensors to translate real-world movements into digital realms. For this project, we'll use one of the most commonly used types: a flex (or bend) sensor. This type of sensor was designed by Abrams/Gentile Entertainment and used in the infamous Mattel PowerGlove (a game controller for the Nintendo Entertainment System), as well as in the more recent Essential Reality P5 gaming glove.

The concept behind a bend sensor is simple: its electrical resistance changes as it bends. This lets it work in a circuit just like other resistive sensors, such as light sensors or potentiometers. Variations in voltage that result from bending can be converted to a digital value and transmitted as data to a computer or other device. The typical

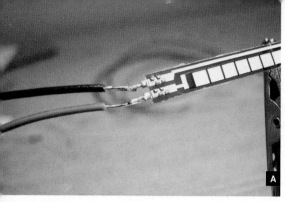

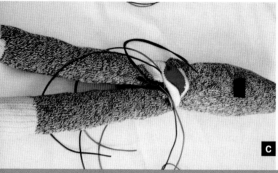

Fig. A: Two leads soldered to the end of the bend sensor.
Fig. B: Bend sensors slip into muslin sleeves sewn onto the insides of the sock monkey's limbs.

Fig. C: Partially assembled sock monkey, with sensor wires and reinforcement sewn at base of monkey's back.
Fig. D: Sensor wires connected to MIDIsense board.

bend sensor application is a data glove, which uses one bend sensor for each finger. Here, we'll try something different.

Monkey See, Monkey Do MIDI

Constructing sock monkeys is well-documented and easy to do. All we need to add are the bend sensors, some wires, and a way of translating the movements into MIDI. There are several off-the-shelf micro-controller boards that could perform this translation, but Limor Fried's open source MIDIsense board, which we also used in the MIDI Control primer, is ideal for our monkey. It's small, inexpensive, and designed to read sensors and nothing more. The MIDIsense takes input from six resistive sensors and has a MIDI jack output rather than serial or USB. This makes it easy to use the monkey with standard standalone music and video hardware, rather than only with computers.

For this design, I collaborated with my sister Anne Kirn, who has a knack for designing and making durable plush toys. Instructions for making sock monkeys are included in the packaging for Rockford Red Heel socks and are also widely available online, so I'll focus on the elements that are unique to this project.

Make the Monkey

1. Cut up the socks. Follow standard sock monkey instructions for cutting up the 2 socks. Basically, 1 sock forms the legs and body, and pieces of the other are used to form the tail, arms, and mouth.

2. Prep the sensors. For each sensor, cut an 8" length of a two-contact ribbon cable or a pair of hookup wires. Strip the ends and solder 1 end to the sensor contact prongs (Fig. A).

3. Reinforce the connection. Place a small piece of heat-shrink tubing over each connection between the wire and sensor. Make sure you have a size that fits loosely. Use a heat gun or hair dryer to shrink the tubing around each joint, and then reinforce the end of the sensor itself with electrical tape. (We found that all the heat caused some of the backing to peel away from the sensor; the tape made sure it continued to bend properly.)

4. Prepare the appendages and sensor sleeves. Sew the edges of the arms and legs per sock monkey instructions, so you have empty, inside-out appendages. Cut 5 sensor-length sleeves out of muslin, and sew them to the outsides of the inside-out legs, arms, and torso. For

MATERIALS

Rockford Red Heel socks 1 pair, men's extra large
Polyester fiberfill Or other stuffing
Muslin fabric For the "sensor sleeves"
Felt To reinforce attachment points
Thread To match sock color
Buttons (2) For eyes
Fleece or other non-raveling fabric, narrow double-fold bias tape, mini anorak/jacket snaps, and snap setter tool For backpack
Bend/flex sensors (6) Jameco part #150551 jameco.com
MIDIsense board Available at ladyada.net/make/midisense
Heat-shrink tubing RadioShack sells an assortment
Electrical tape
Solder Rosin-core/lead-free, any finer size
Insulated copper hook-up wire Or two-contact ribbon

TOOLS

Wire stripper Jameco #159290 is a good size
Wire cutter
"Helping hands" Or vise (optional)
Heat gun Or hair dryer
Soldering iron with a fine tip Weller is a good brand
Soldering stand and de-soldering tool or braid
"Micro" flat-head screwdriver
Computer or other MIDI-to-USB interface Such as the Edirol UM-1EX or M-Audio Uno
Sewing machine (optional)
Needle and scissors

the torso, turn the uncut sock inside out and attach the sleeve midway between what will become the head (the sock's toe) and waist (just above the heel). You won't need a sleeve for the tail sensor, since the tail is smaller than the other appendages.

5. Insert the sensors. Turn the various parts of the monkey parts outside out as per the sock monkey instructions. Then gently push the sensors and wires into their sleeves (Fig. B). To make it easier later, you can label by body part the wire ends of each sensor.

6. Add the stuffing. Stuff the monkey with the fiberfill. Make sure you add enough so that the monkey feels firm and not floppy, but not so much that he refuses to bend easily.

7. Assemble the parts. At this point, you'll have some stuffed limbs and a lot of loose wires. Follow the sock monkey instructions to finish the assembly. Then cut a hole just large enough for all the wires at the base of the monkey's back, just above the tail. Reinforce the hole by sewing on a small piece of felt (Fig. C).

8. Feed the wires. Label the sensor wires, if you haven't already. Feed the wires out of the base of the monkey's back so they can connect to the sensor board. Pull the wires through the body so they all meet on the way out of the hole, and then pull them out together.

9. Make the sensor-board backpack. Assemble the MIDIsense board according to the enclosed instructions. Then sew together a monkey-sized backpack with a board-sized pouch and a hole in the bottom to allow the wires to feed into the board from the monkey. Small straps let the monkey wear the board snugly. We made a pack with a snap closure out of fleece trimmed with double-fold bias tape.

10. Attach the wires to the sensor board. Strip the wires from the sensors and attach them to the 6 screw terminals on the MIDIsense board. Each terminal has 2 holes, one for each wire. Insert the wires, and screw the terminals down tightly using the micro screwdriver (Fig. D). We found that the wires would sometimes pop out when torquing the screws, so we added a bit of solder to make the connections more permanent. Carefully put the board and wires into the backpack.

11. Add some finishing touches. Eyes give the monkey personality, and using 2 buttons of slightly different size give it even more character. From there, accessorize at will. Now the monkey is ready for action!

Configure the Board

1. Set up the sensors. Connect the monkey to your computer via the MIDI-to-USB interface. Use the MIDIsense software to enable all 6 sensors, and select Control Change (CC) type messages for each of them to transmit. There are many MIDI CC messages, but they all transmit continuous values rather than on/off, which is what we want. For best results, use one of the message types with a range of 0 to 127.

If you're planning to use your monkey with external MIDI hardware rather than computer software, make sure you choose messages that your hardware can interpret; most hardware documentation includes a MIDI Implementation Chart that you can check for this.

2. Calibrate the sensors. Prep each sensor so it will transmit the full range of data it's sensing. To do this, use the MIDIsense application to calibrate the low, middle, and high end of the sensor's values. Bend each sensor to its maximum, then partway, and then fully straighten while tracking the corresponding slider position onscreen. Then set the green, yellow, and red arrows to those locations. You may need to invert the high and low values so that the sensor sends 127 when bent and 0 when unbent.

Sample Application

Now our monkey MIDI is all set up. But what can you do with it? The answer, as we saw in the MIDI Primer, is anything you like; you can control music and visuals software as well as any hardware with MIDI ports. Once MIDIsense has configured the ports, you don't even need a computer; you could hook your monkey up to a hardware beatbox and use the monkey in place of the box's own knobs and sliders.

Using the free Quartz Composer tool included in the developer tools that ship with Mac OS X 10.4 and later, I rigged up a simple MIDI-controlled, animated monkey for testing and demonstration purposes.

The Quartz Composer patch I assembled takes an image of a banana and assigns it to a Particle System, which is a way of animating lots of objects onscreen with similar but independent movements. (In this case, the bananas are the "particles," spurting out in clouds of floating fruit.) Quartz Composer translates MIDI input from each of the monkey's limbs into the size of the bananas in the particle system; if the sensors are straightened, the bananas are hidden, and as you flex the sensor, the clusters of floating bananas appear and grow. You could add sound for an audiovisual interactive sock-monkey installation. For more on this patch and the calibration and programming procedures for the monkey, with videos, see makezine.com/08/diycircuits_monkey.

Of course, you can also ignore this monkey business and do your own thing, putting the sensors into another object, or using them to control something else. That's the whole fun of this. But the next time you see a stuffed monkey in a DJ booth, you'll know why he's there.

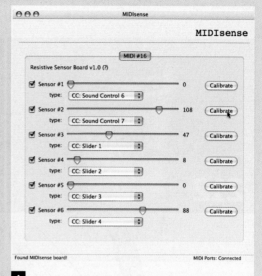

A

B

Fig. A: The MIDIsense board supports 6 sensors, which you configure and calibrate through a simple window. Set sensors as MIDI continuous controllers so that they generate a constant, reactive stream of message data.

Fig. B: A patch built in Apple's free Quartz Composer lets the MIDI monkey trigger on-screen bursts of bananas, which are animated via a particle system — delicious!

Composer and media artist Peter Kirn is the author of *Real World Digital Audio* (Peachpit, 2005), and editor of createdigitalmusic.com.

212

222

232

244

254

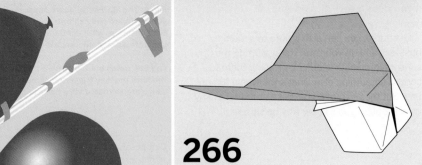

265

266

THE BEST OF
FLIGHT &
PROJECTILES
PROJECTS

from the pages of MAKE

>>

ACME PRODUCTS AND
THE ROADRUNNER HAVE
NOTHING ON US

Call it Big Bang Envy, call it pyromania, call it whatever you like, but large numbers of us in the human herd love to see and hear things explode and fly apart. When this is on a battlefield or due to some unfortunate Act of God? Not so much. Under controlled and reasonably safe conditions? Cover our ears and bring it on! Here are some of our favorite jet, rocket, and cannon projects from the pages of MAKE. And for those who like their aeronautics a little more peaceful, we offer a blimp and a rubber band-powered ornithopter project.

THE JAM JAR JET

By William Gurstelle

Don't think you can build a jet engine at home? Here's a simple jet engine — a pulsejet — that you can make out of a jam jar in an afternoon. All it takes is bending some wire and punching a few holes. >>

Set up: p.215 **Make it:** p.216 **Use it:** p.219

JOIN THE JET SET

Turbojets and fanjets contain hundreds of rotating parts. But the ancestors of these designs, called pulsejets, convert fuel and air into propulsive force by using a fixed geometry of chambers and ducts, with no moving parts. The simplest pulsejet is the Reynst combustor, which uses one opening for both air intake and exhaust.

The pioneering Swiss jet engineer Francois Reynst discovered this combustor as a pyromaniac child. He perforated the lid of a glass jar, put a small amount of alcohol inside, and lit the top. Flames shot out of the hole and then were sucked back into the bottle before being ejected again. This almost-magical process repeated until all of the fuel was expended. Reynst had discovered a jar that literally breathed fire, like St. George's dragon. Our jam jar jet is based on Reynst's discovery.

William Gurstelle enjoys making interesting things that go whoosh then splat. He is the author of *Backyard Ballistics* (2001), *Building Bots* (2002), and *The Art of the Catapult* (2004). Visit backyard-ballistics.com for more information.

HOW IT WORKS

When the fuel and air inside the jar first ignite, the jam jar jet generates a burst of hot gas, raising the internal pressure and pushing the gas out. The exiting gas leaves a slight vacuum behind, and fresh air rushes back into the jar to fill the void. More methanol mixes with the fresh air in the still-hot jar, triggering another combustion. Scientists call the resulting cycle relaxation oscillation.

The V-1

The pulsejet engine is simple, cheap, and powerful, but isn't used in commercial aviation because large versions are incredibly noisy, and they vibrate like gigantic, unbalanced chain saws. Invented in the early 20th century, pulsejet engines had no practical use until German scientist Paul Schmidt developed a no-frills but dependable pulsejet-powered cruise missile. This was the notorious V-1 rocket, a.k.a. the "buzz bomb," which terrorized Britain during World War II.

The V-1 was a 25-foot tube with two stubby wings. Its simple engine gave the missile a range of about 150 miles with an explosive payload of nearly a ton. The first V-1 hit London on June 12, 1943. At the height of their use, 190 were launched daily. The V-1 attacks ended only when the Allies marched back through Europe, and seized the missiles' launch sites, which were located across the English Channel.

INTAKE AND EXHAUST

A ½" diameter hole drilled in the center of the Mason jar lid serves as both the air intake and exhaust port. Most functional pulsejet engine designs use two separate ports, but because the combustion cycle's intake and exhaust stages are not simultaneous, pulsejets can also use a single port. The continuous combustion cycles of more advanced jet engines, such as turbojets, require separate intake and exhaust ports.

HEAT DIFFUSION

The copper diffuser also conducts heat and transfers it out to the four wires that it hangs from. The long wires radiate heat to the air outside, which takes some thermal expansion strain off the jar, reducing the risk of cracking the glass.

AIR FLOW

Inside the jar, a conical copper diffuser ring guides the flow of the gases so that they follow a simple whirl pattern. This improves the efficiency of the combustion cycle.

Illustration by Tim Lillis

SET UP.

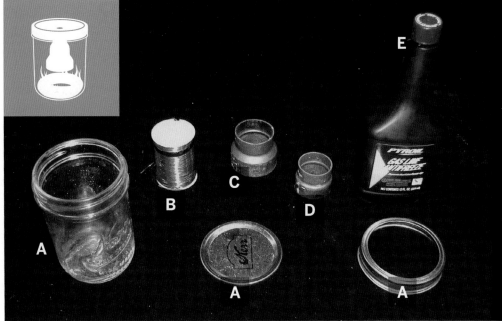

MATERIALS

[A] Pint-sized Mason jar with extra screw caps and lids

[B] 22- to 26-gauge magnet wire (thin enamel-coated copper)

[C] 1½" to 1¼" copper drain/waste/vent (DWV) reducing fitting

[D] 1¼" to 1" copper DWV reducing fitting Available at home centers or hardware stores. These two pipe fittings are for the conical air diffuser.

[E] Small bottle of methanol Available at auto supply stores as gas-line antifreeze; common brands include Heet and Pyroil. Methanol absorbs water readily, which is why it works well as gas-line antifreeze. But this property also causes it to go bad quickly, so you should always use fresh methanol.

[NOT SHOWN]
Package of long fireplace matches, or a long-handled barbeque lighter

Table salt (optional)

Boric acid crystals (optional)

1' long, 1" diameter plastic or metal pipe (optional) **Optional items are for experimental variations.**

TOOLS

Electric drill with ½" and ⅛" drill bits

Wire cutters

File or sandpaper

Teaspoon measure

Cookie sheet

Refrigerator/freezer

Safety glasses

Gloves

SAFETY GUIDELINES

This is a jet engine you're building, a tempest in a tea-pot. I've never had any problems with this design, but no one — not me, not this magazine — can guarantee your safety. If you do choose to go forward with this project, here are some important safety measures.
1. Do not experiment with different sized jars and openings. A too-large jar with a too-small opening might result in an explosion of glass shards.
2. Use no more and no less fuel than directed. Wipe up any spilled fuel immediately.
3. Use only the parts listed in the directions. These are proven to work safely, and I haven't tried or analyzed all the substitutions that people might think of.
4. Wear gloves and safety glasses or goggles.
5. Do not handle the jar for 5 minutes after a successful run, and then be sure to tap it first to make sure it is cool enough. The Reynst combustor is an extremely efficient heating device, and it gets hot enough to burn skin after just a few seconds of run time.
6. After a long run, the glass jar may crack. If so, carefully sweep the entire assembly into a bag without touching it. Seal the bag and throw it in the trash. Jars are cheap enough, so just get another one.
7. Keep spectators at a safe distance.
8. Always ignite the engine with a long-handled match or barbeque lighter, to avoid getting burned by the pulse of hot gas that immediately follows ignition.
9. Examine all parts for wear before and after use. Discard any worn parts.
10. Always use common sense before, during, and after running the jam jar jet.
11. This is an outdoor project only.

MAKE IT.

CONSTRUCT YOUR JAM JAR JET

START >>   Time: **An Afternoon** Complexity: **Low**

1. DRILL THE PORT

Drill a ½" diameter hole in the lid of the jar. Use a file or sandpaper to completely remove the burr. If the hole is so jagged that it cannot be made smooth and round, discard the lid and re-drill another one.

2. DRILL THE DIFFUSER HOLES

Drill four ⅛" diameter holes in the small copper adapter. The holes should be located about ¼" down from the smaller, 1" diameter end. Space the holes evenly around the perimeter at 90, 180, and 270 degrees from the first hole.

3. ASSEMBLE THE DIFFUSER

Insert the large end of the small copper adapter into the small end of the large copper adapter. Press-fit them together firmly. This forms the conically shaped jet diffuser and heat sink.

4. CUT THE DIFFUSER WIRES

Cut four 4" long wires from the spool of magnet wire.

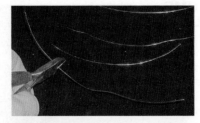

5. ATTACH THE WIRES

Loop one wire through each one of the holes you just drilled, and tie a knot. Extend the other end of the wires outward, radially, from the diffuser cone.

6. SUSPEND THE DIFFUSER

Center the copper diffuser in the middle of the jar. Crimp the wires over the edge of the jar so that the cone hangs suspended close to the top of the jar, with a gap of about ¼" between the diffuser and the top.

7. ADD THE FUEL

Carefully pour or use an eye-dropper to measure and add 5 to 10ml (roughly 1-2 teaspoons) of methanol into the bottom of the jar. You can vary the amount of methanol by a small amount to improve performance. At most, the methanol should just cover the bottom of the jar.

8. CLOSE THE JAR

Screw the Mason jar lid down onto the jar and over the copper wires. The lid will hold the diffuser cone securely in place at the top of the jar.

9. VAPORIZE SOME OF THE FUEL

Prepare the jar by letting it sit in the freezer for two minutes. Hold your thumb over the opening in the lid. Vigorously swirl and shake the methanol inside the jar. Place the jam jar jet on a cookie sheet and place the cookie sheet on a secure surface, away from any flammable objects.

NOTE: When you remove your finger from the hole, you should notice a slight pressure release, and the jar should make a very faint "pffft" sound. If you feel no slight pressure and hear no sound, shake the jar again. If there is still no pressure, there is a leak in the seal of the jar that you'll need to fix.

10. FIRE IT UP

Wearing safety glasses and gloves, hold a flame over the opening in the jar's lid.

The fuel will ignite, and for the next 5 to 15 seconds, the jam jar jet will cycle, pulse, and buzz, running at a low but audible frequency of about 20Hz, depending on conditions in the jar and in the surrounding air. With the lights down low, you'll enjoy a noisy, deep blue pulse of flame that grows and shrinks under the lid as the jar breathes fire. It's an amazing effect.

Pint-Sized Fireworks

During the air-intake part of the cycle, the bottom of the jam jar jet glows brightly. The photo on page 210 shows the blue flame you'll get from burning straight methanol, and this photo (at right) shows the yellow variant that comes from adding a little salt to the fuel. By adding salt or boric acid crystals, you can color your flames in a variety of attractive, retina-burning hues, as described on the next page.

TIPS AND TRICKS FOR YOUR JAM JAR JET

FLAMBÉ RECIPES

Here are some interesting variations on the jam jar jet that you can experiment with:

1. For a bright yellow flame instead of the blue, add a pinch of table salt to the methanol.

2. For green-colored flame, add a pinch of boric acid crystals to the methanol.

3. To amplify the sound of the jet, hold a tube a half inch or so above the hole. You can use a metal or plastic pipe, and even the cardboard from a roll of paper towels will last a little while.

Use pliers or a gloved hand to hold the tube in position. Experiment with the length and diameter of the tube. When the size is right, you'll be rewarded with an unmistakably loud, deep, resonant buzz.

4. Some enthusiasts make Reynst combustors with metal jars instead of glass, and outfit them with resonator tubes permanently attached above the hole. These are sometimes termed "snorkelers."

The most advanced snorkelers also have fuel-feed systems that drip methanol into the combustion chamber, which allows them to sustain combustion for long periods of time.

TROUBLESHOOTING

If the methanol burns with a single big whoosh instead of pulsing:

» Check the size of the hole and make sure it is accurately drilled to a ½" diameter.

» Be sure to place the jar in the freezer for two minutes before lighting. In my experience, slightly cooling the fuel and the jar improves performance.

» Make certain the jar is charged with the recommended amount and type of fuel.

If you hold the long match over the opening and it doesn't ignite, or it does ignite but the pulse is weak:

» Make sure the methanol is fresh.

» Cool down the jar in the freezer for two minutes.

» Start with just one teaspoonful of fuel in the bottom, and vary the amount slightly until you get better performance results.

» Check the seal by listening for the "pffft" when you remove your finger from the hole. If necessary, rejigger the lid to get a good seal.

» Reposition the diffuser by adjusting the support wires, or try shortening the diffuser by removing the bottom section.

If the jar cracks:

» Carefully dispose of the broken jar and replace it with another one of the same size. The Reynst combustor/pulsejet is a very efficient burner and therefore extracts a lot of heat from the fuel very quickly. If the jar you're using cannot handle the rapid expansion, it will crack.

RESOURCES

Pulsating Combustion: The Collected Works of F.H. Reynst, Pergamon Press, 1961

Homemade pulsejets webpage and discussion forum: pulse-jets.com

Larry Cottrill's *jetZILLA*, an online magazine of amateur jet propulsion: jetzilla.com

SODA BOTTLE ROCKET

By Steve Lodefink

You don't have to be Burt Rutan to start your own rocket program. With a few empty soda bottles and some PVC pipe, you can build a high-performance water rocket. **>>**

Set up: p.224 Make it: p.225 Use it: p.230

LIQUID FUEL ALTERNATIVE

I've been a big fan of model rocketry since I built my first Estes Alpha back in third grade. Nothing is more exciting to a 9-year-old proto-geek than launching a homemade rocket. But flying those one-shot solid-fuel rockets can burn a hole through a young hobbyist's wallet faster than they burn through the atmosphere, and with today's larger, high-powered rockets, locating and traveling to a safe and suitable launch site can require substantial planning and effort.

Instead, you can use 2-liter carbonated drink bottles to build an inexpensive, reusable water rocket. The thrill factor is surprisingly high, and you can fly them all day long for the cost of a little air and water. It's the perfect thing for those times when you just want to head down to the local soccer field and shoot off some rockets!

Steve Lodefink works as an interactive designer and web producer for The Walt Disney Internet Group in Seattle.

(WATER) ROCKET SCIENCE

The soda-bottle rocket works the same way as those little red and white plastic rocket toys you had as a kid.

The parachute is packed in the nose and wants to expand, but the nose stays on the rocket during ascent thanks to the upward acceleration, which pushes the rocket up against the nose as the nose gets pushed back from wind resistance.

In this design, the launch tube extends fully into the bottle, which boosts performance by acting as a sort of piston, letting the rocket shoot up some distance before it starts releasing water and losing pressure. The tube also acts as a launch guide, helping to keep the rocket headed straight.

O-ring creates a seal, so the pressure can build.

The simple release mechanism, triggered by pulling a wire retaining pin off a grooved section of PVC, is robust and reliable.

Compressed air forces a jet of water out through the exhaust nozzle, producing thrust and sending the rocket skyward.

PRESSURE TESTED

Two-liter carbonated drink bottles are made to withstand high internal pressures, so they're natural water-rocketry material.

PSHHHT
PSHHHT

SET UP.

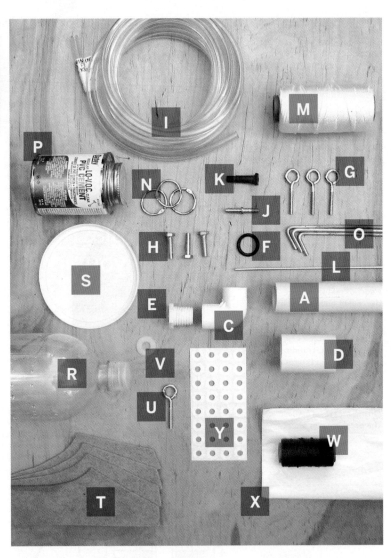

LAUNCHER PARTS

[A] 4" length of 1" PVC pipe **For the release body**

[B] 50" length of ½" Schedule 40 PVC pipe **For the launch tube (not shown)**

[C] ½" PVC elbow pipe **For the end cap**

[D] 1" PVC pipe coupler **For the release collar**

[E] ½" PVC plug cap

[F] Rubber O-ring, 22mm outside diameter (OD)

[G] Eyebolts (3)

[H] Hex bolts (3)

[I] 15' length of ⁵⁄₁₆" OD x ³⁄₁₆" inside diameter (ID) flexible vinyl tubing

[J] ³⁄₁₆" hose barb

[K] Tire air valve

[L] ⅛" music wire **For the release spring**

[M] Nylon cord

[N] Small binder rings (3) **For stay clips**

[O] Small tent stakes (3) **For stays**

[P] PVC cement

[Q] Bicycle pump with pressure gauge **(not shown)**

ROCKET PARTS

[R] 2-liter carbonated drink bottles (3)

[S] 4" deli cup lid

[T] Fin material, such as balsa, thin plywood, or Plastruct sheeting

[U] 2" eyebolt

[V] Medium nylon washer

[W] Kite string

[X] Large garbage bag **For parachute material collar**

[Y] Round hole reinforcement labels

[Z] Quick-set epoxy **(not shown)**

TOOLS

Hacksaw

Utility knife

⅛" file

Drill

Locking pliers

120-grit sandpaper

Thread-cutting taps and dies (optional)

Photograph by Kirk von Rohr

MAKE IT.

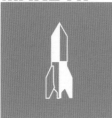

BUILD YOUR SODA-BOTTLE ROCKET

START »   **Time: An Afternoon Complexity: Low**

1. BUILD THE LAUNCH TUBE

1a. Cut the launch tube. Use a hacksaw to cut the ½" PVC pipe to length. A 50" tube will make a launcher that's a convenient height for most adults to load from a standing position. The ½" Schedule 40 PVC pipe fits perfectly into the neck of a standard 2-liter soda bottle.

1b. Install the O-ring. Mark the O-ring position by fully inserting the launch tube into the type of bottle that you plan to use for your rockets. Locate the O-ring roughly in the middle of the bottle's neck. Use the edge of a file to cut a channel for the O-ring to occupy. Rotate the launch tube often while you work to maintain an even depth of cut, and be careful not to go too deep. Then slip the O-ring over the launch tube and seat it in the groove.

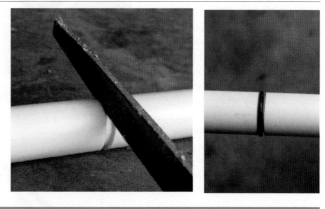

2. BUILD THE RELEASE MECHANISM

2a. Assemble the release body. Cut a 4" length of 1" PVC pipe and press-fit it into the 1" coupler. Cut squarely and deburr all PVC cuts with 120-grit sandpaper.

2b. Cut the release spring slots.
Insert your bottle's neck into the
release assembly and determine the
distance of the bottle's neck flange
from the end of the bottle. Mark the
flange location on the 1" pipe coupler
and use the hacksaw to cut a ³⁄₁₆"
long slot on each side. These slots
will hold the retainer/release spring.

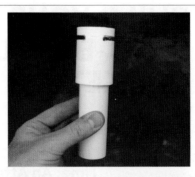

2c. Attach bolts. Drill three evenly
spaced holes through the release
collar and release body together, and
thread the three eyebolts into these
holes. Similarly, drill three holes in
the lower release body tube to accept
the three hex bolts.

Optional: Cut threads in these holes
with a tap to accept the hex/eyebolts.
If you won't be tapping them, drill the
holes just undersized, and the bolts
will cut through the PVC just fine. Be
careful not to strip these holes.

**2d. Make the retainer/release
spring.** Bend a piece of ⅛" music wire
1½ turns around a piece of scrap ½"
pipe clamped into a vise. The spring
should be roughly V-shaped.

Make a retainer clip for the spring by
drilling two holes in a scrap of ½" pipe.
The ends of the compressed spring
will fit into these holes. This keeps the
spring closed above the bottle's neck
flange, holding the bottle in place.

Tie a 15' trigger line to the clip. At
launch time, you pull the clip off with
this trigger line, which allows the spring
to open and the rocket to take off.

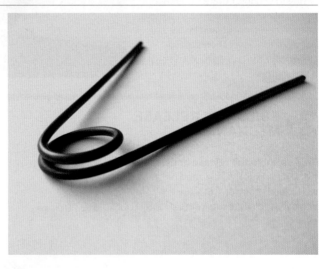

3. MAKE THE AIR HOSE

3a. Drill a ³⁄₁₆" hole in the center of the threaded ½" end cap, and press in the ³⁄₁₆" barb fitting.

3b. Thread the end cap into the elbow fitting and tighten it with a wrench. Using PVC cement, solvent-weld the elbow to the bottom end of the launch tube. The end cap is tapered, so it should require no Teflon tape or adhesive.

3c. Use a utility knife to strip the rubber from the tire valve to one inch from the end. Insert the valve into one end of the ³⁄₁₆" flexible tubing.

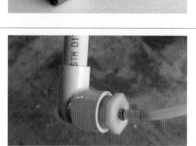

Optional: You can use a die to cut threads into the plain end of the valve stem, and then twist it into the tube.

3d. Push the other end of the air tube onto the barb fitting.

4. SET UP AND TEST THE LAUNCHER

4a. Stake down the stays. The launcher is installed in the field using three stays, each consisting of a 72" length of light nylon cord. Stake one end of each line to the ground, and clip the other end of each stay to the eyebolts on the launcher.

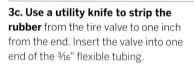

Photograph by Topher Lucas (step 4a)

4b. Pressure-test the launcher. Now is a good time to ensure that all the launcher's connections are airtight. Fill a bottle to the top with water (this way, if the bottle fails this pressure test, it will not explode). Quickly invert the bottle and slip it onto the launcher. A little Vaseline inside the neck will help the bottle make a seal against the O-ring. Squeeze the release spring into the slots in the release collar and clip it in place. Use the bicycle pump to pressurize the system to 70psi. If the pressure holds steady, all is well. Otherwise, fix any leaks and test again.

5. ASSEMBLE THE ROCKET

Water rocket designs range from a simple finned bottle to elaborate six-stage systems with rocket-deployed parachute recovery and on-board video cameras. Ours is a painted single bottle affair with wood fins and parachute recovery. Chute deployment is by the passive "nose cone falls off at apogee" method.

5a. Cut 3 or 4 fins from a light, stiff material such as balsa, thin plywood, or Plastruct sheeting. Roughen the surface of the bottle, where the fins will attach, with some sandpaper and then glue the fins to the bottle with epoxy, or a polyurethane adhesive such as PL Premium. Sand the leading edges smooth.

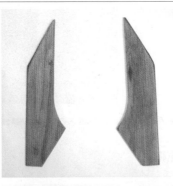

TIP: Gluing on the fins at a slight angle will cause the rocket to spiral as it flies, adding stability to the flight.

5b. Make the nose section by cutting off the neck and base of another bottle. Cut a 6" circle of material from a third bottle. Make a radial slit on the circle, fashion it into a nose cone, and cement it in place atop the nose section.

5c. Outfit the nose cone. When the cement is dry, turn the nose over and epoxy the 2" eyebolt to the inside tip of the nose cone. This bolt serves as a place to anchor the parachute shock cord. It also adds extra mass to the nose section, which will help to pull this section off as the rocket decelerates, exposing the parachute.

5d. Make the nose-stop. Cut the center from a 4" deli container lid, leaving only the outer rim. Cement the rim onto the rocket's lower "motor" section such that it allows the nose to sit loosely and straight on the rocket. This "nose-stop" will prevent the nose from being jammed on too tightly by the force of the launch, which ensures that the nose will separate off and deploy the parachute during descent.

5e. Make a parachute canopy from a 36" or so circle cut from a large trash can liner. For best results, use 12 or more shrouds made from kite string. Apply paper reinforcement labels to both sides of the chute, where the shrouds attach, to keep the chute from tearing. Tie the loose ends of the shrouds to a nylon washer or ring to make the chute easy to manage.

TIP: Ideally, a parachute's shrouds should be a bit longer than the diameter of the chute canopy.

5f. Epoxy a parachute-anchoring ring to the top of the rocket base and tie the parachute to the ring with a short cord. Cut a 4' connecting cord and tie it between the nose cone eyebolt and the parachute-anchoring ring. This cord will keep both halves of the rocket together during descent.

TIP: Make sure the connecting cord is long enough to allow the parachute to completely pull out from the nose cone.

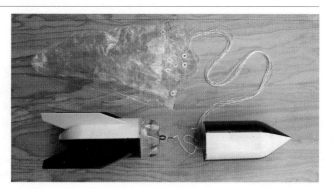

USE IT.

THREE, TWO, ONE, LIFTOFF!

SAFETY

Water rockets produce a considerable amount of thrust, and getting in the way of one could cause severe injury. Take the same common-sense precautions that you would when launching any type of rocket. Make sure that everyone in the area is clear of the rocket and aware that it is about to launch. Do a verbal countdown, or yell something alarming, such as "Fire in the hole!" just before you launch the rocket.

SELECTING A SITE

A well-built, single-stage water rocket is capable of flying several hundred feet into the air and drifting a considerable distance during descent. Less well-built rockets may choose to travel several hundred feet to the side. In any case, you need to choose a launch site that is large and open enough to allow your rocket to wander a bit without getting lost in a tree, or on the roof of some Rottweiler's doghouse.

Big sports fields are the logical site choice for most of us, but if you are in a rural area, any wide-open space will work as a rocketry range. Be sure to take wind direction into consideration when deciding on which side of the field to set up the launcher.

OPERATION

1. Set up the launcher by clipping the three support stays to the launcher's eyebolts. Take up any slack in the lines and stake the other ends to the ground, evenly spaced. Uncoil the air hose and tuck it under one of the tent stakes to keep it from coiling. Attach the bicycle pump to the air hose.

2. Pack the parachute. Grab the center of the parachute canopy between your thumb and fore-finger, and let it hang. Draw the chute through your closed hand to gather it, and then fold it into thirds, zigzag style. Lay down the parachute shrouds on the ground, and accordion-fold them back on themselves. Don't wrap the shrouds around the canopy; just slide the whole thing into the nose section and bring the two halves of the rocket together.

TIP: Line the nose section with parchment paper or Teflon baking sheet liner to help the parachute deploy smoothly. A light dusting of talcum powder will also help keep the chute from sticking.

3. Fill and set up the rocket. While holding the nose in place, turn the rocket over and fill it one-third full of water. Apply petroleum jelly to both the launch tube O-ring and the inside of the bottle's mouth to help it slip onto the O-ring. Hold the mouth of the rocket up to the launch tube. In one smooth motion, pivot the rocket up, slide it down

onto the launch tube, and twist it back and forth, if necessary, to help it engage the O-ring seal.

4. Compress the release spring into the slots of the release collar, locking the bottle flange in place. Install the spring-retaining clip on the ends of the spring, and carefully run the trigger line back to your "ground control" area.

TIP: Tie the release spring to the launcher with some string to keep it from flying across the field and getting lost every time you launch.

5. Launch! Jump over to the bicycle pump and bring the pressure up to about 70psi. When you are ready, clear the area, count down to zero, and pull the trigger line, releasing the spring and freeing the rocket.

If all goes as planned, your rocket will shoot upward, dispensing with its entire fuel load in less than half a second. Then it will begin to decelerate, and the nose will want to separate. As the rocket reaches apogee, the two halves will come apart, deploying the recovery chute and bringing the craft gently back to Earth, much to the excitement of the assembled crowd.

Experiment with different amounts of water and air pressure until you find the sweet spot that sends your rocket the highest. Don't exceed the amount of air pressure that your bottle is designed to withstand; 70psi seems to be about right for a standard 2-liter soda bottle.

The author kindly (and we'd say somewhat naively) allowed the MAKE team to borrow his rocket. After making the rounds through numerous photo shoots, kids' birthday parties, and a high-decibel, code-violating Thanksgiving affair, the rocket sustained the damage you see here. We deeply regret the mishap and promise not to do it to the next rocket we've asked Steve to build for us.

ADVANCED DEVELOPMENT

Once you've tasted the joys of basic water rocketry, you will inevitably want to improve and refine your rocket designs. If you want your rockets to fly higher, the best improvement you can make is to increase the volume of the rocket "motor." This is usually done by splicing or otherwise coupling two or more bottles into a single pressure chamber. There are also various schemes for building multi-stage rockets, as well as more elaborate parachute deployment setups.

There is an abundance of water rocket information available online. Here are a few sources to get you started:

Antigravity Research Corporation – ready-made water rocket components: antigravityresearch.com

Water rocket links: ourworld.compuserve.com/ homepages/pagrosse/h2orocketlinksi.htm

The Martinet Launcher, the basis for this project's launcher design: martinet.nl/articles/20050101.html

Rocket Boy

Watch a video clip of author Steve Lodefink and his 4-year-old son Ivan launching their soda-bottle rocket at makezine.com/05/rocket.

ROCKET-LAUNCHED CAMCORDER

By John Maushammer

EYES ON THE RISE

Hack a $30, single-use camcorder to make it reusable, then launch it up in a model rocket and capture thrilling astronaut's-view footage of high-speed neighborhood escape and re-entry.

You can build this project over a weekend and the results are fantastic. The idea goes back to 1929, when Robert Goddard launched the first scientific payload on a rocket: a still camera and a barometer. During the height of the space race, model rocketry supplier Estes offered a tiny Super 8 film kit that recorded about 10 seconds of rocket POV action. Today, Estes sells a launchable DV recorder called The Oracle, but it costs $120 and its image quality is lacking.

We'll do much better with a new camcorder that costs only $30, is very light, and has enough memory for several flights. It was designed for single-use only, but we'll make it reusable.

NOTE: Pure Digital has changed the firmware in later versions of the camera, which may make your images impossible to download. Check the current states of hackability in the forums at camerahacking.com, or check other options at makezine.com/07/camerarocket.

Set up: p.235 Make it: p.236 Use it: p.243

John Maushammer reverse-engineered the firmware in all three Pure Digital disposable cameras and figured out how to connect them to home computers. While not technically a rocket scientist, he has designed hardware and software for satellites.

ROCKETCAM SCIENCE

The CVS Camcorder is a $30 unit that's cheap enough and light enough to fly on a model rocket, after some modifications.

The camcorder is in the nosecone, with its lens in a peephole cut in the side. An optional rearview mirror redirects the view downward during the rocket's ascent.

Wiring a USB cable's 4 leads to 4 connectors on the edge of the mainboard gives the camera a working USB interface.

By stripping down the camera and substituting one CR2 lithium battery, we can cut the total weight down from 94 grams to 32 grams: 21 for the camera essentials, and 11 for the battery.

Downloader software available at camerahacking. com lets your computer control and copy video from the camera through this new USB port. You can even upload modified firmware that quadruples the camera's resolution from 320×240 to 640×480.

A C-size motor can easily accelerate a rocket at 13g. Larger motors and hard landings can generate even more. The Space Shuttle, for comparison, experiences a maximum acceleration of 3g — this is limited to protect the astronauts.

Illustration by Timmy Kucynda

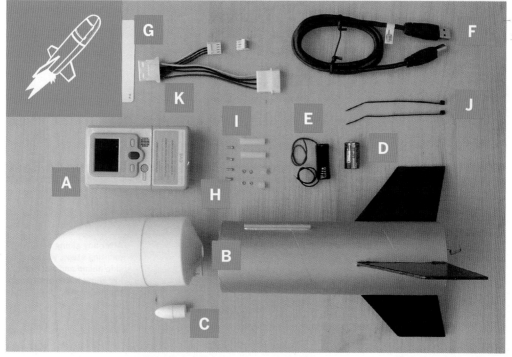

MATERIALS

[A] CVS One-Time-Use Video Camcorder $30 at CVS drugstores. Rite Aid sells a similar camera; check camerahacking.com for compatible drivers. Target carries a reusable Point and Shoot Video Camcorder (no hacking required), but it costs $130.

[B] Model rocket kit The body tube must be at least 2½" in diameter. Also, check the Estes Engine Chart (*see Resources*) to make sure the engine can lift 41 grams of extra weight. Two recommended rockets are the Fat Boy and the Canadian Arrow, both from Estes.

[C] Small rocket nose cone (optional) Estes sells spare nosecones in packs.

[D] Small 3-volt lithium battery Such as a CR2

[E] Battery holder You can adapt an N-sized holder to fit a CR2.

[F] USB cable

[G] Credit card plastic An old credit or gift card, or fake card from junk mail

[H] Small machine screws and matching nuts (4) No. 1 or M1.8 metric

[I] Nylon standoffs big enough to glue the nuts into (4) Or threaded standoffs that match the screws, if available

[J] Small plastic cable ties (2)

[K] A mating pair of light-weight, 4-pin free-hanging power connectors I used connectors from inside an old floppy disk.

Thin wire

Small front-surface mirror (optional) You can find a good one inside a View-Master toy ($5), but be sure to get the kind that looks like binoculars.

Masking tape

Engines and igniters With the smaller Fat Boy rocket, you'll need to upgrade to a C11 or D-sized engine and use a larger motor mount. The Canadian Arrow's standard D or E engine has enough power in stock configuration.

Launch controller Such as Estes' E controller (30' cable) or Electron Beam controller (17' cable)

Parachute recovery wadding Heat-resistant paper that prevents the parachute from melting.

Launch pad (stand, blast shield, and guide rod) Also from Estes (or see MAKE, Volume 05, page 141, for a DIY version)

TOOLS

Windows XP computer

Small Phillips screwdriver

Keyhole saw or hacksaw

Hobby knife, scissors

Soldering equipment

Wire cutters and stripper

Polyurethane glue, such as Elmer's ProBond or Gorilla Glue This glue is tough and foams up to fill gaps.

C-clamps (2)

Vice Helpful to hold parts while cutting or sawing.

MAKE IT.

BUILD YOUR CAMCORDER ROCKET

START ≫   Time: **A Weekend**   Complexity: **Medium to Easy**

1. BUILD THE ROCKET BODY

1a. Follow the instructions to assemble your model rocket kit.

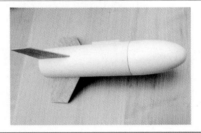

After any gluing and painting steps, you can skip ahead and work on installing the camcorder in the nosecone.

2. DISASSEMBLE AND STRIP DOWN THE CAMCORDER

The CVS camcorder is already small, but it's still too heavy to fly in most rockets. By removing everything that isn't essential, we can cut its weight down from 140 grams to 21 grams (without batteries).

2a. Remove the camcorder's battery cover by inserting something pointy into the opening on the bottom while sliding the cover off.

2b. Unlock the grey battery holder and remove it.

2c. Unpeel the sticker on the back and use a small Phillips screwdriver to remove 4 screws, one at each corner.

2d. Snap apart the case. Inside is the main circuit board.

2e. Remove the 2 screws at the 2 points shown here to release the circuit board.

2f. Find the small 4-pin connector that connects the circuit board to the batteries. Pull straight up to remove the board.

2g. Optional: Remove the speaker, which is encased in a vibration-resistant rubber housing glued to the circuit board. Just pull it to remove, and cut or unsolder the speaker wires.

Removing the speaker saves 2 grams, which doesn't seem like much, but every bit counts on smaller rockets. Leave the microphone in place, though, to record the roar of takeoff.

3. RIG THE CAMERA INTERFACE

The camera connects to a computer using USB protocol, but instead of a standard USB port, it has its own card-edge connector that we'll need to wire into. We'll save flying weight by using 2 USB cables, a short one that's attached to the camera and ends in a lightweight connector, and a longer, second cable that connects the lightweight connector to your computer's USB port.

3a. Cut and strip a short section of the USB cable. Solder the red, black, green, and white leads to the mainboard's edge connector, contacts 6-9, respectively. (If the computer indicates a problem later, your cable may not have the standard color-coding, and you should try swapping the green and white wires.)

This cable makes the same connections as the cable for the Dakota Digital camera (MAKE, Volume 03, page 130).

3b. Solder or crimp the 4 leads to one of your power connector pairs, and solder or crimp the other connector to the computer end of the USB cable, preserving the wire ordering.

My connectors weren't designed to be taken apart that often, but shaving down the locking tabs made them disconnect more easily.

3c. Plug the cable into the computer. Windows XP should identify it as a "Saturn" and prompt you to install drivers. These drivers don't exist yet, so hit "Cancel."

If the computer gives an error message and cannot identify the camera as a "Saturn," swap the green and white USB connections.

3d. Browse to camerahacking.com. Under "CVS One-Time-Use Camcorder," click "FAQs & Links," and choose some up-to-date driver software. Carpespasm, BillW, and Corscaria have good wares, among others (and I wrote a Mac downloader that works with the oldest version of the camera, 3.40). Install per instructions.

3e. Record a video and test it. It will be in XVID 1.0 format, 320x240 resolution, 30 frames/sec. After you're able to download videos from the camera, you may need to install a video codec to play them back; I recommend MPlayer from mplayerhq.hu.

4. MOUNT THE CAMERA IN THE NOSECONE

Now we'll put the camcorder in the nosecone. This is the most protective part of the rocket, and adding the weight in the front will help stability during flight. There are many ways to secure a camera so it won't come off during liftoff, including styrofoam and glue, but this method is reversible, which I prefer.

4a. Cut a hatch in the side of the nosecone, for inserting the camera. I taped a 3"×2¾" piece of paper to the cone to use as a template. We will be replacing this hatch, so don't cut any more than necessary.

The hatch should be big enough to let you operate the camera. Don't cut all the way to the bottom of the cone; leave a ring above the base to keep it strong. If you mess up, you can buy a replacement cone.

4b. Make 4 threaded standoffs by gluing M1.8 nuts into the unthreaded standoffs. (The circuit board's mounting holes are so small that I couldn't find threaded nylon standoffs that would fit.) Screw the standoffs onto the board.

Score the inside of each standoff to give the glue something to attach to. Make sure the glue is fully cured before screwing the standoffs on.

4c. Cut the standoffs so that they match the curve of the nosecone. Start with 2 short and 2 long. Use the rocket body to mark them, and cut them more precisely, angled to fit.

4d. Solder the battery holder's negative lead (black) across the 2 pins closest to the lower-left corner, near the on/ off button. Solder the positive (red) lead to bridge the adjacent 2 pins, near the mystery-chip blob. If you're using an N battery holder with a CR2 battery, cut the sides off of the holder to let the battery seat. CR2s are the same length as Ns, but wider.

4e. Cut a peephole in the side of the nosecone and verify it with the camera's viewfinder. The closer you can mount the lens to this hole, the smaller it can be. But don't worry about the size yet — you can enlarge it after the camera is glued into place.

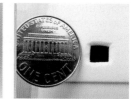

4f. Verify that everything fits, the camera lens aligns with the peephole, and there's enough wire to reach the battery holder, which you'll mount below. Mark where the standoffs will glue to inside the nosecone, and rough these spots up. I used a knife to cut a crosshatched pattern.

4g. Glue in the standoffs and double-check that the camera can still see out the hole. Use C-clamps to hold the board in place while the glue foams up. Allow to cure completely. Polyurethane glue requires moisture to cure, so I put a small wet paper towel in the cone.

4h. After the initial application of glue cures, add some more to reinforce the standoffs.

4i. Optional: If you want a view of the ground as the rocket lifts off, install a rearview mirror on a small nosecone, attached in front of the peephole. The reflection will make the body of the rocket appear at the top of the picture. If you want the body at the bottom of the picture, you can mount the camera upside-down, or correct the videos digitally later. Cut a plastic View-Master mirror down to size with wire cutters. After the glue has set, trim around the mirror with a hobby knife and file.

View-Masters (and SLR cameras) use front surface mirrors, where the coating is on the front of the plastic (or glass). This prevents image ghosting, but the shiny coating is easier to scratch.

4j. Optional: To save even more weight, remove the LCD viewfinder. Remove the camera circuit board from the nosecone and unscrew the 2 screws that hold the screen on the back. Slide the gray plastic catch on the screen's ribbon cable (I used 2 small screwdrivers to help), and pull to completely disconnect the screen and cable.

5. INSTALL THE BATTERY

Battery holders have no springs on the positive terminal, so a good whack upon landing can pop the battery out, interrupting the current. If this happens before you press the Stop Record button, you may write garbage to the camera's memory chip and lose your entire recording, or even render the camera inoperable. To prevent this, I strung a small wire through the positive terminal of the battery and connected it to the positive terminal of the battery holder. That way, if the battery is knocked loose, the wire still keeps it connected electrically.

5a. Thread some thin wire through the vent holes on the positive terminal of the battery. If you don't have wire small enough, use a few strands from a larger, stranded wire.

5b. Load the battery into the holder and connect the wire to its positive terminal. Glue the holder into the bottom of the nosecone, below the hatch opening.

5c. Cut two small holes through the bottom of the cone and thread a cable tie through the holes and over the battery, to hold the battery in place. Because the battery is mounted perpendicular to the direction of flight, the cable tie will absorb most of the shocks.

6. FINISH THE NOSECONE

6a. Referring to the camera's original housing, label the Power and Record buttons on the bare circuit board so they are easy to use.

6b. Add some tabs to be able to close the hatch securely. Glue 2 thin strips of credit card plastic inside the cone along the sides of the hatch. Glue 2 smaller pieces to the front of the hatch itself, and a third small piece inside the nosecone, to fit in between. When done, the hatch should sit flush with the nosecone.

Make sure you reinforce the hatch at the bottom, to keep it securely in the rocket, and at the leading edge, where it will bear the most pressure.

6c. Plug any holes in the bottom of the nosecone by gluing more pieces of credit card plastic.

At the top of the rocket's flight, a small explosive charge fires in the rocket motor to pop the nose off and deploy the parachute. Holes in the bottom of the nosecone can vent pressure without popping the cone, which will send your rocket crashing into the ground!

FINISH ☒

NOW GO USE IT »

Rocket Stability

Did you ever wonder why rocket fins stick down past the tail? The reason has to do with stability, which you need to understand if you are modifying a rocket kit.

Two imaginary points determine whether a rocket will fly straight or corkscrew hopelessly out of control. The first point is the *center of gravity* (COG). If you were to place your rocket on a razor's edge, this is the point where it would balance. Add weight to the nosecone and the center of gravity moves toward the nose; add bigger engines and it moves rearward.

The second point is called the *center of pressure* (COP), and it is a little harder to explain. This is similar to center of gravity, except it involves the aerodynamic forces balancing out. If you were to trace the 2-dimensional side-view of the rocket onto cardboard and then cut it out, it would balance near the center

of pressure. Enlarging the fins and extending them down moves this point rearward.

Once you know these two points, you can estimate the rocket's stability. The general rule for stable flight is that the COG must be at least one body diameter in front of the COP. During flight, aerodynamic forces will push rearward on the COP. If it isn't already behind the COG, then the rocket will attempt to turn around, which is bad news. If the COP is behind the COG by just a little, the flight will be marginally stable and probably corkscrew.

For this project, we won't be modifying the outside too much, so the COP will remain the same. We will be adding the weight of the camcorder to the nosecone, moving the COG forward. Aerodynamically, this will only make the flight more stable.

USE IT.

NOW GIVE IT
A SHOT

LAUNCH SEQUENCE

If you are using a smaller engine, you may want to do a final weigh-in before going ahead with a launch. With everything installed, verify that the total weight is within the capability of the rocket motor. If so, follow the rocket kit's launch instructions. Here's the basic sequence:

1. Check to make sure that the nose fits securely, but not too tightly. Then tie up the parachute, engine, and wadding.

2. Set up the launch pad per instructions. Make sure the field you are launching in is large enough; otherwise you'll lose your precious payload. For a rocket powered by a D-sized engine, the field should be at least 500 feet in diameter.

3. Ensure that the rocket launcher is not armed (usually this means the key is removed) and then set up the rocket on the pad.

4. Turn on the camera and press the record button. The red record light, just under the lens, should come on.

5. Close the hatch and tape it closed. Although not typically rated for space-faring use, ordinary masking tape works fine.

6. Start the countdown and launch.

7. Recover the rocket. Open the hatch and press the Record button to stop recording. The recording light should turn off. Then turn the camera off by pressing the On/Off switch. If you forget to turn the camera off, it will switch off automatically after a few minutes of non-use. Do not turn the camera off by removing the battery.

8. Back at your computer, hook up the USB cable, download the video, and enjoy.

WATCH IT

Watch a high-flying video captured by John Maushammer's Rocket-Launched Camcorder at makezine.com/07/camerarocket.

UPGRADES

Downward View During Descent

If you installed the downward-facing mirror, you'll get a whole new view of the launch. One drawback, though, is that the descent will typically have views of the parachute and sky. You can change this by attaching the parachute to the tip of the nosecone instead of the base. Add a small eyehook to the nosecone's tip, and run the parachute cord along the side of the nose and into the body.

Improved Resolution

The CVS camera (which is manufactured by Pure Digital, along with the Rite Aid and Target cameras) actually has a 640×480 sensor. But in order to extend recording time, it is configured to record at only one quarter of this resolution. Recording time isn't a problem for rocket flights that last only a few seconds, so you can set the camera to record at the full resolution. You can do this by uploading a modified version of the binary file *USP.BIN* into the *P3* directory of the camera. See Resources, below.

RESOURCES

Estes Engine Chart
hobbylinc.com/rockets/
info/rockets_enginemountchart.htm
Determining Center of Pressure
makezine.com/go/cop
Downloader software
makezine.com/go/software
Improving image resolution
makezine.com/go/resolution

BUILDING AN ORNITHOPTER

By William Gurstelle

FLIP, FLAP, FLY

For millennia, men and women have studied birds, bats, and beetles, observing and experimenting, attempting to determine what humans must do to fly by flapping.

But people can't fly by flapping: not with wings covering their arms; not with pedaled, chain-driven wings; and, so far, not with internal-combustion engines, either. Nonetheless, the concept of manned ornithopters continues to hover on the periphery of aeronautical engineering. This project shows you how to build a small, rubber band-powered ornithopter we call Orly.

There are many types of ornithopter designs. Orly is a simple monoplane, meaning there is a single wing mounted above the motor-stick, and its motion is similar to a bird in flight.

Set up: p.248 Make it: p.249 Use it: p.253

William Gurstelle enjoys making interesting things that go whoosh then splat. He is the author of *Backyard Ballistics* (2001), *Building Bots* (2002), and *The Art of the Catapult* (2004). Visit backyard-ballistics.com for more information.

Photograph by Gerry Arrington

HOW DO ORNITHOPTERS FLY?

Get up on the downstroke.

How do ornithopters fly? According to Nathan Chronister of the online Ornithopter Zone, "The ornithopter wing is attached to the body at a slight angle, which is called the angle of attack. The downward stroke of the wing deflects air downward and backward, generating lift and thrust.

"Also, the wing surface is flexible. This causes the wing to flex to the correct angle of attack we need in order to produce the forces that we want to achieve flight."

The mechanics of flapping flight are far more complicated than that of fixed-wing flight. For an aircraft with fixed wings, only forward motion is necessary to induce aerodynamic lift. But for flapping flight, the wing not only has to have a forward motion, but also must travel up and down. This additional dimension means the wing constantly changes shape during flight.

Illustration by Timmy Kucynda

FROM ICARUS TO ORLY: A SHORT HISTORY OF ORNITHOPTERS

Without doubt, even the earliest humans watched birds fly past and felt, well, rather envious. Thus when Thag, a Pleistocene caveman, looked up and saw flocks of ducks and geese soaring above, he might have gathered together a few palm fronds, lashed them around his arms with a vine, and leapt off a tree. Poor Thag never got airborne, or at least he didn't live to record the episode in petroglyphs on his cave wall.

Later, from ancient Greece, comes the legend of Daedalus and Icarus. Daedalus was a skilled engineer who angered King Minos. Minos ordered him imprisoned in a tower.

According to *Bulfinch's Mythology*, "Daedalus contrived to make his escape from his prison, but could not leave the island by sea, as the king kept strict watch on all the vessels.

"So he set to work to fabricate wings for himself and his young son Icarus. He wrought feathers together, beginning with the smallest and adding larger, so as to form an increasing surface. The larger ones he secured with thread and the smaller with wax, and gave the whole a gentle curvature like the wings of a bird."

Unfortunately for Daedalus, the attempt at flight didn't entirely work. His son, Icarus, flew too near the sun, melting the wax that held the wings together. Icarus fell out of the sky and drowned in the ocean.

To a large extent, that's been the typical outcome of human flapping flight experiments, right up to modern times.

The New York Times has run many stories over the years:
» "Ornithopter Somersaults — Captain White Hurt in Crash" (June 1928)
» "Inventor Tries to Soar Like a Bird; Narrowly Escapes Drowning" (March 1932)
» "100,000 See French Birdman Die in 9,000 Foot Fall" (May 1956)
» "[University researchers] named their ornithopter 'Mr. Bill' after the perpetually maimed character on the television show *Saturday Night Live*." (May 1992)

Around 1490, Leonardo da Vinci was carefully studying the mechanics of avian flight. From his bird-watching came perhaps the first blueprint for a human-carrying ornithopter. More of a theoretician than a true maker, Leonardo never got his flying machine off the paper in his notebooks. Had it been built, it likely would not have flown. Still, experts say his design is clever, and embodies modern aerodynamic principles developed hundreds of years later.

Interest in flapping flight took off again in the 1870s. Building model ornithopters became fashionable in Europe, and a number of enthusiasts — among them Alphonse Penaud, Hureau de Villeneuve, and Gustave Trouvé — built internally powered birds that soared over the fields of France and Flanders. Soon, ornithopters powered by rubber bands, gasoline, electricity, and even gunpowder were flapping away, but as scale models, not people-carrying aircraft.

Since then, many people have tried to build a manned ornithopter, but none have yet succeeded. There are unconfirmed reports that the Germans made one during World War II and that the Soviets flew one during the Cold War, but solid evidence is lacking. Today the University of Toronto is making a game attempt.

Why bother with ornithopters at all? Because flappers can do things other aircraft cannot. They probably have the best maneuverability of any aircraft. Unlike fixed-wing drones, ornithopters, at least in theory, can stop and hover like a hummingbird, which makes them extremely versatile, and they need less space to maneuver than a helicopter. Couple all that with their ability to fly at very slow speeds, and ornithopters may be the perfect surveillance vehicles. The military applications for unmanned ornithopters are numerous.

Ornithopters have practical applications in civilian life, as well. For instance, the Colorado Division of Wildlife uses an ornithopter to research a hard-to-capture endangered species called the Gunnison sage grouse. This skittish bird flies away at the first sign of danger but will stay on the ground if it sees a hawk flying above. So state biologists use a motorized, radio-controlled ornithopter painted like a hawk to keep the flighty grouse on the ground long enough for them to capture it.

SET UP.

MATERIALS

[A] Balsa rod ⁵⁄₁₆"×¹⁄₈",
7 inches **Cut into lengths of:**
   5" motor stick
   1½" front vertical
   connector

[B] Balsa rod ³⁄₁₆"×¹⁄₁₆",
6 inches **Cut into lengths of:**
   2¾" connecting rods (2)

[C] Balsa rod ¹⁄₈" square,
24 inches
Cut into lengths of:
   8" wing spars (2)
   5" top wing attachment
   member
   1½" back vertical
   connector
   ½" crank standoff

[D] Balsa rod ³⁄₃₂"
square, 14 inches
Cut into lengths of:
   7" tail members (2)

[E] Music wire .032"
diameter, 10 inches
**Thicker wire is too heavy
and can adversely affect
Orly's performance.**
Cut into lengths of:
   3" tail/rear motor
   attachment wire
   2½" crank/front motor
   attachment wire
   2" wing spar wires (2)

[F] Sheet of tissue paper,
about 18"×18" **Made spe-
cifically for modeling, about
.04 ounces per 100 square**

inches. **Regular tissue paper
is comparatively heavy.**

[G] Square of 16 lb. paper,
2"×2"

[H] Heat-shrinkable tub-
ing, ¹⁄₁₆" diameter **Cut into
3 pieces each ¹⁄₈" long**

[I] Small beads (2) **With
inside diameter just large
enough to accommodate
the .032" diameter music
wire. From bead or craft
stores.**

[J] Model airplane rubber
12" long **Tied in a loop to
make a big rubber band.
From hobby shops.**

[K] Cyanoacrylate (CA)
glue

[L] Cyanoacrylate drying
accelerator

[M] Glue stick

Vegetable oil (not shown)

TOOLS
(not shown)

Needle-nose pliers (2)

Utility knife **To cut the
balsa wood to size**

Ruler

Scissors

Photography by Matthew Dalton and Ty Nowotny

MAKE IT.

BUILDING ORLY THE ORNITHOPTER

START >> Time: **A Day** Complexity: **Medium**

1. MAKE THE FUSELAGE

1a. Form a hook in the tail/rear motor attachment wire as shown. Carefully push the wire through the center of the motor stick at a point ⅜" from the tall end. Then make two 90-degree bends in the wire as shown, and glue into place using CA adhesive. Reinforce the wire-to-balsa joint by placing a tissue paper cover over it. Dab the joint with a thin layer of CA. Spraying CA drying accelerator on the joint makes the process faster and less messy.

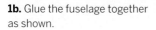

1b. Glue the fuselage together as shown.

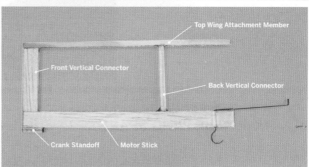

Top Wing Attachment Member
Front Vertical Connector
Back Vertical Connector
Crank Standoff Motor Stick

1c. Roll the 2"×2" paper into a narrow tube using the music wire for a mandrel. Remove the music wire and carefully daub the tube with CA, taking care to maintain the tube's openings. Spray with CA accelerator. Cut into three ½" long tubes and discard the remainder.

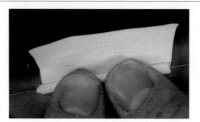

1d. Attach the 3 tubes to the fuselage as shown, with CA and accelerator. Make certain the tubes are aligned with the long axis of the fuselage.

1e. Use the needlenose pliers to bend the wire so the crank appears as shown. Insert the crank wire through the paper tube glued to the crank standoff. Place the 2 beads on the wire. Create a bend in the back end of the wire to serve as the motor hook.

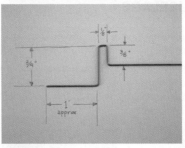

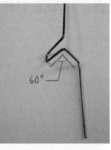

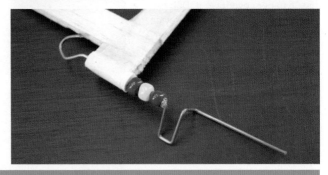

2. MAKE THE WING SPARS

2a. Bend the music wire as shown. Carefully push the wire through the wing spar at a point ¾" from one end. Glue into place using CA. Reinforce the joint by wrapping a layer of tissue paper around the joint and coating with CA.

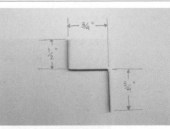

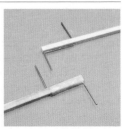

3. MAKE THE TAIL

3a. Use CA adhesive to glue the balsa rods into a T shape. Reinforce the joint by covering it with tissue paper soaked with a thin layer of CA.

3b. Poke the end of the fuselage tail attachment wire into the balsa tail member; then glue the assembly with CA. Reinforce by wrapping with tissue and CA.

4. MAKE THE CONNECTING RODS (CONRODS)

4a. The conrods undergo considerable stress. Harden the rods by coating the last ½" of each end with CA.

4b. Make 2 holes for the music wire in each conrod, ¼" from each end.

5. FINAL ASSEMBLY

5a. Cut out the tissue paper wings and tail using the templates.

NOTE: Full-scale templates are posted at makezine.com/08/orly.

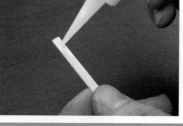

7"

16½"

5b. Glue the tissue paper wing to the wing spars and the top wing attachment member. Glue the tissue paper tail to the balsa T frame with a glue stick.

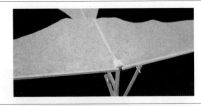

5c. Connect the conrods to the wing-spar attachment wires and the crank. Adjust the spacing of the conrods so the crank turns smoothly. Place heat-shrinkable tubing over the crank and wing spar wires to maintain alignment, and carefully heat with a match to shrink the tubing.

WARNING: The tissue paper, the balsa wood, and the CA catch fire easily! Use great care in this step.

5d. Bend the tail up so it is at about 15 degrees from the plane of the motor stick.

6. SENDING ORLY AIRBORNE

6a. Double the rubber band and place it over the front and rear motor attachment hooks.

NOTE: To accommodate a longer rubber band, double it into 2 loops and place it over the front and rear attachment hooks.

6b. Rub the band with a little vegetable oil for lubrication.

6c. Wind up the rubber band motor by turning the crank at least 35 turns.

6d. Angle Orly's nose up slightly and release gently.

FINISH ☒

NOW GO USE IT »

USE IT.

FLY, BE FREE!

TROUBLESHOOTING

Ornithopters can be difficult to fly. Common problems include stalling, nosediving, and veering, in various combinations. If your ornithopter doesn't fly well, try the following:

1. If the ornithopter dives and veers, winding the motor in the opposite direction may help.

2. Balance is important. Make certain the action of each wing is the same. Make the conrods and crank carefully to ensure balanced wing operation.

3. If Orly has a tendency to flip or roll in flight, you need to improve your craft's stability. Try lengthening the distance between the motor stick and the tail, or adding a rudder (a vertical stabilizing surface on the tail).

4. If the ornithopter veers consistently in one direction and then nose-dives, add a small wire weight to the end of the wing tip on the side opposite the direction of the veer.

5. The angle of the tail is important. Bend it slightly up if the flapper nose-dives, and bend it down if it stalls.

6. If Orly goes through a series of stalls before ultimately diving into the ground, your tail may be mismatched to the rest of the aircraft. Fix this by decreasing the size of the tail. If that doesn't help, extend the length of the tail boom — that is, increase the distance between the wing flappers and the tail.

7. A direct head first plunge to the ground may be a signal to increase the size of the tail stabilizer.

8. If your ornithopter flaps vigorously but won't gain altitude but lowers into the ground tail first, try to move the center of gravity forward. It is best to make the rear lighter instead of making the front heavier.

9. Bank and spiraling problems are common in ornithopters, and can be tough to correct. If your ornithopter starts out with few good looking flaps, but suddenly banks or rotates around its longitudinal axis and then spirals down, try the following:

• Reapply the tissue paper to the wings, making sure the paper is not applied too loosely or too tightly stretched. Both wings should have the same amount of tension.

• Bend a small rotation in the tail plane relative to the longitudinal axis of the ornithopter.

• Add a small weight to the outside of the wingspar opposite the direction of the bank.

Fixing the bank and spiral problem can be difficult. You may need to try a number of fixes in combination before the problem clears.

EXPERIMENT!

You may be able to extend the duration of flight by making a few changes to Orly's design. For instance, you can experiment with rubber bands of differing lengths and thickness. You can also vary the shape, sweep angle, and size of the wing.

Sometimes, adding wing gussets made from a thin piece of transparent tape improves performance. Finally, experiment with the length and width of the tail.

RESOURCES

There is an active community of ornithopter enthusiasts, with a lively online forum at ornithopter.org. This excellent site also provides plans, kits, motorized models, and advice.

For the latest news on the University of Toronto manned ornithopter project, visit ornithopter.net.

THE NIGHT LIGHTER 36

By William Gurstelle

Launch potato projectiles 200+ yards with this stun-gun triggered, high-powered potato cannon with see-thru action. (Good thing potatoes are biodegradable.) »

Set up: p.258 Make it: p.260 Use it: p.264

POTATOES, BEWARE

The potato cannon, a.k.a. the spud gun, is a popular and very entertaining amateur science project. It's simple to make, and few devices offer such bang for the buck. You can use the Night Lighter both day and night, but when it's dark, the clear PVC provides an excellent view of the interior ballistics. Also, the stun gun gives better performance than weaker sparks from piezoelectric or flint/steel igniters. It's fun both to fire and simply to watch in action.

A basic spud gun can be built with plain, white PVC for less than $25. The Night Lighter 36 costs more, but I scrounged leftovers from plastics suppliers and built mine for less than $50. After mastering basic gun construction, the intrepid potato cannoneer may want to design and assemble more complex and artistic devices.

William Gurstelle enjoys making interesting things that go whoosh then splat. He is the author of *Backyard Ballistics* (2001), *Building Bots* (2002), and *The Art of the Catapult* (2004). Visit backyard-ballistics.com for more information.

PRINCIPLES OF SPUD GUNNERY:

Serious spud gun designers tinker with the ardor of hot-rod builders. Our NL-36 improves upon the basic potato cannon by substituting transparent tubing and a stun gun igniter.

BEVELED EDGE Load a spud, and the sharpened front edge cuts a plug that seals airtight against the barrel.

BARREL The three-foot barrel guides the potato plug along its trajectory as it picks up speed from the explosion.

IGNITION CHAMBER A spark from a stun gun ignites hydrocarbon-rich aerosols, causing the internal combustion that sends the spud. You can watch it all through the clear PVC.

CHEMICAL REACTION Fire, everyone's favorite exo-thermal reaction, breaks aerosol propellants into hot, expanding CO_2 and water vapor.

BUTANE

ETHANOL

PROPANE

CARBON DIOXIDE

WATER

ISOBUTANE

OXYGEN

RANGE COMPARISONS WITH OTHER FAMILIAR PROJECTILES
Our spud gun propels a 9-ounce potato plug approximately 200 yards. Here's how this compares with some other launch events.

- Shot put: 16 pounds, 25 yards (Olympic record)
- Cell phone toss: 4-5 ounces, 90 yards (Savonlinna record)
- Football punt: 14-15 ounces, 98 yards (NFL record)
- Baseball throw: 5.25 ounces, 149 yards (Guinness record, 1957)
- Civil War cannon: 6-pound ball, 1,500 yards
- SCUD-B missile carrying 10-ounce potato payload: 186 miles

186 miles

| 25 | 90 | 98 | 149 | 1,500 |

Not to scale.

Illustration by Tim Lillis

SET UP.

Visit makezine.com/03/spudgun for source list.

MATERIALS

36-inch length of 2-inch diameter transparent Schedule 40 PVC pipe (1)

14-inch length of 3-inch transparent schedule 40 PVC pipe (1)
Used mainly in the food-processing industry, transparent PVC is available from industrial plastic suppliers such as Harrington Plastics or Ryan Herco. Check your local business listings. You can also order it from McMaster-Carr (www.mcmaster.com). Transparent PVC is generally expensive, but you might find reduced-price remnants at plastics suppliers or food-processing companies.

Large potatoes

3- to 2-inch diameter PVC reducing fitting socket a.k.a. reducer coupling or bell reducer (1)

3-inch PVC female adapter (1)

3-inch PVC end plug (1)
Use Schedule 40 grade for all PVC fittings. Source from large home or hardware stores or plastics suppliers.

PVC primer and cement
Available wherever PVC is sold.

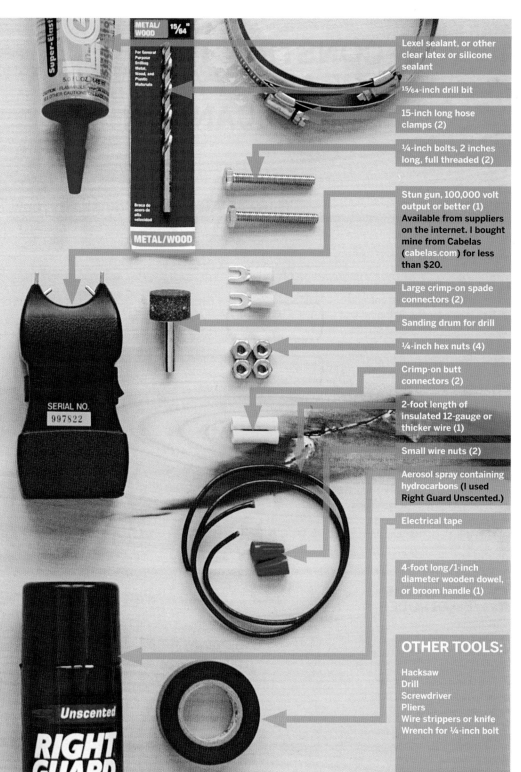

Lexel sealant, or other clear latex or silicone sealant

15/64-inch drill bit

15-inch long hose clamps (2)

1/4-inch bolts, 2 inches long, full threaded (2)

Stun gun, 100,000 volt output or better (1) **Available from suppliers on the internet. I bought mine from Cabelas (cabelas.com) for less than $20.**

Large crimp-on spade connectors (2)

Sanding drum for drill

1/4-inch hex nuts (4)

Crimp-on butt connectors (2)

2-foot length of insulated 12-gauge or thicker wire (1)

Small wire nuts (2)

Aerosol spray containing hydrocarbons (I used **Right Guard Unscented.**)

Electrical tape

4-foot long/1-inch diameter wooden dowel, or broom handle (1)

OTHER TOOLS:

Hacksaw
Drill
Screwdriver
Pliers
Wire strippers or knife
Wrench for 1/4-inch bolt

MAKE IT.

CONSTRUCTING THE MIGHTY POTATO CANNON

START ❯❯

Time: **An Afternoon** Complexity: **Low**

1. PREPARE THE PVC

1a. Cut pipes. Measure and mark a cutting line 14 inches from one end of the 3-inch diameter PVC pipe. Use the hacksaw to cleanly and squarely cut the pipe. This will be the cannon's combustion chamber. Then measure, mark, and cut a 36-inch length of the 2-inch diameter PVC pipe. This will be the cannon's barrel.

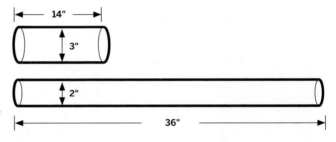

1b. Taper end of gun. Use a file or a drill and sanding attachment to taper one end of the long 2-inch diameter pipe, so that it forms a sharp edge. A clean, sharp edge is important, since it should cut the perfect-sized potato plug projectile as you ram the potato into the muzzle of the gun.

When PVC gets hot, it releases poisonous chlorine gas. Perform this step in a well-ventilated area.

2. ATTACH THE ELECTRODES

2a. Drill electrode holes. Four inches from one end of the 3-inch diameter pipe, drill a slightly undersized hole for the ¼-inch bolt. Drill a second hole directly opposite the first hole, four inches from the end.

The 3-inch pipe will contain the fuel and the spark, and act as the combustion chamber.

Photography by David Albertson

2b. Attach electrodes. Screw in the 2-inch long bolts, with nuts attached (two per bolt), into the holes in the 3-inch pipe. The nuts go outside the barrel. The bolts should tap themselves into the softer plastic, but don't over-tighten or you'll strip the PVC. Position and adjust the nuts as needed so there is a ¼-inch gap between the bolt ends inside the barrel.

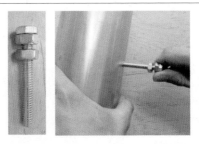
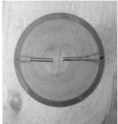

This is the spark gap that will ignite the fuel, firing the cannon.

3. SOLVENT-WELD THE PVC

The spud gun is composed of PVC pipes and fittings that are solvent-welded in place using PVC cement and primer. To prevent leaks and weak spots where the parts are joined, the solvent welding must be done properly. Meanwhile, the primer and cement are toxic and flammable, so you need to work in a well-ventilated area, keep the chemicals away from open flames, and follow all safety precautions on the labels. First, we'll solvent-weld the reducing connector to the front of the combustion chamber. Then we'll follow the same procedure to attach the threaded adapter to the back of the chamber and to connect the barrel.

3a. Inspect parts. Check the 3-inch pipe ends and 3- to 2-inch reducing connector for cracks, dirt, and abrasion, and remove any plastic burrs with a knife. Don't use damaged PVC pipe or fittings.

3b. Weld parts. Following the procedure at right, solvent-weld the 3- to 2-inch reducing connector to the end of the 3-inch pipe closest to the electrode bolts. Then join the unthreaded side of the female adapter to the other end of the 3-inch pipe, and attach the 2-inch barrel to the narrow end of the 3- to 2-inch reducing connector.

3c. Let the cannon dry for several hours in a well-ventilated area before using. You don't want to fire it while the solvents are wet and flammable.

3d. Screw the 3-inch PVC end plug into the back of the chamber after drying.

How to Solvent Weld

1. Clean the weld surfaces with PVC primer. Apply the primer with a dauber or brush (usually inside the cap). The primer cleans and softens the PVC and allows the cement to penetrate the surface.

2. Brush on a thick coat of PVC solvent, first to the end of the pipe, and then to the fitting socket. Leave no bare spots.

3. Immediately join the pipe and the fitting socket, pushing the pipe to its full depth and making sure it's seated squarely with a slight twist. If you've used enough solvent cement, you should see a small, continuous ooze of cement around the fitting. Once joined, you can't reposition the pipes or otherwise fix errors. If you accidentally put the wrong fitting on a pipe, you need to trim it off and start over.

4. WIRE THE IGNITION

4a. Test-fit butt connectors. Using a sharp utility knife, remove excess insulation from each crimp-on butt connector. With the stun gun turned off, test-fit the trimmed ends of the connectors over the gun's main electrodes. These are the twin electrodes that point forward, rather than toward each other, and we're hooking these up to our ignition wires, in order to bring the spark into the combustion chamber.

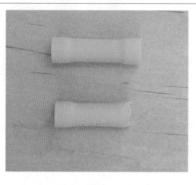

(Depending on the make and model of the stun gun, you may need to modify these directions and connect the wires in other ways, such as with wire nuts or soldering.)

4b. Prepare ignition wires. Cut the wire into two, 1-foot wires. These are the ignition wires. For each, attach a crimp-on spade to one end and the untrimmed end of a butt connector to the other end.

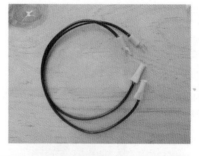

4c. Attach ignition wires to stun gun electrodes by crimping on the modified butt connectors.

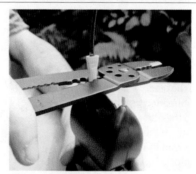

4d. Cover the stun gun test leads (inboard electrodes) with wire nuts cut down to size, or other high-voltage insulators. Insulate all exposed metal areas of the ignition path on the stun gun and bolt electrodes, with electrical tape or silicone glue. It's easy for electricity to find its way underneath any insulation gap at the base of the electrodes.

Wire nuts need to be trimmed to fit onto the test electrodes.

5. ATTACH THE IGNITER

5a. Attach the stun gun body to the rear of the chamber using two hose clamps. Do not over-tighten. Position the stun gun body at a 90-degree angle to the axis of the electrode bolts.

5b. Attach ignition wires to electrode bolts, securing the spade connectors underneath the bolt head or between the nuts. You may have to bend the spades to widen them enough to fit around the bolt.

6. FINAL INSULATION

6a. Cover bolt connections with globs of silicone sealant. To further insulate, wrap the whole ignition area with bubble wrap, and tape down. The stun gun operates at such high voltage that the wrap still may not completely prevent shocks. Avoid contacting electrodes when operating the cannon. Don't be the path of least resistance!

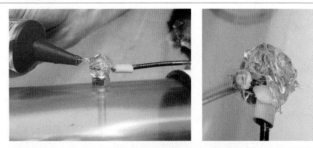

Congratulations! Your Night Lighter 36 potato cannon is complete.

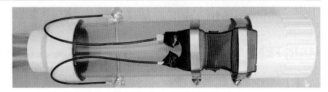

FINISH ☒

NOW GO USE IT »

USE IT.

HOW TO USE YOUR SPUD GUN RESPONSIBLY

FIRING THE NIGHT LIGHTER 36 POTATO CANNON

1. Remove the end plug.

2. Center and push a potato into the cannon, keeping your hand clear of the edge. You may want to wear a leather glove. The muzzle's sharp edge will cut the potato into a plug that should fit snugly on all sides. Any gaps will reduce performance.

3. Use the stick to push the potato plug 30 inches down into the barrel.

4. Direct a stream of aerosol into the firing chamber. Unscented deodorant works well, but check the label to make sure your choice contains hydrocarbons such as alcohol, propane, butane, or isobutane. Start out with a one- to two-second burst, and determine the optimal amount by trial and error.

5. Immediately replace the end plug and screw it in securely.

6. Turn the stun gun on, and double-check that the firing area is clear.

7. Press the stun gun's ignition button. Enjoy your work. For a tracer, stick a glowstick into the spud.

If you have a misfire, and the projectile is not ejected, carefully remove the end cap, and ventilate the combustion chamber thoroughly.

MAINTENANCE

Aerosol chemicals can gum up the inside of the cannon. Every few shots, clean it out with a rag and cleaner. The residue can also make the end plug hard to unscrew. If necessary, use pliers.

SAFETY AND LEGALITY

During construction, don't take shortcuts or substitute inferior materials. The vapors from PVC cement are flammable, so allow all joints to dry fully before exposing the gun to ignition sources.

When using the potato cannon, exercise extreme caution. Wear eye protection, and check the area in front of you before firing. Never look down the gun's barrel, or point it at anything you don't want to hit. Excess fluid stays in the chamber and evaporates slowly, so you should always treat the gun as if it can fire. Check frequently for signs of wear, and never operate a damaged gun. Avoid contact with (or proximity to) the ignition path. Stun guns hurt.

PVC is more brittle in cold weather, so don't use the cannon in temperatures below 60°F.

Neither the author nor this magazine assumes liability for your spud gun or your actions.

Potato cannons may not be legal in your area (even if it is legal to tote a 12-gauge down Main Street). Check with local law enforcement regarding the rules in your area, and obey them. Also, check the laws regarding stun gun usage.

Editor's note: Author William Gurstelle uses PVC for his Night Lighter 36 and other designs, but some spud gunners believe this is unsafe, since PVC can shatter and is not recommended for piping compressed gases. They advise using materials made out of ABS (with ABS cement), which is more flexible than PVC, but not available in transparent.

Alternatively, you can use Schedule 80 transparent PVC, which is thicker and stronger than Schedule 40, but more expensive.

Fun for about a dollar.

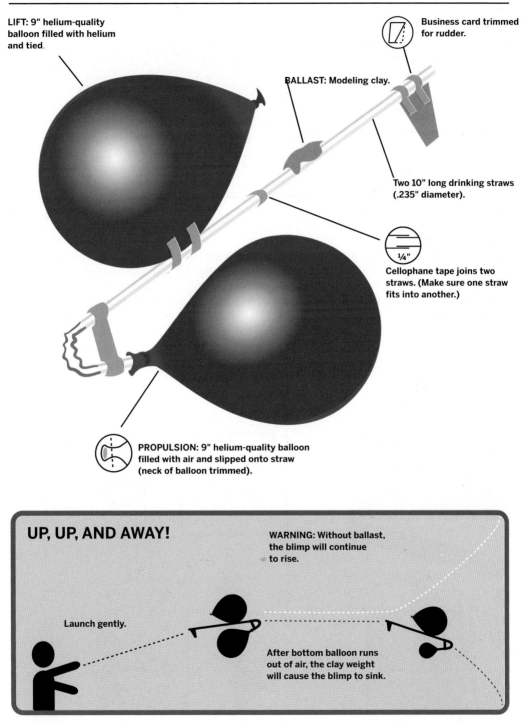

LIFT: 9" helium-quality balloon filled with helium and tied.

Business card trimmed for rudder.

BALLAST: Modeling clay.

Two 10" long drinking straws (.235" diameter).

¼"

Cellophane tape joins two straws. (Make sure one straw fits into another.)

PROPULSION: 9" helium-quality balloon filled with air and slipped onto straw (neck of balloon trimmed).

UP, UP, AND AWAY!

WARNING: Without ballast, the blimp will continue to rise.

Launch gently.

After bottom balloon runs out of air, the clay weight will cause the blimp to sink.

Illustration by John Perez

Paper Water Bomber

Winged origami missile with front-load tank delivers wet payload.

You will need:
Piece of A4 or letter-size paper, knife or scissors, water, target

Key:

Mountain fold	Valley fold	Crease line

1. Make the body.

a. Cut the paper to make a sheet with a length/width ratio of about 2:1.

b. At one end, fold each corner down to the opposite edge, then unfold, to make a square X crease. Turn over and fold the top edge down to bisect the X. Turn over again, push the edges in, and flatten the paper so there's a point at the top.

c. Turn over, and fold the triangle down along its bottom edge. Fold each entire top corner down along the centerline and unfold just the back corners. Leave the front corners folded down.

d. Invert the crease you just made on the back corners, so they lie underneath and inside, making the top pointy again.

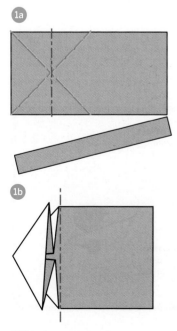

1a

1b

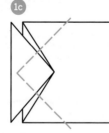

1c

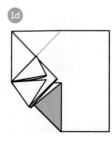

1d

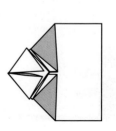

Illustrations by Gerry Arrington

2. Fold the cargo bay.

a. Fold the side of each flap in front to touch the centerline.

b. Fold the bottom point of each flap up to touch the same point.

c. The tricky part: Fold the bottom points up again diagonally, and tuck them inside the pockets you made in the previous step (see 2d).

d. Fold each side corner of the diamond back to touch the centerline behind. Fold the top and bottom corners in front to touch the centerline. Unfold all four folds.

e. Crease the wings to line up with the edges of the cargo section.

f. Blow into the nose of your plane while gently pulling up to inflate the cargo section into a cube.

3. Bombs away!

a. Fill the cargo section with water, pouring carefully into the hole in front.

b. (Optional) Wait until no one is looking.

c. Chuck the bomb at your target or victim of choice.

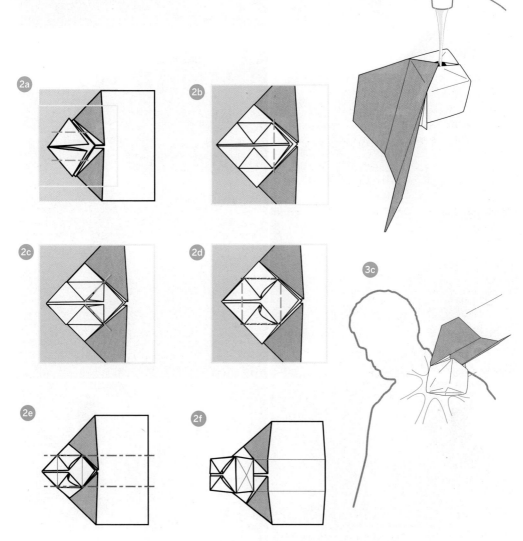

Ewan Spence (ewanspence.com) is a podcaster, blogger, reporter, new media junkie ... generally a hoopy frood who's fun to be with.

270

282

286

292

295

298

301

THE BEST OF PHOTOGRAPHY & VIDEO PROJECTS

from the pages of MAKE

DON'T JUST SIT THERE AND LET BIG BROTHER HAVE ALL THE FUN

Pocket digital cameras, cellphone cams, webcams, CCTV/surveillance cams, Google's satellite and street-level views, Flickr, YouTube. There's never been a more photographed time in history. We now live inside of media, and likely will, forever more. The Maker response is to hack the signal and the machinery that captures it. In this section, we cover projects as diverse as flying disposable digicams on kites to creating your own stop-motion animation to compositing yourself into TV shows.

PINHOLE PANORAMIC CAMERA

By Ross Orr

PIN-O-RAMA

Lensless and low-tech, pinhole cameras have always been maker-friendly. But forget the Quaker Oats carton, and go wide with this roll-film, panorama design.

I bought a new scanner recently, and soon found myself spelunking through drawers of old photos from my many misspent years in photography. Some of the most interesting shots were the pinhole camera experiments I had done as a teenager. With ghostly outlines from multi-minute exposures, and shapes warped into boomerangs by curved film, these otherworldly images got me dreaming about pinhole cameras again.

So I headed to the workshop to build a new one. And then another, and another. I eventually made more than a dozen, and this "Pin-o-rama" design is my favorite. Unlike simpler pinholes, it uses standard 120 format roll film, which means you don't have to open the camera and reload after each exposure, and you don't need a darkroom to process the results — just take the rolls to a photo lab. Also, it's built entirely from scratch, rather than hijacking the film-transport from an existing camera.

Set up: p.273   Make it: p.274   Use it: p.279

Ross Orr keeps the analog alive in Ann Arbor, Mich. A frequent contributor to MAKE, Ross hacks low-tech gadgets and invasive plants in his spare time.

Photograph by Sam Murphy

THE HOLE PICTURE

A pinhole camera is a light-tight box with a piece of film (or photo paper) on one side and a tiny hole in the other. An image forms because each point on the film can only "see" the one patch of the outside world that's lined up with the pinhole, whether it's light, dark, blue, red, etc. Because the pinhole does not focus light like a lens, the film can be any shape and a flexible distance from the hole, and the enclosure can be made from any lightproof material.

With construction so forgiving, pinhole tinkerers have produced a riot of camera creations, from mint tins to airplane hangars — even animal skulls and hollowed-out vegetables. Pinholes are the most hackable camera type ever devised.

> By curving the film, you can uniformly illuminate very wide angles, while getting a groovy, bulged perspective. The Pin-o-rama's 105° horizontal coverage matches that of an expensive, superwide 16mm lens on a 35mm camera.

> To capture detail, we use 120 format film, which is nearly 2½" wide and has no sprocket holes. Frame numbers on the backing paper let you put a peep sight on the back of the camera, and wind until the next number shows. To shoot panoramas, we count off every other number, yielding 6 shots per roll.

PROPER PINHOLE SIZE

TOO LARGE: The pinhole's image comes from overlapping pinhole-sized blobs of light. A large pinhole loses all detail smaller than its own diameter.

TOO SMALL: With a too-small pinhole, light diffraction smears the image. A smaller hole also decreases brightness, requiring longer exposures.

JUST RIGHT: Calculating proper pinhole size has absorbed much scholarly brainpower. For cameras with typical focal lengths, it's 0.2mm to 0.5mm.

Illustration by Nik Schulz

SET UP.

MATERIALS

[A] Scrap piece of ½" MDO plywood at least 6" square. Medium density overlay (MDO) has a smooth finish that looks nice.

[B] ½"×¾" pine strips (2) about 6" long each

[C] Aluminum sheet 0.01-0.02" thick (e.g. roof flashing), about 1' square

[D] Tinsnips

[E] Sewing needle

[F] Utility knife

[G] Jigsaw **or scroll saw or band saw**

[H] Drill and assorted bits

[I] Drawing compass

[J] English and metric ruler **or calipers**

[K] Flat black spray paint

[L] Roll of 120 format film and cheap, expired film rolls for testing (2) **or one old roll and a spool**

[M] Light meter **or camera with a built-in meter (optional)**

[N] Slide projector **or slide scanner**

[O] Tuna can **or other source of springy steel**

[P] Black silicone sealant

[Q] Pop rivet tool and pop rivets **(rivets not pictured)**

[R] Ball-peen hammer

[S] ⅜" threaded eye bolt **(or thumbscrew) with matching nuts (2) and washers (2)**

[T] #10×1" bolt **with assorted matching nuts and washers**

[U] ¼"×20 nut

[NOT SHOWN]
Epoxy or wood glue

Scrap piece of ¼" plywood **at least 9" square**

⁵⁄₁₆" fender washer

¾" round-head wood screws (4) and 1½" round-head wood screws (4)

Cereal box cardboard

Staples or tacks

Electrical tape

320-grit sandpaper

Carpenter's square

Hacksaw

Metal file

Screwdriver **any type**

Block of scrap wood

Photography by Ross Orr

MAKE IT.

BUILD YOUR PINHOLE PANORAMA CAMERA

START »»   Time: **A Day or Two** Complexity: **Medium**

1. MAKE THE BODY AND WINDER

1a. Draw a paper template for the top and bottom pieces. We're going to cut 2 identical D-shaped pieces of plywood; the curved side is a 105° chord of a circle with radius 65mm (about 2.56"), which continues along tangent flat planes to form 2 corners that extend ¼" past the circle's center. Download the template I used at makezine.com/09/pinhole.

1b. Use a jigsaw or band saw to cut the ½" plywood into 2 D-shaped pieces, following your template.

1c. Clamp the 2 Ds together, and sand until the cut perimeters are smooth and matching.

1d. Use the template to mark the spool centers on the top piece. Drill a ⅜" hole for the film take-up spool, which will be on the left as you face the camera's curved back. Drill a 3⁄16" hole for the supply spool on the right. In the center of the bottom piece, on the underside, drill a ½" hole partway through.

1e. Epoxy a ¼"×20 nut into the partial hole in the bottom piece, to make the camera's tripod mount.

1f. Fit the ⅜" eye bolt through the top piece with nuts and washers on each side, then measure the distance that the nut opposite the winder eyelet will stick into the camera. Find a combination of washers and nuts that fits onto the #10 bolt (for the supply reel) and matches this height.

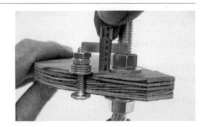

1g. Cut both bolt ends so each pro-trudes ⅛" beyond the nut face. Get an empty 120 format film spool from a friendly camera lab, or untape one from a cheap, expired roll of 120 format film; ends of each bolt flat on 2 sides, so that they engage the slot of the spool. Don't file flats on the #10 bolt; the film supply spool will just spin freely around it.

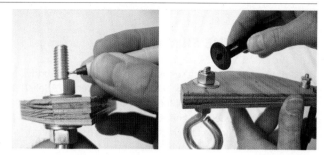

1h. Use thread-locking compound or mash the bolt's threads slightly to keep the nuts in place, so that the bolt assembly spins easily through the wood without loosening or tightening.

1i. Calculate the inside height of the camera by measuring the spool height and adding twice the height of the nut stacks; using this dimension will center the film vertically. Cut a scrap block of wood down to this dimension.

1j. Calculate the outside height by adding 1" (twice the plywood thick-ness) to the inside height, and cut 2 side rails to this length from the ½"×¾" wood. Clamp the top and bot-tom pieces around the scrap block, and glue on the side rails after mak-ing sure that all 4 pieces fit together evenly and are perfectly square.

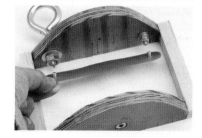

1k. After the glue has set, cut a 1"×5" strip of springy steel — I used the sidewall of a tuna-fish can. Drill ⅛" holes that exactly match the spacing of the bolts on the top, and then attach pop rivets through these holes. The rivet nubs will catch the bottoms of the film spools.

1l. Cut a small wooden block the same thickness as the nut stacks, and staple or tack the spring to it, centered. My block was ⅜"×1"×1½". Apply glue to the block and position it on the floor of the camera. Load the film and take-up spools between the spring nubs and the bolts, and slide the spring block around until the film spools are exactly vertical. Clamp, and let the glue set.

2. MAKE THE FILM GATE

2a. Cut aluminum flashing to the height of the camera body, and 8" wide. Draw a 2¼"×4¾" rectangle in the center. At each corner, draw lines that parallel the top and bottom edges, ½" in (to match the plywood thickness), and extending to the same longitudes as the sides of the rectangle. Measure the camera's curved plywood edge, and symmetrically mark the corners of the aluminum where the edge overlaps this distance. For example, my camera back measured 7¾" around, so I marked in ⅛" from each side.

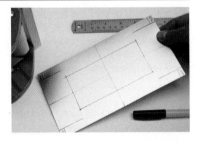

2b. Use tinsnips to cut along the 4 lines that extend in from the sides, and then cut the excess-overlap corners off of each piece, as marked. Then score the center rectangle lines with a sharp utility knife, and flex along the score lines to snap through the aluminum.

2c. Fold the 2 center flaps inward from each side, using a screwdriver shaft as a brake to form smooth, 90-degree bends. Sand any rough edges, especially around the opening in the center.

2d. Fit the gate around the back of the camera. Cut notches in the flaps so that they clear the spring strip, and carefully curl them around so they fit around the film spools.

2e. Spread a thin bead of black silicone sealant along the plywood edges, position the film gate onto the back, and tape it into position until it cures.

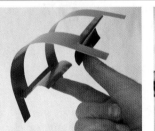

2f. Wrap your test film in position around the back of the gate (if it isn't centered, add or remove washers on the film-winder bolts). Mark the edges of the film on the gate.

2g. Cut 2 straight strips of cardboard from a cereal box, to use as guide rails for the film. Glue them in place along your marks on the gate with a thin layer of silicone. The film should easily slide between these guides with a little wiggle room.

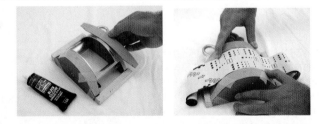

3. MAKE THE PINHOLE

3a. Cut several 1⅞" squares of sheet metal; we'll put a pinhole in each of these, and load them into a 35mm slide projector or slide scanner later to choose the best ones.

3b. For each square, use a ball-peen hammer to tap a small bump into the center of the metal.

3c. Thin the bump with #320 sandpaper. Stop sanding when light pressure with a needle telegraphs a tiny dent through the other side.

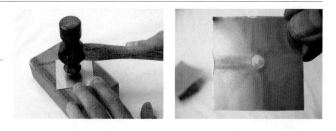

3d. Back the bump against a phone book, and lightly press the needle until a hole shows through. Sand away any raised burr, blow through the hole to remove gunk, and check the diameter.

3e. Use a slide scanner or projector to check each hole's diameter and roundness. For our camera's 65mm focal length, we want a pinhole diameter of 0.33mm, plus or minus 20%. Focal length divided by hole diameter yields the equivalent f-stop, and we're aiming for about f/200.

With a scanner: Set the scanner to its highest resolution, scan the hole, and read its size by setting your image software's units to millimeters.

With a projector: Set the projector up so that a full 35mm slide image measures 52"×35". Load the pinhole into the projector, and look for the light spot to measure about ½" in diameter.

4. MAKE THE FRONT AND SHUTTER

With long exposure times, pinhole shutters can be very low-tech — a piece of black gaffer's tape will work fine but I like this simple "cigar cutter" design. The lip that's formed by the inner and outer panel edges deflects any light that leaks through the sides, making it disappear before it gets a direct line to the film.

4a. Cut 2 rectangles of ¼" plywood, one to the camera front's inside dimensions and one to its outside dimensions. Mine measured about 4"×5¾" and 3½"×5¼".

4b. Drill a ⅞" diameter hole in the exact center of the smaller (inner) piece, and cut a 1"×1½" rectangle in the center of the larger (outer) piece. If you want to use a simple piece of tape as a shutter, glue these pieces together and skip ahead to step 4e.

4c. Use paper and a thumbtack to make templates for a pivoting shutter and the chamber it moves within. The pivot point will sit at the right edge of the front panel (if you're right-handed). The pivot for the shutter will be about 2" or 2½" to the side of the pinhole, and it will take some trial and error with paper templates to find a shape that fits. When you have shapes that work, cut the shutter out of aluminum flashing and the chamber out of cardboard.

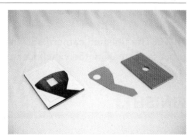

4d. Glue the cardboard chamber to the outer panel, thumbtack the metal shutter into place at the pivot point, and position the small panel in back. Test the shutter to make sure the alignment works.

4e. Disassemble the pieces, and spray-paint black both sides of the inner panel, and the back sides of the shutter and outer panel. Reattach the shutter and glue everything back together.

4f. Center the pinhole behind the rear opening and tape it in place. Hold the panel against the camera, point the pinhole toward some light, and sight along the edges of the film gate to check for any obstructions. The plywood shouldn't block any light from reaching the corners of the film.

4g. Drill clearance and pilot holes, and use two ¾" wood screws to attach the front panel to the side rails, in place.

5. MAKE THE BACK

5a. Cut a piece of flashing to match the height and width of the camera back, bending ½" flanges for the side rails. Drill a ½"-diameter peep sight for the frame counter, offset 1¼" left of center. Put a small square of black tape over this hole; you'll keep it there to cover the hole whenever you aren't winding the film.

5b. Spray-paint black the inside of the back piece you just cut, and the interior of the camera itself. But don't paint the wooden side rails, which you'll be gluing, or the gate where it touches the film. Mask these areas with tape before spraying.

5c. Glue down the back with a sparing amount of silicone, to avoid squeeze-out into the film path. Tack and clamp the side flanges to the wood rails, and hold the back against the D's with tape until the silicone cures.

5d. That's it! Your camera is now ready to use.

NOTE: The finished camera shown above includes an optional viewfinder. To see the step-by-step process for making the viewfinder, please go to makezine.com/09/pinhole.

FINISH ⊠

NOW GO USE IT »

JOIN THE PINHOLE PAPARAZZI

120 FORMAT FILM

Once you remove 120 format film from its foil pouch, the only thing that keeps ambient light from exposing it is the backing paper at each end of the roll, and the fact that it's wrapped so tightly around the spool. So be careful, and don't reload in the sun.

Choose 100- or 200-speed negative film. Anything faster makes sunny-day exposures too brief to time accurately. I recommend Fuji's Superia color and Acros black-and-white films.

LOADING THE CAMERA

When the roll is done, you don't rewind; instead, wind it all onto the take-up spool. The original inner spool becomes the take-up for your next roll.

1. Unscrew the front of the camera, unspool a few inches of backing paper, and use your (moistened) fingertips to fish the paper through the film gate.

2. Tape the paper onto the take-up spool, wind it one turn, and then close up the camera.

3. Wind until the "1" shows in the peep sight. You're on a roll, and ready to shoot! (After each winding, stick the black tape back down over the hole.)

TAKING PHOTOS

With the viewfinder, you will need to move your eye around to see all the edges of the frame.

After every shot, wind to the next odd number, so you don't accidentally double-expose. This yields 6 double-width panoramas per roll.

With 100-speed film, try 2-second exposures in sun, or 10 seconds in bright shade. At this slow shutter speed and camera height, you'll need a tripod to keep your images sharp. In a pinch, resting the camera on a flat surface will work, too, but be careful not to jiggle the camera while exposing the film. For other light conditions, use a light meter or another camera to get exposure times. Your Pin-o-rama has an f-stop of about f/200. Few light meters give readings for f-stops this high, but you can take the shutter speed indicated for f/16 and

multiply the time by 150. Don't hesitate to double or triple the calculated time. Negative films are tolerant of overexposure. To time exposures without looking away from the subject, hold a watch to your ear and count ticks.

FINISHING A ROLL

After exposing shot #11 (the sixth and final shot), keep winding until the paper backing disappears from the peep sight. Open the camera in dim light, and use the adhesive band to tape the roll tight. (If you find the take-up reel has wound too loosely, add more spring pressure on the supply spool.)

Any lab or camera shop that caters to professionals can develop your 120 format negatives for about $5. The 6×12 format is nonstandard, so they should develop the negatives only, and return them uncut. Then you can scan them yourself with a flatbed film scanner. The generous negative size means that even a budget scanner will do a decent job — resolution is a non-issue given the softness of pinhole images. (Darkroom prints are possible too, using a 4×5 format enlarger.)

For pinhole photography resources and additional information go to makezine.com/09/pinhole**.**

PIN-O-RAMA PHOTO GALLERY

The Pin-o-rama camera design was a brainwave from early 2006. Since then, I've loved exploring how it transforms familiar scenes from my hometown into fresh and surprising images.

Big Boy The limitless depth of field of a pinhole lets you move as close as you want to your subject, while leaving distant details sharp — making all kinds of playful juxtapositions possible. The Pin-o-rama's curved perspective, so obvious when manmade straight lines are in the frame, is much less noticeable when they're absent. With this camera's wide, cinematic framing, I like compositions where the subject is strongly off-center.

Cafe Scene This one is a sentimental favorite from my first test roll with the Pin-o-rama prototype. During the 9-minute exposure, people got up and sat down, unaware that the strange object resting on my table was taking a photograph. The clock's minute hand also has blurred into nothingness. At the time, I had no idea what this camera's images would look like, but after pulling the film out of my developing tank, I was delighted.

PINHOLE PORTFOLIOS

📷 See more of Ross Orr's pinhole photographs at flickr.com/photos/vox/tags/pinorama

➕ Homemade Pinhole Flickr Group: flickr.com/groups/homemadepinhole

Liberty Plaza Tree Lights
I started this 15-minute exposure just as daylight was fading from the sky — slightly self-conscious about loitering in a darkened park with my unidentifiable apparatus. I was surprised that on the negative, it's quite noticeable that the pinpoints of light shatter into tiny ripples, from the diffraction of light waves passing through the pinhole. In the background, the taillights of cars creeping by in stop-and-go traffic add their own sparkle to the scene.

Liberty Block Darting into the street between passing cars, I set down my tripod to make this exposure of a nice old commercial block in downtown Ann Arbor, Mich. The couple waiting at the crosswalk stood motionless enough during the 3-second exposure to register on film, but another shopper walking through the scene dematerialized completely. The glint off the window was pure serendipity; pinhole photography seems to invite fun surprises and accidents.

Peaches Pinhole photos can evoke mysterious, even somber moods. But I enjoy working against that stereotype, seeking out subjects that are vibrant and colorful. Strong crossing light can provide extra punch and contrast. I also sometimes use Photoshop's Unsharp Mask filter in an unconventional way: with a very large radius and the amount set to about 12%, to lift the "fog" from pinhole images and regain the brilliance of the original scene.

Yellow Bakery Vertical I've always enjoyed how the afternoon sun lights up this brightly painted neighborhood bakery. This was shot using a Kodak pro film 3 years past its expiration date; any minor color shifts are easily corrected when scanning. Historically, the biggest users of 120 format film were wedding photographers and other professionals. But these shooters migrated to digital so abruptly that camera stores often have excess out-of-date rolls for sale, at half price or lower.

Digital Kirlian Photography

Shoot "auras" without film. By John Iovine

Kirlian photography records a high-voltage corona discharge around objects. Some people call this discharge an "aura" and attribute metaphysical and paranormal factors to its varying parameters. Originally, Kirlian photography was a contact print process that used film. This article illustrates another method of shooting Kirlian photographs, using a digital camera.

The technique can be traced back to the late 1700s, when Georg Christoph Lichtenberg first created "electro-photographs" in dust using static electricity and sparks. Nikola Tesla photographed corona discharges using his famous Tesla coil in the 1880s and in the early 1900s. Others followed, and in 1939, the Russian husband-wife research team, Semyon and Valentina Kirlian, began their 30-year investigation into electro-photography techniques.

In 1970, the landmark fringe book, *Psychic Discoveries Behind the Iron Curtain* by Sheila Ostrander and Lynn Schroeder, popularized the Kirlians' work in the English-speaking world, and electro-photography has been known as Kirlian photography ever since.

Many claims have been made regarding Kirlian images, but most phenomena have conventional explanations. For example, changes in the "aura" of an honest individual who is lying result from the same stress-related increase in galvanic skin resistance that a polygraph lie detector measures. Changes in skin resistance can also be caused by illness, fatigue, drug or alcohol consumption, and other factors, which make these conditions observable using Kirlian photography with no paranormal "bio-plasma" explanation required. The Kirlians themselves believed that their photography could diagnose illnesses before noticeable symptoms manifested, an idea that generated interest, but was never verified by scientific investigation.

One famous Kirlian phenomenon, the "phantom leaf," cannot be explained by known physical laws — but it may be a fake. The experiment is easy to perform: you take a Kirlian photograph of a leaf after cutting off a small portion. Phantom leaf photos from Soviet-era proponents showed the removed portion of the leaf appearing as a ghostly apparition, suggesting an ethereal "bio-plasma

body" persisting where the physical leaf no longer existed. Many researchers and experimenters have been unable to replicate this effect (myself included) although it is easy to fake with a simple double exposure: take a short exposure of the entire leaf, then cut off a piece and continue the exposure.

Although I have never observed any paranormal phenomena with Kirlian photography, I like Kirlian photographs, which are unique and often beautiful. I also like exploring, so I've continued to look for the phantom leaf effect over the years, taking sporadic excursions back into Kirlian photography when I think of new twists to incorporate. If the phantom leaf effect exists and someone finds an exact combination of voltage, frequency, pressure, ambient humidity, exposure time, and other experimental parameters that can reproduce it consistently, it will be the starting point of a new paradigm.

FILM AND DIGITAL TECHNIQUES

The process for traditional, film-based Kirlian photography is simple. In complete darkness, place sheet film on top of a metal plate, and then an object to photograph on top of the film. If the object is inanimate, ground it. Then apply high voltage to the plate momentarily. The corona discharge between the object and high voltage plate is recorded onto the film as a contact print. Develop, and you have a Kirlian photograph of the object.

To shoot Kirlian photographs with a digital camera requires a slightly different technique that uses a transparent discharge plate instead of a metal plate. This discharge plate has a coating of tin oxide on one side that is so thin it is visually transparent, but still electrically conductive.

Place the object on one side of the plate and the camera on the other. As with the film process, connect the object to an earth ground if it's inanimate.

With all the room lights turned off, open the camera's shutter and apply high-voltage power to the transparent discharge plate. The camera captures the corona discharge between the object and the transparent plate.

With film Kirlians, the electrical discharge has to travel through the film's 3 primary color dye layers and activate the silver crystals in each layer at different depths; whereas with digital, it all happens at one level on the plate. This makes Kirlian photographs with film more colorful around the edges, but less accurate than digital Kirlians.

KIRLIAN PHOTOGRAPHY WITH A DIGITAL CAMERA

Digital camera (Options: macro lens, multi-second exposure)

Connect to HV

High-voltage source

Specimen connect to ground

Transparent discharge plate   **Specimen**   **Black paper or background**

MATERIALS

Digital camera with manual focus and a shutter capable of multi-second exposure times. It also helps to have a macro lens setting for close-ups.

High-voltage power supply You can use a Tesla coil, induction coil circuit, high-voltage flyback transformer circuit, or any other source that puts out 5,000+ volts, between 1-5 milliamps, at any frequency (or DC). Variable frequency lets you experiment. The PG13 supply I sell is great for this; see makezine.com/go/highvoltage.

Transparent discharge plate You can buy one of these, or make one out of conductive glass (tin oxide coated), clear plastic, thin copper plate, silver epoxy, and some HV wire. For sources and instructions, see makezine.com/09/kirlian.

Stand or tripod or another way of holding the camera still

Black paper or foamcore

High-voltage wire, Teflon-coated or other, sufficient for your HV power supply

READY, STEADY, SHOOT

Long exposures are required to capture the corona discharge, so you will need to keep your camera still and the object steady. I use a camera copy stand that holds my camera pointing downward, and I set up the object and plate horizontally underneath it. Once you have a steady arrangement, here's the procedure.

1. A black background works best, so put down a piece of black paper or foamcore (must be nonconductive) on your work surface. If you're shooting an inanimate object such as a leaf or coin, you need to

Illustrations by Damien Scogin

ground it. To do this, cut a small hole in the center of the black background, and attach a length of HV wire to an earth ground (a pipe will work). Then strip the other end of the wire, run it through the hole in the background, and place the object on top, making sure it has contact with the wire. You can also use a small grounded copper plate under the background instead of a wire, so long as the object you are shooting makes contact with the plate through the hole.

2. Attach the high voltage lead from the HV power source to the discharge plate's electrode, and place the plate on top of the object.

3. Position your camera to frame the object, and focus it manually. If you can set the camera's f-stop, open the aperture (reduce the f-stop) as much as possible. You will usually shoot a flat object, which doesn't require much depth of field. Set the shutter speed to 15 seconds or more.

4. Turn off all room lights. Turn on the high-voltage, and adjust its frequency to provide the brightest discharge. Hit the camera's shutter, and keep the power going for the full duration of the exposure.

5. Evaluate your picture, adjust, and reshoot. Raise the f-stop and reduce the exposure time if the image is washed out; do the reverse if it's too dim.

SHOOTING PEOPLE

Sooner or later you are going to want to shoot people. The best place to start is with a person's fingertip. To avoid electric shock, make sure that the subject doesn't touch ground. They also need to hold still, of course, and they shouldn't put too much skin surface in contact with the discharge plate. Touching an entire hand to the discharge plate can dissipate the corona discharge so much that it won't photograph, even with a long exposure.

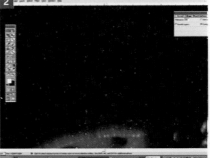

WARNING: When shooting people, make sure the subject doesn't touch ground.

DIGITAL MANIPULATION

My digital camera has a maximum exposure time of 16 seconds, during which the image accumulates some digital noise — mostly gray, red, and yellow pixels of various shades sprinkled throughout the resulting photograph. I remove this in Adobe Photoshop as follows:

1. Starting with the base image of the leaf, select Image » Adjust » Brightness/Contrast, and increase the contrast.

2. Zoom picture to 500% so you can clearly see the background pixel noise. Using the Magic Wand (set to 30 tolerance), click on a noise pixel. Then go to Select » Similar. Now all the pixels with that color in the photograph will be selected.

3. Set the foreground color to black and hit the Delete key to turn all the selected noise pixels black. Repeat this process by selecting and deleting various other noise pixels, but don't select any noise pixel that has a similar shade to the main image you are trying to enhance.

Kirlian photographs are mostly blue-violet with a little white. This is because the corona discharge in nitrogen (air is 78% nitrogen) generates light in

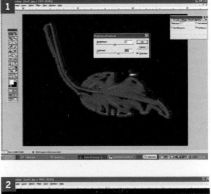

the blue-to-ultraviolet range. You can use Image » Adjust » Color Balance in Photoshop to change your image's color for some beautiful effects.

GOING FURTHER

Using digital cameras to capture Kirlian photographs is relatively new. Because it's a low-light process, we can borrow techniques from the astronomers who use digital cameras, as well as astronomy software such as AstroStack (astrostack.com), which stacks multiple images of the same object to create a brighter, more detailed image.

Shooting Kirlian photographs in atmospheres with other mixes of inert gases, such as helium, neon, and carbon dioxide, will create corona discharges in different colors. Just make sure the gases are inert.

To learn more about Kirlian photography using film, refer to my book, *Kirlian Photography: A Hands-On Guide* (Images Publishing, 2000).

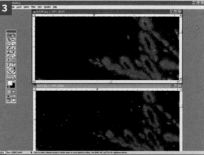

John Iovine is a science and electronics tinkerer who has published a few books and articles and owns and operates Images SI Inc. (imagesco.com). He lives in Staten Island, N.Y.

FILM-BASED KIRLIANS

The light you see during a Kirlian discharge comes from an electric field ionizing the air and parts of your Kirlian subject. The charged air gives off intense heat and ultraviolet radiation as well as visible light. Photographic emulsions interact with all of these, plus the electrostatic field itself, while digital sensors mainly just capture visible light. As a result, traditional film-based Kirlian images require shorter exposure times than digital methods, and can show the lightning-fast discharge arcs in greater detail.

For example, my Kirlian photograph of the quarter on this issue's cover took only a ¼-second exposure. With faster photo papers and films it might take even less. In addition to improving detail, such short exposures also decrease the chances of vaporizing or burning your subject. And when photographing a human being, short exposure times are less painful!

Open-film Kirlian requires a fully darkened room, tens of thousands of volts, and expensive sheet film. All of this can be daunting and, well, dangerous — or at least expensive. But you can start out and get wonderful results using black and white photo paper instead of film, and working under a darkroom safelight or red LED flashlight to help you steer clear of the power supply. Later, you can graduate to a color image, and the digital method is also worth a try! *—Jasper Nance*

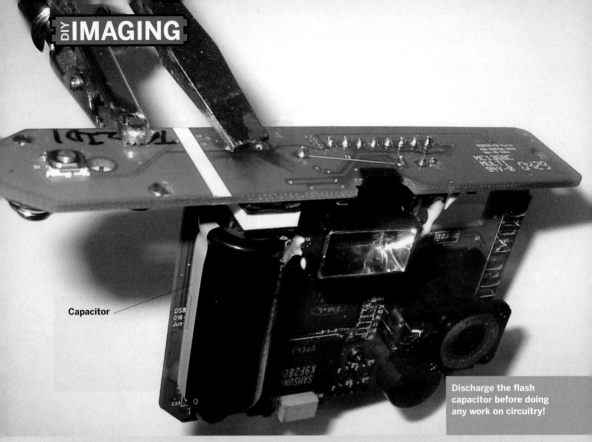

Capacitor

Discharge the flash capacitor before doing any work on circuitry!

SINGLE-USE DIGICAM FOR KITE AERIAL PHOTOGRAPHY

Simple, lightweight timer circuit triggers a shot every minute. By Limor Fried

Kite aerial photography (KAP) is a fun hobby for engineers and artists alike, and the resulting images can be beautiful. However, one unlucky gust can send your camera tumbling to the ground. That's why disposable cameras are a good choice for KAP beginners. Their low cost means there's no big loss if they crash. They're also lightweight, and most are pre-focused to infinity, which means they're good for taking scenery shots.

Many KAP rigs have been designed to take film-based disposable cameras aloft. (See MAKE, Volume 01, page 50, for complete instructions on how to build your own kite aerial photography rig.) But film disposables require winding between

shots, so you can only take one picture per launch.

Fortunately, there are digital equivalents that use electronic switches and need no winding. I use the Dakota Digital PV2; it's $19 at Ritz Camera (*ritzcamera.com*) and Wolf Camera (*wolfcamera.com*), with a $10, non-LCD version available in CVS stores. The PV2 holds 25 photos. As with a film disposable, you send it away for processing.

Here's how to make a kite-ready timer circuit that triggers the PV2 once per minute, after an initial delay. You can also adapt it to trigger other, non-disposable digicams.

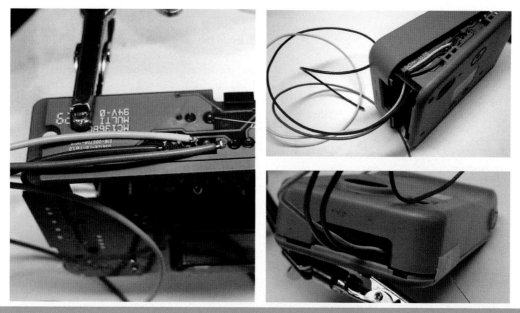

To run the Dakota PV2 from our timer circuit, we need to connect them with three wires: positive power (red wire, VCC in the schematic below), negative power/ground (black, GND in schematic), and the shutter trigger (yellow, OUTPUT in schematic). Conveniently, contacts for all three are arranged in a row on the PV2's top PC board, across from the shutter button. Our timer will piggyback off the camera's power source.

Designing the Circuit

Since it takes time to launch the kite, we'll want to delay the first shot. We'll implement this with a simple resistor-capacitor network. The capacitor slowly charges through a resistor until its voltage is high enough to turn on the timer chip, via the chip's reset pin.

The PV2 automatically shuts off after 3½ minutes idle, so our delay must be less than this. The time depends on the relative values for the resistor and capacitor, and I calculated that a 470µF capacitor and 680kOhm resistor (C2 and R3 in schematic at right) would produce about a 3-minute delay. (See *ladyada.net/make/sudc4kap* for full derivation.) Try this combination, and if the camera shuts off before having a chance to fire, reduce the resistor value.

For the main part of our circuit, we'll use a TLC555 or LMC555 timer chip. Configured with two resistors and a capacitor, 555 chips can generate square waves of almost any frequency up to 2MHz. We want to make ours oscillate every minute, firing a shot once per cycle.

Timing a 555 also uses an RC (resistor-capacitor) network. In our configuration, the 555 discharges a capacitor through one resistor and recharges it through two resistors in series. While the capacitor charges, voltage on the timer's output pin is high (3V) — and it's low (0V) while it discharges. I determined that a 47µF capacitor (C1 in schematic) and two 680kOhm resistors would tune the 555 to produce a 60-second square wave. (See website noted earlier for derivation.)

Other components in the circuit are an indicator LED that flashes once per cycle (for testing), and a diode between the Output and Reset lines. This keeps the Reset line voltage high, which protects against noise from the camera's flash that can trigger the shutter accidentally.

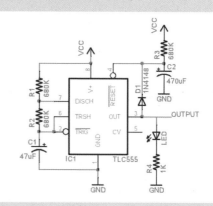

Fabrication

1. Disassemble the camera.

Take the batteries out, remove the three screws that hold the body together, and take off the back cover, being careful not to touch any of the electronics inside (they may be at high voltage).

MATERIALS	Red LED
Single-use digital camera	Tape
Small perf board	EQUIPMENT
TLC555 or LMC555 timer chip	Soldering equipment
470µF capacitor	Multimeter
47µF capacitor	Large flathead screwdriver
680kOhm resistors (3)	Precision Phillips screwdriver
1kOhm resistor	Needlenose pliers
Switching diode (1N914 or 1N4148)	Angle cutters
Small toggle or slide switch	"Helping hands" or mini vise
Hook-up wire, preferably 22-gauge stranded	Solderless breadboard (recommended)

2. Discharge the flash capacitor.

Carefully hold the main PC board as shown on page 286, making sure the vise or jaws don't short any exposed components. The huge capacitor carries the flash bulb charge, and it *must* be discharged before any hacking. Hold a flathead screwdriver by the plastic handle, and touch its tip to both capacitor leads simultaneously. There will be a large pop and flash as it discharges. (Do this again after any time the board has sat idle.)

3. Attach the control and power wires.

Cut three wires in red, black, and yellow (or other color) for the shutter wire. Solder these to the pins located where the top shutter board connects to the front flash board. Counting over from the shutter button, connect black to Pin 4, yellow/shutter to Pin 6, and red to Pin 8.

4. Reassemble the camera.

Thread the wires through the small hole in the flash board, and put the plastic case back on, so the wires come out the side. Then insert the two screws near the bottom of the case, and use a piece of tape to secure the top. Don't try to close the case all the way.

5. Build the controller circuit.

Before soldering your circuit onto the perf board, you should test it out on a solderless breadboard. Build the circuit following the schematic on page 287, making sure to place all the capacitors, diodes, and LEDs in the proper directions.

6. Test the circuit.

Connect the red and black wires from the camera to the VCC and GND points on the circuit, respectively. The LED should light up after two minutes or so, and then cycle on and off about every minute. Use a multimeter to confirm that there are 3V between the red and black wires, and check the voltage across the 470µF capacitor (C2) to make sure it rises slowly to 1V or so, then jumps up to 3V. Check the 47µF cap (C1) and look for slowly oscillating voltage between 1V and 2V. Finally, check the output voltage, which should flip between 0V and 2V, following the LED. If you're prototyping your circuit on a solderless breadboard and it all checks out, you should rebuild it, soldering onto the perf board, and test it again.

7. Test the full system.

With the camera off, discharge the two capacitors (C1 and C2) with the screwdriver. Connect the trigger wire to the OUTPUT line from the 555. Turn on the circuit, and then the camera. When the LED turns off for the first time, the camera should take a picture. (If it doesn't take a picture, try connecting the shutter wire to the black wire.)

8. Take flight!

Mount your camera and control board to your kite, and let it fly.

9. Disconnect the circuit and develop images.

After your flight, check the "Pictures remaining" counter to verify that the circuit worked. After all the pictures are taken, open the camera, discharge the flash capacitor, and desolder the wires. Then reassemble the camera and have it developed normally.

Limor Fried is a science genius girl (with apologies to Freezepop).

With iStopMotion
software, creating stop
motion animations
could not get any easier.

STOP MOTION ANIMATION, THE EASY WAY

With iStopMotion, making *Gumby* is less pokey. By Phillip Torrone

Stop motion animation, one of the oldest special effects, makes the impossible seem real. But the tedious process — move the model, take a picture, repeat thousands of times — discourages citizen filmmakers with non-obsessive patience levels.

I recently discovered iStopMotion, Mac software that automates the process and works with any camera that can capture QuickTime. iStopMotion features a transparent preview that lets you super-impose the previous frame over the current one before you shoot. This alone helps enormously. The Time Lapse feature shoots frames continuous-ly at a specified interval, and Speech Recognition lets you say "Capture" instead of having to click. These two features saved me thousands of trips

back to the keyboard and mouse, and in just a few minutes of experimenting, I made a fairly impres-sive little flick.

You can even shoot frames simultaneously from multiple cameras, to create seamless cuts between different angles, just like the pros.

For inspiration, go watch *King Kong* and visit the example pages at *istopmotion.com/example*. I'll post my animation on the media feed of *makezine.com*. If you create one, let us know!

iStopMotion: free demo version, $39.95 license, *istopmotion.com*

Phillip Torrone is senior editor of MAKE.

Study Andy's attire carefully. Imitation, like flattery, will get you everywhere.

The Fauxlance Photographer

How to get VIP treatment by dressing the part of a pro photographer. By Andy Ihnatko

Being a columnist for a Great Metropolitan Newspaper has its perks, first among them being the fact that the *Sun-Times* and I have worked out a little agreement by which they pay me money for my columns.

But there's another benefit. My idea of a fun time is to walk around with an SLR and make some pictures. Sometimes, getting Just The Right Shot means climbing up on a balcony or asking a street performer to turn toward the sunlight or standing rock-still with my SLR against my face, waiting for the scene to finish composing itself. And being able to flash a business card or a recent clipping is often the difference between being told "I'm sorry, but I'm really going to have to ask you to leave" and being told "You'll get much better shots of the parade from the top of the reviewing stand ... just follow me."

If you're not a columnist, well, simply dressing the part of a working journalist can go a long way. The way you appear and the way you behave subconsciously communicate to people that you're not some creep with a camera or just a tourist getting in the way. It's not about misrepresenting yourself; you're just subtly encouraging people to treat you with the same amount of respect that you invest in your photography.

Here's how you can get treated like a pro:

Overall Dress

Don't look like a bum, unless you actually are a photographer for *Sports Illustrated*.

As a freelance journo, you don't have the money for expensive clothes and they'd just get ruined as you travel from event to event anyway. So the

Photograph by Andy Ihnatko

effect to aim for is "as comfortable as you can get away with while still looking professional and presentable." Meaning: buttoned work shirt (no tees or pullovers), casual slacks, and sneakers that can pass for shoes.

Camera (general)

Yup, it's bigotry, but civilians instantly respect the Big Black SLR. A line of servers at a Chowderfest will not patiently pose for you if you're packing an $80 point-and-shoot.

Pros almost always wear two cameras (one with a short lens, one with a tele). Finally, a reason to get your old film Nikon out of mothballs! Always wear them across the shoulder on a long and comfortable strap.

Camera Lens

Put a lens shade on your camera, like every working photographer — not to combat sun flares, but to protect the glass and filter threads against bumps.

Camera Flash

An external flash is a must, even when shooting outdoors. For extra credibility, rubber-band a white index card around the head as shown. It's a cheap bounce-diffuser for better lighting.

Bag and Waistpack

Photographers need to get their hands on another lens, memory card, or filter without any fiddling around. So: A waistpack is essential, and if you wear a bag, it shouldn't be a backpack.

Necktie (in bag)

Handy to have, in case you discover that most of the other media are wearing 'em. A rare problem, but I've gotten into at least one event because I was able to put a tie on.

Microphone (in bag)

If you're recording audio for a podcast, get a mic that is, or looks, impressive. And, the shabbier the windscreen, the better.

Old Press Pass (in bag)

OK, it's a cheap trick, but I might have a press pass in my bag, too. If I'm talking to the media coordinator of an event and I can sense that they're on the fence about giving me a pass, I'll rum-mage through my bag for something and whoops, a genuine laminated badge to a previous event fell out! Sorry about that. Hmm? Oh, yes, that's from a space launch that I covered, thanks.

More important than a pass is a batch of business cards. No need to lie: just use your name and put the simple word "Photographer" underneath. Include your address, phone number, and email, and a link to your photo site. Photo subjects and media coordinators want to be reassured that you're for real and not some creep; 500 cards for $10 goes a very, very long way.

Breast Pocket

Keep it filled with random crap at all times.

Notepad

Hands-down the most important prop in the picture. All working media need to have a notepad handy at all times. Not a PDA — a notepad. You're always scribbling down names of people you've spoken to or photographed and millions of other notes, and it also serves as documentation that you actually were where you said you were, in case lawyers get involved later on.

Plastic Bag

If you're still shooting film, you gotta keep it in Ziploc baggies. It's just the rule.

Important Notes:

So what about getting an actual press pass? Well, it can't hurt to ask (call the event office and ask for Media Relations). Two big warnings, though: First, do not, do not, do not lie. If I'm not attending an event with the express purpose of covering it for my paper, I'm crystal-clear about my intentions. Trust should never be abused. And of course you want to get into the media tent, with its air conditioning and water and hookups. But stay out. For members of the actual media, it's essential. If I have 30 minutes to file my story but you're using the last network connection to play some MMORPG ... well, look, it'll get ugly. Finally, give away your photos and story to your local paper. Remember there's just one difference between an amateur photographer and a pro: a single published clipping.

Andy Ihnatko (andyi.com) is the *Chicago Sun-Times'* technology columnist and the author of a best-selling series of Mac books for Wiley Publishing.

Recursion Alert: View this image through red-cyan glasses to see how to take a 3D photo, in 3D.

3D Photography

Taking stereographs is easier and more fun than ever. By Bill Coderre

3D photography has been around for almost as long as regular photography. But the 1950s brought the fad of 3D horror movies, starting with *Bwana Devil* (1952) and *House of Wax* (1953). Many of these were so bad that 3D soon became an embarrassment. Even 50 years later, 3D still hasn't shaken its reputation as a cheap gimmick that causes headaches — which is a shame, because the medium can produce compelling images worthy of contemplation.

It's time for a 3D renaissance. In the last few years, computers have made it simple to produce color 3D images in just a few minutes. All it takes is a digital camera, some free software, and a pair of red-cyan glasses. I'll explain how to do this, and also suggest some camera projects that can make taking 3D pictures even easier.

If you don't have red-cyan glasses, you can find some at a comics shop, attached to any Ray Zone title. Or, you can buy them from American Paper Optics (*see page 294, "Further References"*).

The Stereo Shuffle: Taking Basic 3D Photos

Grab your digital camera, and head outside. Find a scene with a fun variety of distances to focus on. The closest objects should be at least 5 feet away, and perfectly still. Also, make sure that everything in the scene will be in sharp focus. Sunny conditions will help your depth of field.

Ready? Hold the camera to your eye, and lean very slightly to the left. Snap a picture. Lean very slightly to the right, and snap another picture. That's your basic "stereo shuffle" right there.

Photograph by Bill Coderre

For best results, make sure that the camera moves from left to right about 2 inches, and does not move up or down, toward or away from the scene, or tilt in any direction. Also, do not realign the frontmost object to the same position in the frame. (For the nerds: no y or z translation; no roll, pitch, or yaw; and do not "toe in.")

Now head back to your computer, import the pictures from your camera, grab some free software (see table above), and use it to load the left and right frames of the pair. I recommend AnaBuilder, which has all the functionality you'll need, once you figure out the buttons and menus.

Display the picture as a color anaglyphic image, and before you put the glasses on, correct any twist or vertical offset. The horizontal lines in both views should overlap and be parallel, and objects in each image should occupy the same vertical position in the frame.

One last step: make sure that the frontmost object in the scene has its two views completely superimposed, with no red and blue halos around it. This means that the nearest object in the picture will be at the stereo window depth, not sticking out in front of the frame. This makes the scene easier for most people to view. You can experiment with "eye-poking" stereo later.

Now, put on the glasses and admire your handiwork. Believe it or not, that's it for the basic method. Everything from here on is just enhancements and alternatives.

Near or Far Subjects: Adjust the Interaxial Distance

If your subject is a small flower or a mountain, having it 5 feet away won't work. With a close shot, 2 inches between pictures (called the "interaxial distance") is too much, and the picture will be unviewable, or at least headache-inducing. With a distant subject, 2 inches will not be enough to give any appreciable 3D effect.

For these situations, the basic rule is to divide the distance to the nearest object by 30, and move the camera that amount between shots. That means 176 feet for an object a mile away, and 0.8 inches for an object that's 2 feet away.

These approximations, however, do not take into account important factors such as lens focal length and the distance of the farthest object. For a more rigorous approach, see the Bercovitz Formula article listed under "Further References" on page 294.

More Accurate Alignment: Make a Candy-Tin Slide Bar

A stereo slide bar will ensure that your camera moves only horizontally between pictures. You can easily make one out of a tripod, a bubble level, and a candy tin.

Use a nibbler, a drill, or some other tool to cut a slot in the bottom of the tin. The slot should be a bit wider than ¼", running the full length of the box from side to side, parallel to the hinges, centered halfway between the hinge side and the front. Place the slot onto the ¼-20 head screw (¼" diameter, 20 threads per inch) that sticks up from the top of your tripod, and align the candy tin so that it is parallel to the tripod's handle. Use poster putty (such as Fun-Tak) to attach the camera to the candy tin, and the bubble level to either the camera or the tripod. To take pictures, make sure the tripod is level, then snap one photo, slide the camera over, and snap the second photo. It also helps to add a ruler to the back of the candy tin, so you can control your interaxial distance.

Live Subjects: Link Two Cameras

You can take 3D photographs of subjects that don't hold still by using two cameras. Place them next to each other on a bar, or tack them together by their bottom surfaces and rotate or invert the images so they match later. If the lenses cannot be placed close enough together, you can rig up a mirror system. The other half of the solution is finding a way to trigger both shutters at the same time. This depends on the type of cameras you're using (*see "Further References"*).

"Found" 3D

My friend Eric von Bayer pointed out that you can create stereo pairs by selecting frames from 2D movies. Copyright considerations prevent me from showing this example, but you can download the *Star Wars Episode I* trailer, and extract two frames from the shot that pans horizontally over a landscape of Naboo. Drop them into AnaBuilder, and you end up with a pretty nice 3D scene.

Loreo Cameras

Loreo (loreo.com) makes inexpensive 3D cameras that use mirrors to put the left and right stereo images on one standard 35mm film frame. For viewing, you could either use their included viewer, or get a digital image, and then cut and paste the two sides into the software provided. The Loreo doesn't let you adjust the exposure, but at least it has a built-in flash.

Other Display Options

Red-cyan glasses take some getting used to, and you can see "ghosts" when the disparity between the images is large. Here are some other display technologies.

The **ColorCode** system substitutes blue and amber for red and cyan. This reduces ghosting, but it's also patented, and illegal to distribute software that produces ColorCode images.

Adjacent-image pairs are viewable by crossing or uncrossing your eyes, but this is a tricky skill to learn, and it limits image size. To view bigger images, you need a stereoscope, like the ones from the turn of the 20th century. These devices have two lenses mounted off-center, to allow for close-focus, and a prism that lets the eyes point straight ahead while the view-lines diverge. Fancier viewers use mirrors to accommodate larger stereo pairs. You can make your own stereoscope, but note that the brain usually interprets misaligned left and right eye views as caused by food poisoning, and reacts by the sudden expulsion of the contents of the stomach. Therefore, it's important to align the mirrors precisely.

Time-interlaced images are viewed by LCD shutter glasses, which you can get for $30 from

Razor 3D (razor3donline.com). Unfortunately, you cannot use these glasses with LCD monitors, progressive scan monitors, or any monitors set to refresh faster than 90Hz. As a result, images often seem to flicker, but it helps to turn down the brightness of the monitor, and reduce room lighting. At standard television frequencies (roughly 50-60Hz), movements appear jerky, and fast-moving objects will often "tear."

Horizontally interlaced image sets are displayed by putting the interlaced image behind a lenticular screen, like with 3D postcards. Free software such as Interlace produces interlaced images that you can print on a high-resolution printer, and then display in a sleeve behind a reusable lenticular screen. The big plus of this technology is that it does not require the user to wear glasses or use a stereo viewer. The downsides include low horizontal resolution, and the difficulty of aligning the image with the screen.

Several television companies have hinted at future **3D TV** products based on lenticular screens, but the NTSC format's width of 720 isn't enough to split in half and retain a good picture. With horizontal resolutions of 1,280 and 1,920, HDTV seems like the "silver bullet" that might make it all work. We can only hope.

Further References

General 3D information: dddesign.com/3dbydan/3dlinks

Explanation of the Bercovitz Formula: makezine.com/go/vic3d

Synchronizing twin cameras: makezine.com/go/twincam

American Paper Optics, manufacturer of 3D glasses and experimenter kits: americanpaperoptics.com

Ray Zone, "The King of 3-D Comics": ray3dzone.com

Bill Coderre dredged most of this article from class notes and memories of a seminar by 3D pioneer Stephen A. Benton, professor at the MIT Media Laboratory, to whose memory this article is dedicated. These days, Bill is a programmer for Apple.

Alito Confirmation Hearing
Opening Statements

 C-SPAN

Me and Judge Alito, at
his confirmation hearing.
I'm sitting at far left,
enjoying a cold brau.

HOW TO DRINK BEER ON C-SPAN

 Put yourself into somebody else's video.
By Bill Barminski

Photography by Bill Barminski

OK, you're not really going to drink beer on C-SPAN or *Larry King Live*. But you can make it look like you did on video. I don't know why you'd want to, but let's just say you do. I know I did.

The method used to achieve this effect is called *compositing*. You will need a source video recorded from a television show, a replacement video you will shoot yourself, and a *static matte* — a shape cut out of the source video with Photoshop to hold the new video.

The first step is to watch TV and record the source video. Sounds like fun, right? Don't get too excited. I recommend using video from C-SPAN, which is a good source for two reasons. First, they

repeat everything, over and over and over. So if you see something good, you can catch it later and record it. I use an analog-to-DV converter on my computer to capture stuff live off the TV directly into Final Cut Pro. But you can record onto a DV camera and then later capture it into the editing software of your choice.

The second and more important reason why C-SPAN is a good source is that the network tends to use "locked-down" camera shots. (A *lockdown* is a camera that has been set on a tripod and is not moving or panning.) No matter where you get your source video, you need to look for shows that use locked-down camera angles. A locked-down

A

B

C

D

Fig. A: Still image extracted from source video.
Fig. B: Matte cut from source still in Photoshop, using the Path tool. Fig. C: The checkered area is transparent.
Fig. D: Shooting the fake video.

camera dramatically reduces the amount of work required. If the camera is moving, the shape of the matte you create has to move, too, in every single frame of the video captured, which is 30 frames per second! Do the math. A 10-second shot will require you to cut something like 8 billion mattes. OK, my math may be off, but it's still a lot. With a locked camera shot, you just need one, which you can cut in Photoshop.

MATERIALS
C-SPAN
Analog-to-video capture device
Final Cut Pro and Photoshop
Video camera
Tripod
Beer
Microphone to record belch
Business suit

I recorded some footage of Judge Alito at his con-firmation hearings. In the shot I used, the camera is locked down and Judge Alito is sitting mostly still. More importantly, the woman wearing light pink in the row in back is very still. I want to sit next to her and drink beer.

Once I've found a source video clip, I extract a still image from it, open it in Photoshop, and cut the matte. I use the Path tool to make this selec-tion; a 1-pixel feather on the selection is recom-mended. I place the selection in a new layer and fill it with any color. I turn off the background image, leaving me with a solid shape. I save this as my matte with the alpha/transparency intact. From the File menu, do a Save for Web with PNG-24 selected as the format and the Transparency option checked. The checkered background indicates the transpar-ent area.

Once the matte is cut, it's a good idea to test it to see whether it really holds up before wasting time shooting a replacement video. Using Final Cut Pro, I place my original video into Layer 1. Then I import my PNG matte and place it into a new layer above the original video. It should fit right into place. I render the shot and check to see whether the edge along the woman in pink's silhouette will work. If she sits still and the camera doesn't move, then the matte should sit perfectly along her outline. If she picks up her hand, it ruins this matte because her hand would travel into the orange area and get cut off.

SEN. EDWARD KENNEDY

D-Massachusetts

C-SPAN

Here I am again, seeing if the Alito hearing is any more interesting if I'm sitting behind Ted Kennedy.

Now Comes the Fun Part: Making Your Fake Video

I put on the only suit I own and set up my camera on a tripod in my living room. In order to get the camera angle right, I print out a still image from the Alito footage and make adjustments until I have the same approximate angle. I don't worry about placing myself in the corner of the shot; in fact, I shoot myself central to the frame, knowing I can hand-place the shot in the corner later.

I originally shot myself doing various things such as drinking beer and making faces and hand gestures. But I found that the faces and hand gestures were just too obvious, too over the top. The beer drinking was subtle. In fact, when I showed the finished composite shot to some people, they had to watch it twice before figuring out that something wasn't right. I consider that success.

Once the fake video is shot, I'm ready to composite all the elements together. I use Final Cut Pro, but After Effects or Premiere will also work. I import my three elements — the original video, a PNG matte, and the replacement video — into a project. I place the original video into the bottom video layer, the

PNG matte into a new video layer above it, and the fake video in a third layer above the first two.

The method used to put the fake video into the matte area of the real video is called *travel matte alpha*. I double-click to select the fake clip, open the Modify column, and then, near the bottom, open the category Composite Mode. I then select the Travel Matte Alpha option near the bottom of that subcategory. The fake video now plays only inside of the shape of the matte.

The video may now need to be moved into position to look correct. I had to resize mine a little and move it to the left until I was "sitting" next to the girl in pink. I also did some color correction so that both shots looked consistent.

The last step was to record a loud burp. I used the voice-over option in Final Cut to record that additional sound. I'm done, unless I want to eat chips behind Senator Kennedy.

▇ These clips can be seen at barminski.com.

Bill Barminski is a multimedia artist currently teaching in the School of Theater, Film and Television at UCLA.

TILT-SHIFT PHOTOGRAPHY

Flexible lens makes scenes look miniature.
By Dennison Bertram

One of the fancier lenses in the world of SLR and DSLR photography is the tilt-shift lens. You might not know what these lenses look like, but you proba-bly have seen their effects. Architectural photogra-phers use tilt-shift lenses to eliminate the perspec-tive distortions that sometimes give buildings the appearance of falling over. Aerial photographers use them to make large cities look like toy models. Art and portrait photographers use them to control exactly where the focus falls.

Tilt-shift lenses cost $1,000-plus, which is far beyond what most photographers will pay to experiment. Fortunately, building your own tilt-shift lens is easy, and doing so will open up a remarkable array of creative optical effects.

To build your own tilt-shift lens, you start with a spare lens that's built for a film format larger than that of the camera you'll use the lens on.

For example, I used a 6×6 lens (designed for 6cm film) to make a tilt-shift lens for a 35mm camera body. With 35mm or APS format digital SLR cam-eras, you'll need a lens built for 6×6 film or larger. The oversized lens gives you extra room to move and distort the image that lands on the film or CCD, while still filling the frame. (You could use this hack to mount a 35mm lens on a 35mm camera, but it would only work with a macro lens, for very close objects.)

Assembly

Using a rotary tool or hobby knife, hollow out the mid-dle of the camera body cap (Figure A), then grind or file it down smooth, so there are no rough spots or burrs (Figure B).

The plunger will act as a flexible camera bellows, allowing us to tilt and shift the lens to our heart's desire. Cut a hole in the top of the plunger, where

Photography by Dennison Bertram

Fig. A: Hacked lens begins with a standard plastic body cap. Fig B: Body cap is hollowed out.

Fig. C: Humble plunger becomes camera bellows.
Fig. D: Plunger bellows should fit tightly to lens.

MATERIALS

SLR (single-lens reflex) or DSLR (digital SLR) camera body with interchangeable lens mount

Oversized lens I used an old Carl Zeiss made for the now-obsolete Pentacon 6 camera format. Millions of these were manufactured, and eBay is full of good deals on them.

Rubber plunger with bellows design Any will do, so long as it's flexible and not too large.

Stiff cardboard (non-corrugated) Or stiff, black plastic

Plastic body cap to fit camera body I use a lot of these in my work. They're the perfect way to attach your camera to your own hacked-lens creations, and they're also cheap.

Hot glue and glue gun or other adhesive You need to bond the body cap to the cardboard/plastic and the cardboard/plastic to the plunger rubber. If you want a more robust setup, skip the glue and attach the pieces with small hobby nuts and bolts.

the stick is (Figure C), making it just large enough to stretch around the base of your lens (or make it a bit smaller and enlarge it later).

Go ahead and stick your lens onto the plunger to see if it fits, and trim the rubber as necessary. Don't worry about gluing the lens down yet, but the hole should be tight enough so the lens fits snugly (Figure D). I even cut grooves in the rubber to let me screw the precious lens into place. Keep holding onto the lens, though; don't expect the rubber to hold so tight that you can let go.

If you're really enterprising; you could buy a bayonet adapter for the lens you're using, attach it to the plunger bellows, and then screw your lens onto that when needed.

Next you need to build the backing. To do this, I cut a ring out of cardboard with the inner circle the same circumference as the body cap and the outside matching the bottom of the plunger (Figure E). Although not shown here, it helps to paint one side of the cardboard black, to cut down light refraction inside the bellows.

Then I hot-glued the parts together, body cap into ring, and ring, black side in, onto bellows (Figure F). The hot glue is for expediency; if, after some experimenting, you think you'll use this lens setup often, I recommend finding something sturdier than hot glue and cardboard to hold it together.

You're all done! Attach your lens, and you're ready to shoot.

E

F   **G**

Fig. E: Cardboard backing ring.
Fig. F: Finished assembly.

Fig. G: Bendable lens makes Prague look miniature.

Shooting

This hack works surprisingly well. The image quality of the Zeiss is awesome, and I didn't get any optical interference between it and my Nikon DSLR's CCD chip. But there are a number of things to keep in mind while shooting. First, automatic exposure modes will not work with this lens, so you have to shoot manually. The apertures will still work, but in general you want to shoot with the lens wide open, or it will be far too dark to focus.

Needless to say, auto-focus also won't work. With this lens, you focus (or selectively un-focus) by squeezing the lens and plunger down, bending it, and twirling it around to get the cool effects you want. To give a "miniature" effect to a city-scape, tilt the lens forward or backward so that the only things in focus are in the middle ground, in mid-frame horizontally. The blurred foreground and background simulate the look produced by a macro lens taking a close-up of something small (Figure G).

If you want to use a lens like this in low light conditions, where you'll need to hold the lens in one position for a long time, you might augment this design with an adjustable mechanical frame that controls the lens' range of motion. If you build a tilt-shift lens with a frame, a photographic bellows will be more flexible than a plunger.

A note to digital camera users: dust is a common problem with most digital cameras. Projects such as this one can exacerbate the situation. Before using this lens, be sure to clean it out with some strong puffs of air, to get rid of any loose dust particles that might be inside.

Dennison Bertram is a fashion and beauty photographer who lives in the Czech Republic.

OUTDOOR WEBCAM ENCLOSURE

Capture winter scenes from hanging sewer pipes. By Alek Komarnitsky

I anteed up the big bucks for a wireless security webcam with motorized pan, tilt, and 10x optical zoom — specifically, the D-Link DCS-6620G. Nice webcam, but I wanted to put it outside so people everywhere could view my Halloween decorations and infamous Christmas lights.

The problem is that the webcam is rated only down to 32°F, and here in Colorado, temperatures can drop below zero. Suitable prefab outdoor enclosures cost about $500 and include a blower and heater, so I decided to build my own simpler webcam enclosure. It cost me a whopping $27, and it has successfully stood up to two full seasons of Rocky Mountain rain, cold, and snow.

I installed the webcam at my neighbors' house, hanging it from a 6" can light fixture under an

eave that had good line-of-sight to our house. The basic idea was to attach 2 brackets hanging down from the inside of the can, and build an easily removable enclosure that would hang from a rod running horizontally through the brackets.

For the brackets, I straightened two 5" L-brackets in a vice, and extended their internal cut with a hacksaw. Then I used sheet metal screws to anchor the brackets to the inside walls of the recessed light fixture. I screwed in an adapter to convert the fixture's socket into a power plug for the webcam, which is the only physical connection the webcam needs.

I made the enclosure itself out of 6" inner diameter, foot-long ABS sewer pipe. (Yes, sewer pipe — no expense spared!) I cut a 1' length of

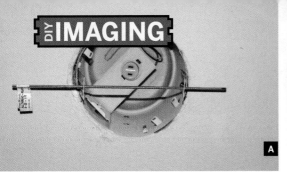

Fig. A: Metal rod "hanger" attaches to can light with metal brackets. Fig. B: D-Link DCS-6620G webcam with cover plate.

Fig. C: Internal view of webcam enclosure. Fig. D: Webcam assembly hangs from friendly neighbor's second floor balcony.

MATERIALS

D-link DCS-6620G with threaded base and transformer
  6" ABS plastic sewer pipe, 1' long cut down from 2'
    section sold at Home Depot
6" ABS end cap
5" round cover plate
5" metal L-brackets (2)
Sheet metal screws (4)
1' threaded metal rods (3) with matching wing
  nuts (6)
¼-20 bolt with washers and nuts
Light socket to 2-prong plug adapter
Electrical tape
Clear plastic CD case cover from free AOL install disc
Outdoor recessed can light fixture already installed
  in good location for observation

pipe and drilled 2 holes at one end to hang it from. A threaded rod runs through the pipe and the hanging brackets inside, and 2 wing nuts hold the rod in place.

Before installing the webcam enclosure at my neighbors' house, I tested the basic concept on a recessed ceiling light in our upstairs hallway. The pipe hung securely, but this indoor can light was 6½" in diameter, which meant that the two brackets stayed visible on either side of the pipe. The light fixture at the neighbors' is just 6" wide, so the pipe fits around the brackets and conceals them, which looks nicer.

Inside the enclosure, the webcam itself hangs from 2 more rods that run through holes drilled side by side farther down the pipe. To suspend the webcam upside down, I ran a ¼-20 bolt through the center of a 5" cover plate and screwed it into the webcam's tripod mount. I added extra nuts onto the bolt to reduce the chance of stripping the webcam threads. Inside the pipe, the cover plate rests on the 2 lower rods, and the webcam hangs underneath.

For the enclosure's view port, I cut a hole in the pipe and made a window by taping on the clear plastic from a free AOL install CD case.

Fig. E: Portable webcam assembly can be hung under eaves or placed freestanding, weather permitting.

Fig. F: Webcam view at 2x optical zoom last fall. During the winter, the aspen leaves drop and open up the visibility. Fig. G. 10x webcam picture of Alek taking picture of it.

I put a flat ABS end cap at the bottom of the enclosure, which is where the webcam's 120VAC-12VDC transformer sits. This provides some warmth for low-temperature operation. I left the end cap vented (there's about a ¼" opening on the top) to prevent fogging in the viewport. The flat end cap also lets me unhook the webcam and set it down to operate anywhere, connected via extension cord, weather permitting.

You can optionally add a cheap wireless temperature sensor inside the PVC, and also one taped to the outside, in order to provide a "delta temperature" reading.

Actual installation was super-duper easy since the entire unit is self-contained, with only a power cord coming out of the top. Plug that in, run the threaded metal rod through the PVC-bracket-bracket-PVC, screw down the wing nut, and you are ready for action!

The installation has worked great, although I did crack some plaster around the light fixture — oops! The webcam works even with the temperature dipping below 0°F a few times. The only issue was that the wireless signal would sometimes drop out, so I installed a Pringles-can antenna at my house pointed toward the webcam, and have had no problems after that.

The webcam enclosure is an integral part of my Controllable Christmas Lights for Celiac Disease website, which lets people remotely view live images of thousands of Christmas lights, and also control them with a click of a mouse. Besides being fun for people around the world, the site has raised over $16,000 for charity.

(Before making my Christmas lights controllable through my website in 2005, I simulated this effect with canned photos and a CGI script. The ruse spread far and wide before I invited the *Wall Street Journal* to reveal it as a hoax. But that's another story.)

Alek Komarnitsky lives in the Republic of Boulder. When not spending time with his wonderful wife Wendy and two sons Dirk and Kyle, he enjoys tinkering with stuff. Read more at komar.org.

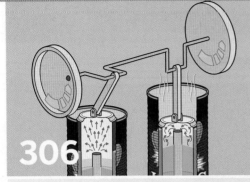

306

318

330

BIODIESEL

THIS SALE

GALLONS

2

Laurelhurst Oil

BIODIESEL

363

E54130

TELSTAR LOGISTICS
PROVIDENCE
SAN FRANCISCO

E54130

CAUTION: DO NOT OVERWIND

2XX7745

366

368

THE BEST OF CARS & ENGINES PROJECTS

from the pages of MAKE

GENTLEMAKERS, HACK YOUR ENGINES!

For many a do-it-yourselfer, the teen car was their first hardware hack. Whether it was putting in a new stereo system, pimping out the interior/exterior, or upgrading/replacing parts of the engine. In today's cars, hardware hackers of the computer kind are right at home, creating diagnostic tools, adding computer control of car systems, embedding PCs in the dash, and turning a car into a mobile wireless hotspot, all of which we've covered in MAKE, and include here. We've also included a piece on building your own wind energy generator and two cool engine experiments.

TWO-CAN STIRLING ENGINE

By William Gurstelle

REDLINING AT 20 RPM

The Stirling engine has long captivated inventors and dreamers. Here are complete plans for building and operating a two-cylinder model that runs on almost any high-temperature heat source.

Stirling engines are *external combustion* engines, which means no combustion takes place inside the engine and there's no need for intake or exhaust valves. As a result, Stirling engines are smooth-running and exceptionally quiet.

Because the Stirling cycle uses an external heat source, it can be run on whatever is available that makes heat — anything from hydrogen to solar energy to gasoline.

Our Stirling engine consists of two pistons immersed in two cans of water. One can contains hot water and the other cold. The temperature difference between the two sides causes the engine to run. The difference in the hot and cold side temperatures creates variations in air pressure and volume inside the engine. These pressure differences rotate a system of inertial weights and mechanical linkages, which in turn control the pressure and volume of the air cylinder.

Set up: p.310   Make it: p.312   Use it: p.317

William Gurstelle enjoys making interesting things that go whoosh then splat. He is the author of *Backyard Ballistics* (2001), *Building Bots* (2002), and *The Art of the Catapult* (2004). Visit backyard-ballistics.com for more information.

THE STIRLING CYCLE

Every heat engine works on a cycle. When heat is applied to a working fluid, the fluid undergoes some sort of change — its pressure, volume, or temperature is increased by the added heat — and in so doing, the fluid does meaningful work on its surroundings. Work could mean making a piston move, or a turbine, or some other mechanical object. The Stirling cycle is a four-step process, using hot air as its working fluid.

Four Steps of the Stirling Cycle

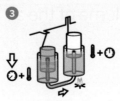

❶

COOLING
Cold piston (left) moves upward by flywheel inertia, drawing hot air over to cold side.

❷

EXPANSION
Hot air is forced to the left cylinder, forcing the cold piston up. This is the power cycle.

❸

COMPRESSION
As air in the cold water contracts, the cold piston moves down.

❹

HEATING
With the cold piston fully down, most air is on hot side and getting reheated.

Illustrations by Nik Schulz/L-Dopa.com

THE STORY OF THE STIRLING ENGINE

All engines run on heat cycles. More properly called thermodynamic cycles, each of these cycles has a name. Cars run on the Otto cycle, trucks on the Diesel cycle. Power plants often run the Rankine, while gas turbines run the Brayton cycle.

One cycle in particular has long captivated inventors and dreamers — the Stirling cycle. The Stirling cycle was among the first of the thermodynamic cycles to be exploited by engineers. Compared to other engine types, it is ancient. When it was patented as a new type of engine by a Scottish cleric in 1816, scientists hadn't even come up with the idea of thermodynamic cycles.

Robert Stirling, a young Scottish Presbyterian assistant minister, had the idea for a new type of heat engine that used hot air for its working fluid. Until then, the steam engines of Watt and Newcomen were the only heat engines in use.

Stirling Engines Go to Work ...
and Are Laid to Rest

Stirling's idea was to alternately heat and cool air in a cylinder using articulated mechanical arms and a flywheel to coax the machine to run in a smooth, endless cycle.

Although complex and expensive for its time, Reverend Stirling made it work. As early as 1818, his engine was in use pumping water from a stone quarry. By 1820, a 45-horsepower Stirling engine was driving equipment in the Scottish foundry where his brother worked.

Auto manufacturers have experimented with the Stirling for years. Its numerous good qualities make the Stirling an attractive candidate to replace or augment internal combustion engines.

Automakers worked closely with the federal government from 1978 to 1987 on Stirling engine programs. The goals were ambitious: low emission levels, smooth operation, a 30% improvement in fuel economy, and successful integration and operation in a representative U.S. automobile.

General Motors placed one in a 1985 Chevrolet Celebrity, and met all of the program's technical goals. But improvements in the efficiency of existing engine types, coupled with the status quo's far less expensive cost structure, doomed the Stirling to automotive irrelevance.

The External Combustion Revival

The Stirling idea was dusted off in the mid-1990s. A prototype Stirling hybrid propulsion system was integrated into a 1995 Chevrolet Lumina. But that test was not particularly successful, as the hybrid vehicle failed to meet several key goals for fuel efficiency and reliability. The program was abandoned. Still, Stirling engine advocates continue to research and apply the technology. The big breakthrough may yet arrive, possibly in a hybrid electric-Stirling engine.

While not terribly complex, the engineering analysis of the engine's thermodynamic cycle goes beyond the scope of this article. Suffice it to say that Stirling engines operate on a four-part cycle in which the air inside the engine is cyclically compressed, heated, expanded, and cooled, and as this occurs, the engine produces useful work.

While most heat engines are fairly understandable to interested amateurs, building one yourself is an altogether different prospect. Most engines require carefully machined metal parts, with close tolerances and tightly sealing clearances for pistons and/or rotating parts. Robert Stirling's heat engine is an exception. Or at the very least, making a working model can be done without any difficult machining.

About MAKE's Stirling Engine

This article provides step-by-step instructions for building a straightforward Stirling external combustion engine.

This engine is simple and cheap, and once you get it going, you really get a feel for how this sort of engine works. It chugs along at a leisurely 20 to 30 rpm, its power output is minuscule, and it makes a delightful squishing/chuffing noise as it operates.

But be forewarned: All engines, even the metal-can Stirling described here, are complex mechanical devices in which myriad mechanical movements must come together in precise fashion in order to attain cyclical operation.

SET UP.

MATERIALS

**Large steel cans (2) At
least 4" in diameter.
Large juice cans or 1lb.
coffee cans work; 13 oz.
coffee cans are too small.**

**Copper gauze Such as
"Chore Boy" pot scrubber**

Aluminum soda cans (2)

**#3 size rubber stopper
To fit middle opening of
the copper tee**

**Plastic spacers, 1" long
(2) The spacer's outside
diameter must match
the inside diameter of
the sheave, while its
inside diameter must
just fit the rod used for
the crank. Look in hard-
ware stores, in the small
parts bins that contain
specialized fasteners.**

¾" copper tee

**¾" copper pipe, about
18" long Cut as follows:
2¾" (2), 5" (2)**

**5"-diameter metal die-
cast sheaves or pulleys
(2) Such as McMaster-
Carr #6245K45**

**Wood 1"×2", 9" long (2)
Pieces A**

**Wood 1"×10", 10" long
Piece B**

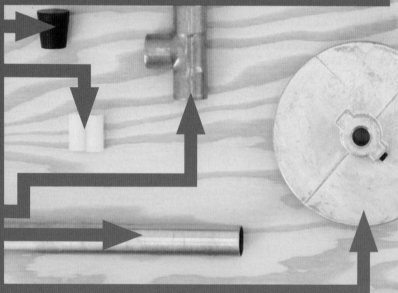

Photograph by Kirk von Rohr

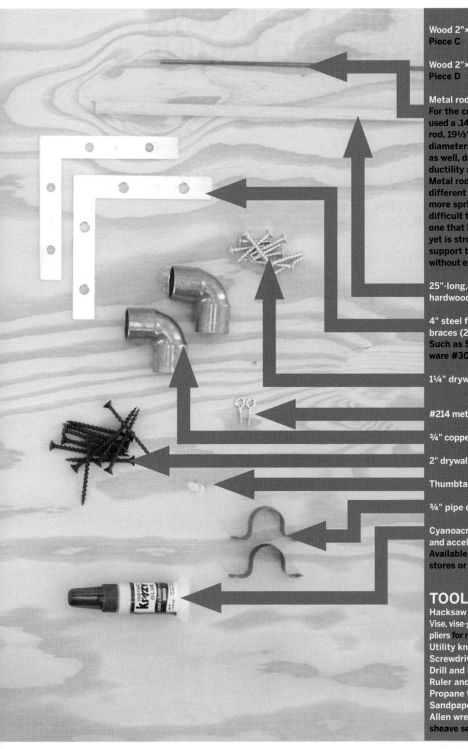

Wood 2"×4", 36" long
Piece C

Wood 2"×4", 4" long
Piece D

Metal rod, about 20"
For the crankshaft. I used a .14"-diameter iron rod, 19½" long. Other diameters may work as well, depending on ductility and strength. Metal rods come in different tempers, some more springy and more difficult to bend. Select one that bends easily, yet is strong enough to support the flywheels without excessive bowing.

25"-long, ⅜"-diameter hardwood dowels (2)

4" steel flat corner braces (2) with screws
Such as Stanley Hardware #306560

1¼" drywall screws (10)

#214 metal screw eyes (2)

¾" copper elbows (2)

2" drywall screws (8)

Thumbtacks (2)

¾" pipe clamps (2)

Cyanoacrylate glue and accelerator spray
Available in hobby stores or online

TOOLS
Hacksaw
Vise, vise-grips, needlenose pliers for rod-bending
Utility knife
Screwdriver
Drill and bits
Ruler and tape measure
Propane torch
Sandpaper
Allen wrench to fit sheave setscrew

MAKE IT.

BUILD YOUR OWN STIRLING ENGINE

Photography by Ty Nowotny and Jake McKenzie

START ⟩⟩⟩   **Time: A Day Complexity: Easy**

1. MAKE THE PISTON SUBASSEMBLIES

There are two pistons in this engine, one for the hot side and one for the cold side.

1a. With a hacksaw, carefully remove the top end of each soda can. Cut the can at the point where the flat side of the can curves to meet the top, resulting in a 4"-long piston. Sand the cut edge to remove burrs, then wash and dry the interior.

1b. Locate the center of the can bottom as accurately as possible. Push the thumbtack through the can bottom at that point. Remove the thumbtack.

1c. From the interior of the can, re-insert the thumbtack through the hole you just made.

It helps to stuff a rag into the can when pushing the tack through. This will stabilize the sides of the can and prevent buckling.

1d. Locate the center on the end of the ⅜"-diameter dowel and push the thumbtack into the wood. Carefully remove the thumbtack and coat the bottom of the dowel and the tack with super glue. Press into place and apply the super glue accelerator spray to hold fast.

1e. Test the can for watertightness. If it leaks, apply more glue.

1f. Locate the center of the opposite end of the ⅜"-diameter dowel, and drill a pilot hole and screw the #214 screw eye into the center. Apply super glue and accelerator spray.

2. FABRICATE THE CRANKSHAFT

The crankshaft consists of a metal rod bent in a precise way that holds the piston connecting rods in alignment.

2a. Lay out bend lines on the rod as accurately as possible using a permanent marker, as shown on the bend diagram.

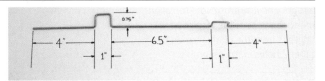

2b. Using a hammer, vise-grips, and vise, bend the metal rod as shown. Use special care when bending the rod to make the bend sizes and shapes correspond closely to the diagram. The 2 bends (the cranks) must be offset by exactly 90 degrees, and the distance from the end of the crank to the centerline of the crankshaft must be ¾".

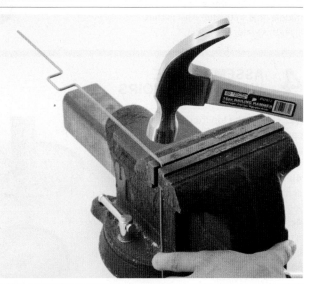

2c. Insert the plastic spacers into the sheaves. Tighten the setscrew inside the collar of the sheave to lock the plastic spacer in place. Do not put flywheels on the crankshaft yet.

3. ASSEMBLE THE AIR CYLINDER

3a. Before soldering or gluing, cut down the 2¾" pipes if necessary, so that the overall distance of the *finished* assembly will be 7½", center-to-center.

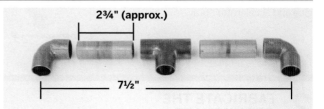

2¾" (approx.)

7½"

3b. Solder or epoxy the copper pipes and fittings together as shown, making certain the connections are airtight and leak-free. Note the alignment: the copper tee is rotated 90 degrees from the plane formed by the other 2 holes in the assembly.

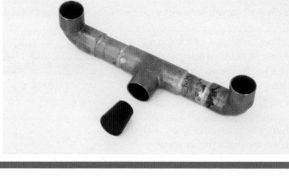

3c. Place the rubber stopper into the middle hole, in the tee. This is the system's water drain.

4. ASSEMBLE THE WATER RESERVOIRS

4a. Remove the top from each steel can, leaving the bottom intact. Sand edges smooth.

4b. Mark a ¾"-diameter circle in the center of the bottom of each can.

4c. With a utility knife, carefully make 8 to 12 radial slits on the bottom of the can, but within the ¾" circle. The slits should form a star shape, radiating out from the center.

If you are soldering the pipe into the can, the bottom of the can should be very heavily scored with a file to provide a toothy surface that the solder can stick to.

4d. Push the 5" copper pipe into the can's bottom, through the hole formed by the slits. Slide the pipe until just 1" of pipe still extends out the bottom.

4e. With the pipe concentric and parallel to the sides of the can, solder the pipe in place. (Alternatively, you can seal the pipe-to-can connection with slow-curing, waterproof epoxy glue, taking care to seal the pipe carefully so it will not leak. Allow to dry completely.)

Do not give up hope on the soldering. It is very difficult to do, but perseverance will pay off.

5. MAKE THE FRAME

5a. Using deck screws or nails, assemble wooden pieces A-D to form a frame, as shown.

6. ASSEMBLE THE STIRLING ENGINE

6a. Insert the water reservoir assemblies into the air cylinder assembly. Fill the reservoir cans with water and check for leaks. Repair leaks with epoxy and let dry.

6b. Measure and then mark a spot on each 1"×2" frame piece, 3¾" from the back edge of the frame. Place the combined water reservoir and air cylinder assembly on the 1"×2" frame pieces at the marked spots. Now place the ¾" copper pipe clamps over the assembly. Screw the pipe clamps into the 1"×2" pieces. The clamps must hold the combined assembly firmly in place.

6c. Slide the screw eyes on the connecting rods onto the crankshaft, so that 1 screw eye is on each of the 2 cranks. Place the soda-can pistons inside each of the water reservoirs so that each soda can rests on copper pipes. Turn the crankshaft so that one of the cranks is pointing downward.

Holding the crankshaft level, lift the crankshaft until the can corresponding to the bottomed crank is about ½" above the top of the copper pipe. This is the desired height for the crankshaft. Mark this height on the upright 2"×4" and attach the angle bracket at this point, making sure that the hole through which the crankshaft will pass is located 3¾" from the back of the 2"×4".

6d. Slide 1 flywheel onto each end of the crankshaft. Position the flywheels so that they are as far inboard as possible without interfering with the cranks or piston rods. Glue the flywheels onto the crankshaft using super glue and accelerator spray.

You're done!

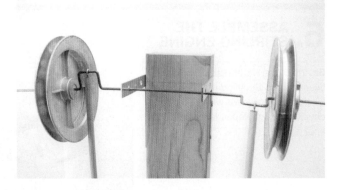

FINISH X

NOW GO USE IT »

MAKER, START YOUR ENGINE

To start your Stirling engine, turn the crankshaft until both cranks are tilted upwards at 45-degree angles to the vertical. With the stopper removed from the drain, fill each side with water, until a trickle runs out the drain. Dry it and replace the stopper.

Designate one side as the hot side, then heat the water on that side to boiling with a propane torch. This takes a while, depending on the heat output of the torch. Be patient.

When the water is ready, start the engine by giving the flywheels a small push. The rotation is determined by this rule: the cold side is 90 degrees behind the hot side.

If built properly, your engine will dip and lift, dip and lift, 20 to 30 times per minute to the chuff-chuff beat of Robert Stirling's ancient idea.

TROUBLESHOOTING

1. Make sure the engine is level. The crankshaft must revolve freely, and the connecting rods should stay in the middle of each crank as it rotates. Use shims or cardboard to level the system. If the connecting rods will not stay centered on the cranks, you can add a small wire loop or small nut to the rod on either side of the eye screw, fastening them into place with super glue.

2. You may have to experiment to find the best flywheel weight. If the flywheels are too heavy, the metal rod will bow, interfering with the crankshaft's rotation. But if the flywheels are too light, there won't be enough inertia to carry the crankshaft past the volume compression phase and into the next expansion stroke. If this happens, the engine will pulse but not run cyclically. You can add weight to the flywheel by simply taping bolts or other weighty objects to its perimeter.

3. Large steel cans full of water take time to heat. Be patient, and let the water heat to 200°F or more.

4. Minimize friction and interference. Friction is your engine's greatest enemy. Minimize rubbing between pistons and water cans, between connecting rods and cranks, and between the crankshaft and the metal support angles that attach it to the wooden frame.

5. Add a regenerator. A regenerator consists of a small piece of heat-conducting metal gauze placed in the air cylinder just behind the rubber stopper. A regenerator will improve cycle efficiency and make the machine turn faster. The copper gauze sold for cleaning kitchen pots ("Chore Boy") works well.

WIND POWERED GENERATOR

By Abe and Josie Connally

With a motor and some piping, it's surprisingly easy to build this inexpensive, efficient wind generator — and enjoy free energy forever. ❯❯

Set up: p.321 Make it: p.322 Use it: p.329

Photography by Abe and Josie Connally

CURRENT FROM CURRENTS

There are no limits to what you can do with wind power. It's abundant, clean, cheap, and easy to harness. We designed this Chispito Wind Generator (that's Spanish for "little spark") for fast and easy construction. Most of the tools and materials you need to build it can be found in your local hardware shop or junk pile. We recommend that you search your local dump or junkyards for the pieces required. Or, if you live in a city, search freecycle.com for salvaged parts, and see if you can install one on your roof.

   We believe that anyone can be in control of where his or her electricity comes from, and there is nothing more rewarding and empowering than making a wind-powered generator from scrap materials. Remember: puro yonke (pure junk) is best!

Abe and Josie Connally are off-grid adventurists based in the remote Big Bend region of Texas, where they experiment and live with sustainable technologies built from puro yonke (velacreations.com).

WIND GENERATOR BLOW-BY-BLOW

The Chispito Wind Generator is a simple little machine that's great for getting started with wind power. In a 30 mph wind, ours gives us about 84 watts, 7 amps at 12 volts.

Field Magnets

Rotors

Axle

When the motor is connected to a load rather than to power, and you turn the rotor, the field magnets will induce an electric current in the rotating electromagnet coils. This is how the motor works as a generator.

The blades for the Chispito's turbine are cut from PVC pipe — strong, lightweight material with a gently curving shape that increases efficiency by scooping up moving air, rather than letting it bounce and blow past.

Pipe and pipe fittings make up the Chispito's tower and mounting hardware. At the base, a short 1¼" pipe inside of a 1½" pipe creates a hinge that allows the tower to be raised and lowered.

A diode between the windmill and the battery ensures that the power only flows in one direction, charging the battery rather than drawing power away from the battery and running the motor. For its diode, this project uses a bridge rectifier, a component that uses three or four diodes to convert AC to DC. You could also use a simple one-way diode, but these usually aren't sealed or protected.

EDITORS NOTE: There was some confusion with this piece because the direction of the windmill blades shown in the step-by-step pictures turn in the opposite direction to the motor listed for this project (some motor hubs screw on to the motor clockwise, others counterclockwise).

More details and discussion about this project and the correct cutting and assembly of the blades can be found at makezine. com/05/windmill and at velacreations.com/chispito.html.

The Chispito charges up batteries through a regulator, which protects them from overcharging. These same back-end components could also store power from a solar cell array, a micro-hydro turbine, or any other off-grid, environmental power source.

BATTERY

Illustration by Tim Lillis

SET UP.

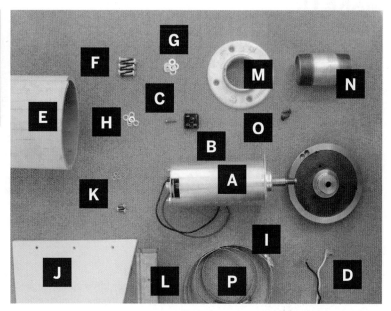

MATERIALS

MOTOR AND ELECTRIC

[A] 260 volts DC, 5 amps, treadmill motor with a 6" threaded flywheel You may use any other simple, permanent-magnet DC motor that returns at least 1V for every 25 rpm and can handle upwards of 10 amps. Our motor is rated at 5A, no load, and we've found that the coils can withstand 15A going through them without heating up.

If you use another motor, change this supply list to match. For example, if the motor lacks a flywheel, you will have to find a hub for it. A circular saw blade with a ⅝" shaft adaptor will work.

[B] 30-50A bridge rectifier with center hole mount surpluscenter.com, item #22-1180

[C] Mounting screw (1)

#8 (or larger) copper wire, red and black, both stranded Enough length for both a red and black piece to run from the top of the tower, down through length of pole, to batteries. We recommend at least #8 wire, but if your tower will be sited a long distance away from your batteries, you may need a heavier gauge.

[D] Spade connectors for wires (4) For bridge rectifier

Heat-shrink tubing or electrical tape

Battery bank We recommend deep-cycle lead-acid storage batteries, and a total battery bank capacity of at least 200 amp-hours.

Ammeter

Regulator or charge controller

Fuse

BLADES

[E] 2' length of 8" Schedule 80 PVC pipe If PVC is UV resistant, you will not need to paint it.

[F] ¼" #20 bolts, ¾" long (6)

[G] #20 washers (9)

[H] Lock washers (6)

[I] Hose clamp (1)

VANE

[J] 1 sq. ft. (approx.) of sheet metal

[K] Mounting screws and lock washers (approx. 9)

MOUNT AT TOP OF TOWER

[L] 36" of 1" square metal tubing or 1" angle iron

[M] 2" floor flange pipe fitting

[N] 2" steel pipe nipple, at least 4" long

[O] Mounting screws (2)

MOTOR MOUNT

[P] #72 hose clamps (2)

TOWER POLE

10'-30' length of 1½" steel pipe, threaded at both ends

TOWER BASE

2'x1¼" steel pipe nipple (2)

6"x1¼" steel pipe nipple

1¼" 90-degree steel pipe elbows (2)

1½" steel pipe T

10 lb. bags of quick-mix concrete (2-3)

¾" #10 sheet metal screws (4)

TOWER STABILITY

Guy wire, galvanized steel With a working load of 200 pounds

1½" U-bolt

Stakes (4)

Turnbuckles (4)

TOOLS

Drill and drill bits (⁵/₃₂", ⁷/₃₂", ¼"), jigsaw, thread-tapping set, pipe wrench, crescent wrench, flathead screwdriver, vise and/or clamp, wire strippers, metal punch or awl, tape measure, level, marker, tape, compass and protractor, shovel, wheelbarrow, several ropes (each at least twice the length of the guy wires), and an extra person or two to help.

MAKE IT.

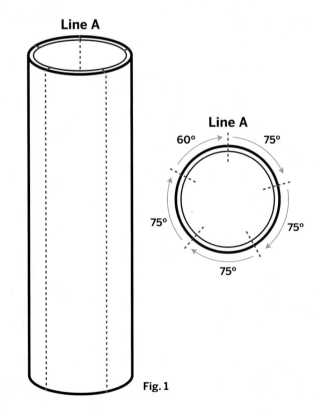

BUILD YOUR WIND-POWERED GENERATOR

START ⟫⟫ **Time: A Couple Weekends Complexity: Medium**

1. CUT THE BLADES

Let's begin by cutting.

1a. Place the 24" length of PVC pipe and square tubing (or other straight edge) side by side on a flat surface. Push the pipe tight against the tubing and mark the line along the length of the tube. This is Line A (see Fig. 1).

1b. Starting from Line A, draw parallel lines at 75-degree intervals along the length of the pipe. You should have a total of five lines on your pipe as shown in left Figure 1. Note that one strip will have an arc width of only 60 degrees. That's OK.

1c. Use a jigsaw to cut along the lines, splitting the tube into five strips. Four will be wider than the fifth (60°) strip. Set the 60° strip aside for now.

1d. Place the four 75° strips concave-side-down. For each one, make a mark 20% of the width of the strip from one corner along the diagonally opposite side as shown (see Fig. 2).

1e. Mark a diagonal line between the two marks you just made on each piece, and use the jigsaw to cut along these lines (see Fig. 3). You should wind up with eight identically shaped trapezoidal blades. You can trim a ninth blade out of the 60° strip left over. You now have enough blades for three generators, or plenty of spares for one generator.

Line A

Line A

60°   75°

75°   75°

75°

Fig. 1

1f. Now you are going to cut one corner from each blade. First, measure the width of the blade (if you are using an 8" diameter PVC pipe as your stock, it should be about 5.75" wide). Call this value W. Then make a mark along the diagonal edge of the blade, a distance of W/2 from the wide end (3" is good enough if you are using 8" PVC). Make another mark on the wide end of the blade at 15% of W from the long straight edge (1" with 8" PVC) (see Fig. 4).

1g. Connect these two marks and cut along the line. Removing this corner prevents the blades from interfering with each other's wind.

1h. The blades should look like the ones shown in Fig. 5. Pick the three best ones of the batch and let's move to the next step, making the tail.

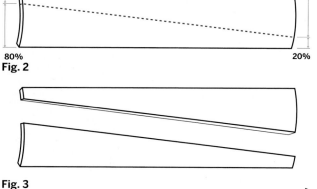

20%   80%

80%   20%

Fig. 2

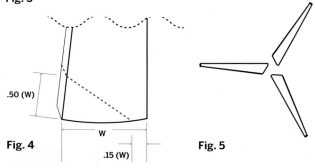

Fig. 3

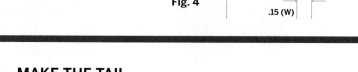

.50 (W)

W

.15 (W)

Fig. 4   **Fig. 5**

2. MAKE THE TAIL

2a. Cut the tail. You can make it any shape you want, as long as the end result is stiff rather than floppy. The exact dimensions of the tail are not important, but you'll want to use about one square foot of lightweight material, preferably metal.

2b. Using the 5/32" drill bit, drill two or three holes, spaced evenly, in the front end of the tail. Then place the tail on one end of the square tubing, noting that it will attach to what will become either the right or left side of the tubing, as the generator sits upright. Mark the tubing through the tail holes.

2c. Drill holes in the square tubing at the marks you just made.

2d. Attach the tail to the tube with sheet metal screws. (Or you can do this later, so it doesn't get in the way.)

3. ATTACH THE BLADES

3a. Take three blades. For each blade, mark two holes along the long, right-angle side of the blade (as opposed to the long diagonal side), at the wide end, next to the cut-off corner. The first hole should be ⅜" from the long side and ½" from the end, and the second hole should be ⅜" from the straight edge and 1¼" from the end.

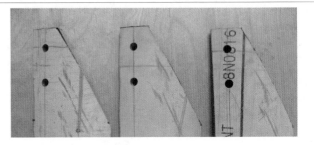

3b. Using the ¼" drill bit, drill these six holes for the three blades.

3c. Detach the hub from the motor shaft. With our motor, we removed the hub by holding the end of the shaft firmly with pliers and turning the hub clockwise. This hub unscrews clockwise, counter to the usual direction, which is why the blades turn counterclockwise.

3d. Using a compass and protractor, make a template of the hub on a piece of paper. Then mark three holes, each of which is 2⅜" from the center of the circle, 120 degrees apart, equidistant from each other.

3e. Place this template over the hub and use a metal punch or awl to punch a starter hole through the paper and onto the hub at each hole.

3f. Drill the holes with the ⁷⁄₃₂" drill bit, then tap them with the ¼" tap.

3g. Attach the blades to the hub using ¼" bolts, running them through the holes closest to the ends of the blades. At this point, the three outer holes on the hub have not been drilled.

3h. Measure the distances between the tips of each blade, and adjust them so that they are all equidistant. Then mark and punch starter holes for the three outer holes on the hub through the empty holes in each blade.

3i. Label the blades and hub so that you can match which blade goes where.

3j. Remove the blades, and drill and tap the three outer holes on the hub.

3k. Position each blade in its place on the hub, so that all the holes line up. Using the ¼" bolts and washers, bolt the blades back onto the hub. For the inner three holes, use two washers per bolt, one on each side of the blade. For the outer holes, just use one washer next to the head of the bolt. Tighten.

4. ASSEMBLE THE GENERATOR

4a. Drill a ⁵/₃₂" hole in the tubing, about 5 inches from the front end of the tube, opposite the tail holes end, on any side. Place the bridge rectifier over the hole, and screw it to the tubing using a #10 sheet metal screw.

4b. Using hose clamps, mount the motor on the end opposite the tail. Do not tighten the clamps, because you will make a balance adjustment later.

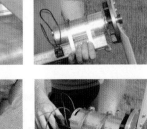

4c. Crimp spade connectors onto the black and red wires from the motor, and connect them to the two AC voltages in terminals on the bridge rectifier, L1 and L2. Insulate connections with heat-shrink tubing or electrical tape.

4d. If you haven't already, attach the tail.

4e. Re-attach the blade assembly on the motor.

4f. Now we'll attach the tower mount. Using a pipe wrench, screw the nipple tightly into the floor flange. Clamp the nipple in a vice so the floor flange faces up and is level.

4g. Set the generator on the flange/nipple and balance it by adjusting the position of the motor, then tighten the hose clamps down. Mark spots in the square tubing that match up with the flange holes.

4h. Drill these two holes using a ⁵/₃₂" drill bit. (You will probably have to take off the hub and tail to do this.)

4i. Attach the square tubing to the floor flange with two sheet metal screws.

MAKE THE TOWER ❖❖

5. PLANT THE TOWER BASE

The tower is one of the most important components in your wind generator system. It must be strong, stable, easily raised and lowered, and well anchored.

5a. Dig a round hole about 1 foot in diameter and 2 feet deep.

5b. Feed the 6"x1¼" steel pipe nipple through the horizontal part of the 1½" steel pipe T.

5c. Screw the pipe elbows onto each end of the nipple, one on either side of the T, so that they both point in the same direction.

5d. Screw the two 2'x1¼" pipe nipples into the free ends of the elbows.

5e. Set this hinged base assembly in the hole, so that the T just clears the ground. Dig around, adjust, and position things so that the 2' nipples point straight down and the horizontal part of the T is perfectly level.

5f. With the base properly positioned, mix some concrete and pour it into the hole.

6. ERECT AND STAY THE TOWER

The higher your tower is, the more wind your generator will catch, and the more power it will produce.

6a. Drill a large hole about 1 foot from the bottom of the 10'-30' pipe, for the copper wires to exit.

6b. Screw the pipe into the vertical part of the base's hinged T.

6c. Make four strong, flexible rings out of guy wire, about 5 inches in diameter. For each ring, loop the wire around several turns, and twist it closed.

6d. Place the 1½" U-bolt around the pipe, 3 feet from the top of the pipe. Thread the four wire loops around the U-bolt, and space them evenly around the pipe. Then tighten the nuts of the U-bolt.

6e. Secure a guy wire to each of the loops on the U-bolt. Also loop the ropes (safety ropes) through loops on opposite sides of the pole.

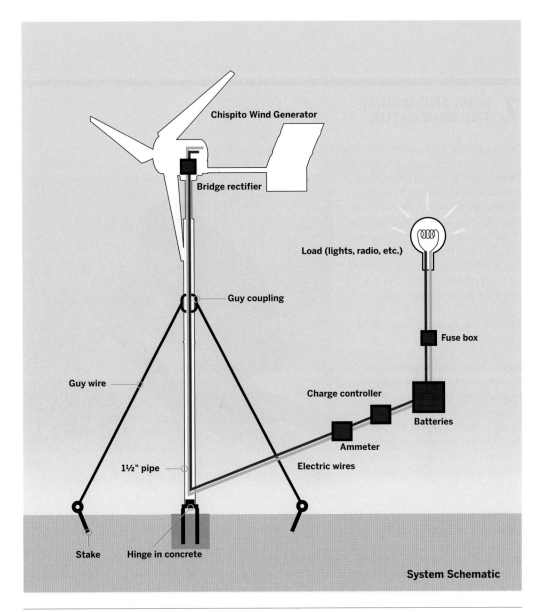

Chispito Wind Generator

Bridge rectifier

Load (lights, radio, etc.)

Guy coupling

Fuse box

Guy wire

Charge controller

Batteries

Ammeter

1½" pipe

Electric wires

Stake **Hinge in concrete**

System Schematic

6f. Position the four stakes, spacing them evenly apart at a distance away from the base that's at least 50% of the tower's height. For our 15-foot-tall pole, we positioned the stakes 12 feet away from the base. Then drive the stakes firmly into the ground, slightly angling them away from the base. Or, for greater strength and permanence, dig holes 2 feet into the ground, and set the stakes in concrete.

6g. Wire a turnbuckle to each stake, using several strands of guy wire.

6h. Raise the pole up and tie each of the safety ropes to something solid, like a truck or a building (this is where having another person or two really helps). Attach the guy wires to the turnbuckles.

6i. Hold the pole straight upright, and tighten all turnbuckles to ensure a secure fit.

6j. Mark the front turnbuckle for future reference, so you know how far you need to screw it back in when you're re-raising the pole.

7. WIRE AND MOUNT THE GENERATOR

7a. Release the front guy wire and lower the pole to the ground.

7b. Feed two lengths of #8 wire, red and black, down through the pole and out through the hole in the bottom of the pipe. Then wrap the bottom ends of the two wires together, to create a closed circuit. This is a safety precaution; it puts a load on the wind generator to prevent it from spinning around fast while you're working on it.

7c. Slide the generator assembly over the top of the pole.

7d. Pull the pole wires up through the mount, strip the ends, and crimp them into spade connectors. Plug the red wire into the DC+ terminal of the spade connector (which will probably be perpendicular to the others), and the black wire into the DC− terminal. Insulate connections with heat-shrink tubing or electrical tape.

7e. Raise the pole by pulling the front guy wire into place, and tighten the turnbuckle to the mark made earlier.

7f. Unwrap the ends of the wires and wire up your system as shown in the schematic (previous page). Connect a regulator, an ammeter, a fuse, and a stop switch on the positive line coming from the generator, between the generator and the battery bank. Refer to the manufacturer's instructions. Then hook up the battery bank, and watch it fill up with free power!

Teach a Man to Make

Give a man a fish, or teach a man to fish? Abe and Josie Connally prefer to teach people how to make their wind generators, rather than providing the finished product. They take their wind generators to some of the poorest rural communities in northern Mexico, where rough terrain and long distances make grid power not a viable option. They also hold workshops in their construction for entire villages and have also set up solar water pumps in a few communities.

"Basically, we focus on getting communities to help themselves," Abe says. "So far, we've been met with outstanding enthusiasm."

For more information about their work, and to get complete how-tos of their projects, visit velacreations.com.

"The simplicity and cost are the driving factors. Self-repair and low-maintenance options are essential. Our goal is simple: help those who want to know how to do it."

USE IT.

FREE WIND POWER IS YOURS

SAFETY

Safety should be your highest priority. Human life is more important than electricity, so please follow any and every safety guideline you come across. Wind generators can be dangerous, with fast-moving parts and high voltage, especially in violent weather conditions. Some things to consider:

» Ground and fuse your electrical system, as well as each component within it.

» Always stand upwind when inspecting the wind generator, to avoid debris in case of failure.

» Always attach safety ropes and/or cables when erecting your tower and/or wind generator.

» Always wire connections securely, with proper insulation such as heat-shrink tubing and/or electrical tape.

» Never touch the positive and negative wires at the same time while they are connected to the battery.

» Never leave your wind generator unconnected to anything, unless it is on the ground. Always connect it to a battery or some other load, or else short it out by crossing the positive and negative wires that lead from the generator, to create a closed circuit. Without one of these precautions, the generator can spin freely and attain dangerous speeds.

» Never expose batteries to heat, sparks, or flames, and do not smoke near them. Batteries can ignite and explode easily, resulting in injury.

SYSTEM CARE

To give your wind generator a longer life span, you should paint the blades, and tape the motor with something like aluminum tape.

Batteries are usually the most expensive component in your system, and it pays to take good care of them. The Chispito Wind Generator is rated at 12V, so your system should be configured to match. We recommend deep-cycle, lead-acid storage batteries, and a total bank capacity of at least 200Ah. Check your batteries' water levels regularly to ensure the longest possible life span. Don't use car or marine-type batteries; they will wear out fast and can be dangerous if used in a battery bank.

For proper system health, you should include a regulator or charge controller, and an ammeter. An ammeter lets you view your wind generator's current-charging rate, and the regulator prevents your batteries from overcharging.

SPECIFIC USES

We use the Chispito Wind Generator, along with a 100-watt solar panel, for all of our electricity needs in our home. This includes a laptop, satellite receiver, TV, VCR, DVD player, printer, stereo, power tools, and lights. Our system consists of a 450Ah battery bank, a 1,000-watt inverter, and the proper fuses and breakers for a home supply.

The wind generator has a wide range of applications depending on your particular needs, from powering your entire house, to running your computer and printer, to recharging the battery packs in your RV. The really nice thing about our model is that you can literally build ten of them for the price of one pre-built wind generator sold in stores.

RESOURCES

Wind generator information and inspiration:
otherpower.com

U.S. Wind Energy Resource Atlas wind maps, for assessing the suitability of a region for wind energy:
rredc.nrel.gov/wind/pubs/atlas

Edwin Lenz's wind generator made from an old microwave:
windstuffnow.com/main/microwave_wind_generator.htm

MOD YOUR ROD

CARS ARE HACKABLE PLATFORMS ON WHEELS

JAMES BOND DEPENDED ON Q TO TRICK OUT HIS
CARS. BUT WITH MAKE'S GUIDE TO CAR HACKING,
YOU'LL LEARN HOW TO TURN YOUR RIDE INTO A
FULLY LOADED, GREASE-EATING, MP3-BLASTING,
WI-FI-TRANSMITTING MONSTER MACHINE.

Illustrations by Nik Schulz

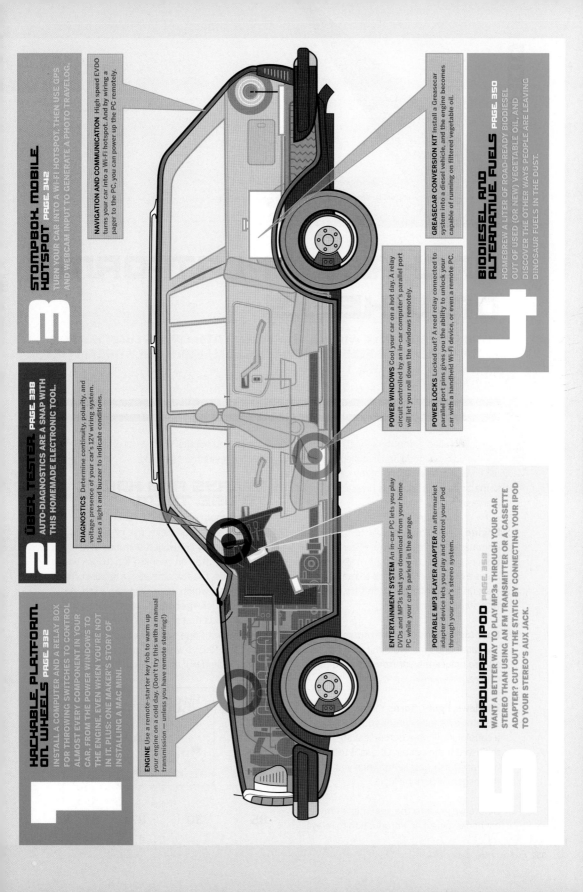

1

HACKABLE PLATFORM ON WHEELS PAGE 332

INSTALL A COMPUTER AND A RELAY BOX FOR THROWING SWITCHES TO CONTROL ALMOST EVERY COMPONENT IN YOUR CAR, FROM THE POWER WINDOWS TO THE ENGINE, EVEN WHEN YOU'RE NOT IN IT. PLUS: ONE MAKER'S STORY OF INSTALLING A MAC MINI.

ENGINE Use a remote-starter key fob to warm up your engine on a cold day. (Don't try this with a manual transmission — unless you have remote steering!)

2

ÜBER TESTER PAGE 338

AUTO-DIAGNOSTICS ARE A SNAP WITH THIS HOMEMADE ELECTRONIC TOOL.

DIAGNOSTICS Determine continuity, polarity, and voltage presence of your car's 12V wiring system. Uses a light and buzzer to indicate conditions.

3

STOMPBOX MOBILE HOTSPOT PAGE 342

TURN YOUR CAR INTO A WI-FI HOTSPOT, THEN USE GPS AND WEBCAM INPUT TO GENERATE A PHOTO TRAVELOG.

NAVIGATION AND COMMUNICATION High speed EVDO turns your car into a Wi-Fi hotspot. And by wiring a pager to the PC, you can power up the PC remotely.

POWER WINDOWS Cool your car on a hot day. A relay circuit controlled by an in-car computer's parallel port will let you roll down the windows remotely.

POWER LOCKS Locked out? A reed relay connected to parallel port pins gives you the ability to unlock your car with a handheld Wi-Fi device, or even a remote PC.

ENTERTAINMENT SYSTEM An in-car PC lets you play DVDs and MP3s that you download from your home PC while your car is parked in the garage.

PORTABLE MP3 PLAYER ADAPTER An aftermarket adapter device lets you play and control your iPod through your car's stereo system.

4

BIODIESEL AND ALTERNATIVE FUELS PAGE 350

HOMEBREW A LITER OF ROAD-READY BIODIESEL OUT OF USED (OR NEW) VEGETABLE OIL, AND DISCOVER THE OTHER WAYS PEOPLE ARE LEAVING DINOSAUR FUELS IN THE DUST.

GREASECAR CONVERSION KIT Install a Greasecar system into a diesel vehicle, and the engine becomes capable of running on filtered vegetable oil.

5

HARDWIRED IPOD PAGE 358

WANT A BETTER WAY TO PLAY MP3s THROUGH YOUR CAR STEREO THAN USING AN FM TRANSMITTER OR A CASSETTE ADAPTER? CUT OUT THE STATIC BY CONNECTING YOUR IPOD TO YOUR STEREO'S AUX JACK.

HACKABLE PLATFORM ON WHEELS

YOUR CAR'S 12-VOLT WIRING IS LOADED WITH OPPORTUNITIES FOR ENHANCING YOUR VEHICLE IN WAYS CARMAKERS ONLY DREAM OF. BY DAMIEN STOLARZ

I've installed computers in all my cars. I've also designed and installed circuits to control different parts of my cars. Here are some of the most obvious hackable access points on a car:

Engine Install a remote starter (such as the Commando EZ-2500 Remote Starter available from commandoalarms.com), and you can press a button on a key fob to make the car start up automatically. (Don't try it on a car with manual transmission!)

Door locks We've all locked ourselves out of our car at one time. Wouldn't it be nice to be able to unlock the car from any computer using a password?

Windows, lights, horns You can make a relay system to remotely unroll your car's windows a couple of minutes before leaving the office to flush out the hot air. Or use it to blast your horn and flash your lights when you're lost in the megamall parking lot.

Before we get into the specifics of my car hacks, here's an introduction to relays,

diodes, and parallel ports — all of which figure into most electric car hacks.

RELAYS AND HOW TO USE THEM

Many components in a car, such as car windows and door locks, use relays. Relays are electrically controlled switches. When you press the button on your key fob to lock your car's doors, the fob activates a small current to the relay, which in turn flips a switch that allows a much stronger current to flow through the solenoids that activate the locks.

Relays with one switch are called single pole (SP). Relays with two switches are called double pole (DP). If the switch has only one

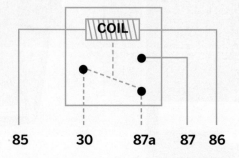

position, i.e., two terminals, on or off, it's a single throw (ST). If the switch has two positions, i.e., three terminals, it's a double throw (DT).

The relay shown here is a SPDT (one switch, two positions). You can see that there are five numbered terminals. Two of them are used to activate the switch, and the other three make up the two-pole (double pole) switch. When you power up the relay (putting 12V across 85 and 86), it activates a magnet that pulls a switch inside the relay, and this switch then connects 30 and 87. If you disconnect the voltage between 85 and 86, the magnet stops, and the spring-loaded switch clicks back to its normal position between 30 and 87a.

You can connect relays to make complex conditional switches ("If the car is on and the window is down and the trunk is open, then honk the horn"). For some examples of interesting car relay applications, visit the12volt.com/relays/relays.asp.

DIODES: A ONE-WAY STREET FOR ELECTRONS

Diodes are electronic components that ensure current only runs one way. They have a stripe indicating the negative terminal. The current can only flow from plus to minus, so you should put the stripe toward the direction you want the current to flow.

Let's say you wanted to add a switch that keeps your radio on, even if your car's keys are removed. You could just run a 12V line from your battery, through a switch, and to the "ignition" input on your radio. If you flipped that switch, your radio would turn right on. The problem is that current would flow back through the ignition wire and back into all the other devices it powered. The solution is to use two diodes: one between your new switch and the radio (stripe toward the radio) and another between the ignition and the radio (stripe toward the radio) to ensure current goes to the radio only.

IN CONTROL: PARALLEL PORTS

Parallel ports are an easy way to control switches from an in-car computer (see "Halloween Haunted House Controller," MAKE, Volume 03, page 86). A parallel port has eight pins that can be set at 5V or 0V. Using a resistor and a small 5V "reed" relay (such as a RadioShack 275-232 or 275-310), you can programmatically activate or deactivate a switch by setting 0 or 1 on a parallel port line.

If you're triggering a light load, such as a small light, you can just connect the 12V through the relay. If you're trying to switch a larger load, such as headlights or the door locks and windows, you'll want to chain the relays together. The tiny 5V relay will pass 12V to the big 12V 30A automotive relay, which will control your heavy load.

You don't have to hack a parallel port. There are many USB, serial, and parallel port-based relay controllers you can purchase and plug into your PC. These are already safety-fused and easy to program, and include example code. For instance, ontrak.net makes the ADR2010 that I used to control my Nash's windows via an onboard computer.

PAGER POWER

I wanted to be able to activate and access my cars' PCs from my house, so that I could connect via Wi-Fi to download music and grab email (via my company's CarBot software) to be read to me as I drive. Instead of keeping the in-car computer on continuously (which would kill the car battery), I bought a pager (service is about $4/month) to connect to the "ignition" switch of the in-car PC. The power supply for the PC (an M1-ATX from mini-box.com) has a mode where it will turn on and then stay on for two hours, giving me time to do what I need to do.

By rigging the pager buzzer up to the ACC/ignition pin on the PC's power supply (isolated by a diode, so that ACC doesn't fry the pager), I can turn on the PC in my car from anywhere in the city.

The PC, running Windows XP Pro, is configured to automatically connect to the internet (via Wi-Fi or EVDO) and run Trillian, a popular instant-messaging client. I can now dial a pager number to make my car computer wake up, go online, and start up IM — at which point I can log in and start controlling the machine.

POWER WINDOWS HACK

I decided to upgrade my old Nash with power windows. The kit I purchased online took the good part of a day to install but was very simple electrically. It's just a motor with two wires. If you apply 12V to one wire and ground to the other, the window goes up. Reverse the connection, and the window goes down. The rocker switches that came with the kit perform this reversal when you manually activate them, but to trigger this up-down with a computer required a more complicated arrangement of relays. You can see from the diagram (next page) that two of the relays apply 12V and ground to make the window go up, and the other two relays apply ground and 12V to make the window go down.

My friend developed an onscreen interface for controlling my windows, and I bought an ADR2010 controller board from ontrak.net. It has simple command language for turning ports on and off and was ideal for this application.

Using my in-dash touch screen, I can control the ADR2010, which activates the relays I've connected to the parallel port, which in turn can switch the bigger relays that control the windows.

DOOR LOCKS HACK.

To unlock the doors via computer, I had to figure out how the switches work in my Caravan. After pulling the right door panel off, I disconnected the lock switch and used an ohmmeter to test all three positions: off, up, and down. I found that it usually has a 20Kohm reading, but that it drops to 2Kohm when I "unlock" the switch and 4Kohm when I "lock" the switch.

I used the voltmeter to measure the voltage of both lines that went to the switch, with respect to ground. One terminal was on the ground of the car, the other on each wire. (The leads on my ohmmeter

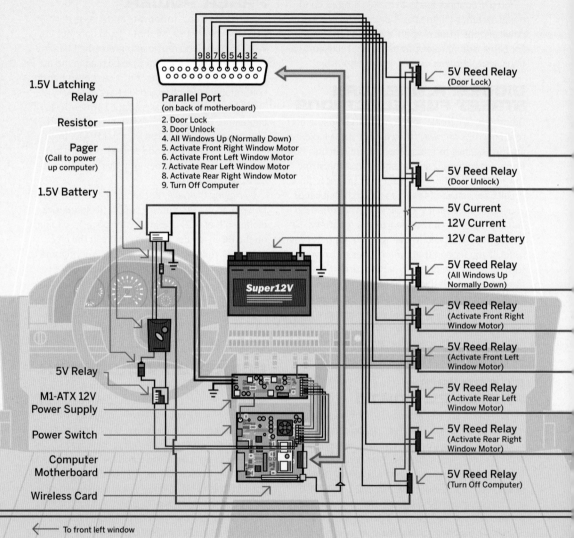

1.5V Latching Relay

Resistor

Pager
(Call to power up computer)

1.5V Battery

5V Relay

M1-ATX 12V Power Supply

Power Switch

Computer Motherboard

Wireless Card

9 8 7 6 5 4 3 2

Parallel Port
(on back of motherboard)
2. Door Lock
3. Door Unlock
4. All Windows Up (Normally Down)
5. Activate Front Right Window Motor
6. Activate Front Left Window Motor
7. Activate Rear Left Window Motor
8. Activate Rear Right Window Motor
9. Turn Off Computer

Super12V

5V Reed Relay
(Door Lock)

5V Reed Relay
(Door Unlock)

5V Current

12V Current

12V Car Battery

5V Reed Relay
(All Windows Up Normally Down)

5V Reed Relay
(Activate Front Right Window Motor)

5V Reed Relay
(Activate Front Left Window Motor)

5V Reed Relay
(Activate Rear Left Window Motor)

5V Reed Relay
(Activate Rear Right Window Motor)

5V Reed Relay
(Turn Off Computer)

← To front left window

have sharp ends, so I just poked a tiny hole into the wire to test the voltage.)

The voltages were both around 0V. I then followed the wires out through the door into the body of the car. One wire, a purple and green one, continued into the passenger footwell area. Using alligator clips and the sharp terminal probe, I connected that purple and green wire through the 2Kohm resistor to ground. Lo and behold, the pleasant "unlock" sound reverberated throughout the minivan.

Since the switches to unlock the doors are merely applying a resistance temporarily, I felt comfortable using a simple reed relay connected to parallel port pins.

Now, if I lock myself out of my car, I can call a friend or relative who can log into my car, and ask them to "unlock" my car.

Further exploration:
the12volt.com, crutchfield.com, ontrak.net.

Inventor Damien Stolarz has spent over half his life making different kinds of computers talk to each other. His book, *Car PC Hacks*, was published by O'Reilly Media, Inc.

HACKING YOUR CAR'S NERVOUS SYSTEM

Every one of author Damien Stolarz's cars are outfitted with a computer and a Wi-Fi card. In addition, he's added a pager to the computer — a CarBot PC, made by his company (carbotpc.com) — so he can call in from any phone and power up the system for two hours (to keep the car's battery from running down). By adding a relay system controlled through the PC's parallel port as shown here, Stolarz can control his windows and locks remotely, as well as fill his in-car PC with MP3s, videos, and email from his home computer.

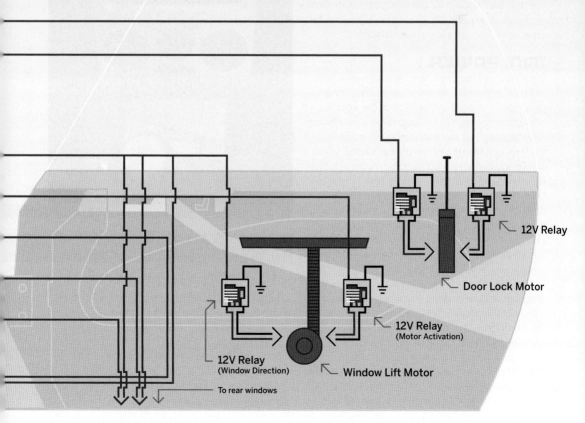

12V Relay

Door Lock Motor

12V Relay
(Motor Activation)

12V Relay
(Window Direction)

Window Lift Motor

To rear windows

MAHING A MACSWAGEN.

ADDING A MAC MINI TO A VW: A PRO TELLS US HOW HE DID IT.
BY PHILLIP TORRONE

Matt Turner is a professional fabricator and installer of mobile electronics. He's been working in the industry for 12 years and has built numerous award-winning show vehicles. His latest project, outfitting a 2001 VW GTi with Apple's Mac mini, brings hope to all car owners looking to Macify their ride.

When the Mac mini was announced, along with its form-factor dimension, Turner spent two weeks planning how he would install the mini into his car. Through careful planning and examination of his VW GTi, he selected over a dozen off-the-shelf parts that would provide a driver/user with the best in-car computer experience from a variety of standpoints.

MINI POWER.

In car computer systems, the first challenge is power. The Mac mini presented a new puzzle. The power button is located on the back of the enclosure — great for homes, but not so great when it's inside a car. You can't let the mini stay in sleep mode all the time (it'll take it's toll on the car's battery), so Turner first contemplated a secondary battery bank with a relay, but "that didn't really seem to me to fall in line with the inherent simplicity of Apple's computers," he says. After taking apart the mini, he discovered a simple, two-conductor momentary pushbutton switch that tells the hardware to fire up, and this actually made the process of relocating it (electrically, at least) pretty simple. "The wiring for the button actually terminates in a connector that plugs into the main board, so I simply unplugged it and cut the original wires, extended them with a Monster Cable 2-channel 3-meter RCA cable and soldered this cable to both ends of the cut switch wires," says Turner. A drilled hole and button relocation solved the problem. Turner proclaims, "There's really nothing quite like hopping in the car, starting it up, and pushing that button to be greeted with

the signature Apple startup sound."

Turner's initial power setup took the electrical system's output (which varies between 12V and 14V DC), converted it to 110V AC, and then stepped it back down to 18V DC through the Mac mini's power supply brick. But the Monster Cable inverter he used would sometimes go into a protection mode and not allow the Mac to power up normally. The lack of ignition-controlled wake and sleep functions

Photograph by Matt Turner

installed in my VW is a pre-production model using the GTi as a test environment." Production models have been available since the end of April, and a Mac mini-specific plug-and-play wiring harness should be available in August 2005.

IT'S IN THE BOX.

Turner first wanted to have the mini mount in-dash, but with all the connectors sticking out of it, the mini stretched out from 6.5 inches deep to over 10.5. The choice was to either rework all the airflow controls and ductwork in the car or find a new home, such as the glove box. Turner says, "I still wanted the Mac mini to look as if it were a factory-installed option, so simply mounting the computer in the glove box itself or in the glove box door wasn't really what I wanted to do. That's not the way that I felt Volkswagen would do it. Instead, I chose to modify the internal storage dividers in the glove box to make an actual compartment that would house the Mac."

The full complement of connections was then routed through the sub-dash of the car, wire-tied along factory harness runs, and plugged into the computer. The Mac mini itself slides into the housing from the front to allow access to the back of the computer if removal of the computer is ever necessary.

DASH FABRICATION.

The most striking feature of the mini installation is the in-dash LCD screen that controls the whole shebang. Turner adds, "The first thing to do when trying to make a large component fit in a small area (like a 7-inch monitor in a double DIN-sized opening) is to take it apart and remove as much stuff as you can that is extraneous or unnecessary for the project at hand. In the case of the Xenarc monitor, this meant opening the case of the monitor, removing the built-in speaker, reversing the orientation of the power lead, removing and relocating the infrared receiver for the remote control, and doing away with the front of the case entirely, including the buttons on the front bezel (all these functions are now performed via remote control). This gave me a flat surface to build trim on top of, as well as reduced the overall size of the monitor significantly."

To make the aluminum panel, Turner made an acrylic version first. This was used as a cutting template for the aluminum. Turner notes, "All of the other aluminum parts in my vehicle (the iPod and

and the quirky workarounds to power the computer and inverter to stay on while pumping gas or running into a store for a quick pickup were solved with new hardware specifically designed for the Mac mini: the CarNetix CNX-P1900 ($90, carnetix.com).

This dual-output, 140-watt intelligent DC to DC power regulator simply replaces the Mac mini's standalone power brick. It accepts 7.5V to 18V of constant input, has an ignition sense and pulse trigger input, and outputs a stable and consistent 18.5V, a secondary 5V or 12V output for powering USB hubs or screens, a delayed 12V amplifier for accessory turn-on lead, and a pulsed ground output for triggering the Mac mini's power button for automated operation of sleep, wake, and startup functions. The wide range of voltage input capability allows the P1900 to never sacrifice its output based on low voltages encountered during engine cranking, when battery voltage can often drop to as low as 7.5V.

When Turner was Slashdotted, the manufacturer leapt at the opportunity to fill the void for in-car power solutions for Macs. Turner says, "The unit I

trackpad plate, the port and flash reader plate, and the ring around the boost gauge) were all made with the same technique of using separate jigs to make a complete, one-piece, acrylic jig that was used to cut the aluminum parts. They were all then hand-sanded, polished, and brushed in the same manner."

Turner then hand-sanded the inside of the beveled aluminum starting with 80-grit sandpaper and working through 120, 180, 220, 280, 320, 400, 600, and 1000-grit sandpaper to reach a polishable

> **THERE'S REALLY NOTHING QUITE LIKE HOPPING IN THE CAR AND PUSHING A BUTTON TO BE GREETED WITH THE SIGNATURE APPLE START-UP SOUND.**

surface. He used a pneumatic die grinder with a polishing head and a block of blue jeweler's rouge to polish the aluminum bevel to a mirror finish.

CONTROL: RUB, ROLL, AND PUSH

The touchscreen isn't the primary method of user interface; in fact, there are three ways the driver or passenger can operate the mini. The first is, of course, the touchscreen, but Turner comments, "The touchscreen doesn't have the resolution to control the OS, nor should it." The Griffin PowerMate USB control knob and the Cirque Easy Cat USB Trackpad control most of the functions.

What's next for Turner? He's interested in a turn-by-turn GPS and would like to see a Mac-based OBDII interface to show engine speed, timing, and vehicle speed. When that happens, Turner proclaims, "The dream will be complete."

For more on Turner's creations, along with extensive details on the Mac mini install, visit: tunertricks.com.

Phillip Torrone is senior editor of MAKE.

2 ÜBER TESTER

MAKE YOUR OWN 4-IN-1 CAR WIRING DIAGNOSTIC TOOL.

BY DAVE MATHEWS

When you're working on your car's 12V wiring system, it helps to have a few special tools to get the job done. The Uber Tester is a 4-in-1 gadget that won't remove door panel clips but does test for most wiring conundrums you might encounter. Plus, with its dual notification (buzzer and light), you'll be able to use the tool when your stereo is blasting or when you're contorted under the dash.

This 9V-powered handheld device will test the following scenarios with just three wires:

1. DC polarity — positive or negative voltage
2. Speaker polarity — "pop test"
3. Connectivity — wire loop back
4. Presence of voltage — fuse tester, constant or ignition switched power

Installing a 12V accessory or stereo properly — that is, with the correct speaker polarity, switched power source, and constant voltage for the clock and preset memory — requires a tester such as this. Tracing wires in today's vehicles is next to impossible with their nearly identical colors and tight wire looms. This gadget will help you find that needle in a haystack.

SET UP.

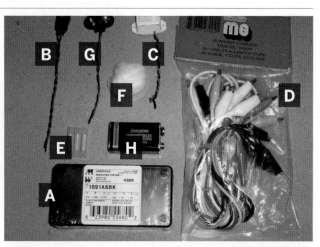

MATERIALS:

[A] Enclosure (gadget box) 3"x2"x1"
(RadioShack #270-1801)

**[B] 12V, snap-in, high-brightness lamp
– blue**
 (RS #272-335)

[C] Mini 12V DC electric buzzer
 (RS #273-055)

**[D] 1" alligator insulated test/jumper
cable set**
(RS #278-001)

**[E] Wire crimps, shrink tubing, or
electrical tape**

[F] Cotton balls
(to keep the innards from moving around)

[G] 9V battery snap connector
(RS #270-325)

[H] 9V battery

TOOLS:

[I] Soldering iron and solder

[J] Drill with large and small bits

[K] Wire cutters

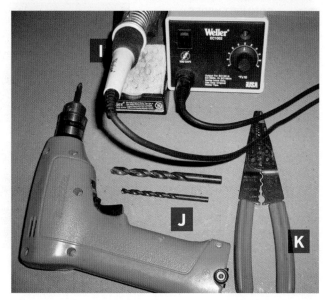

MOD YOUR ROD

1. DRILL HOLES AND INSERT LAMP

Start by drilling a hole for the lamp in the bottom of the gadget box (not on the removable lid) about a quarter of an inch down. Then drill a hole just large enough to let the three jumper wires pass through it. Strip ¾" of insulation off the lamp wires and insert the lamp into the hole you drilled for it.

2. INSERT WIRES

Cut one alligator clip off the end of each of the red, green, and black wires and poke the wires into the gadget box, leaving the alligator clips outside of the box. Tie the three wires into a knot to prevent them from pulling through the hole, leaving about 2" of wire to work with inside the box. Strip the wires back ¾" to prepare for the connections.

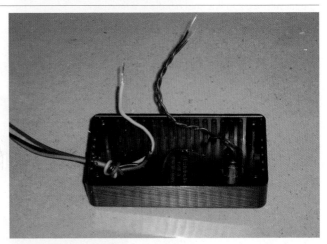

3. CONNECT WIRES

Connect the black wire from the 9V harness to the black alligator clip wire.

Connect the red wire from the 9V harness to the red alligator clip wire, the red buzzer wire, and to one of the lamp wires.

Connect the black wire from the buzzer to the green alligator clip wire and to the second wire on the lamp.

Check wiring. Note that coming from the alligator clip jumpers, the red wire makes three connections, the green wire makes two, and the black one only connects to one. That's it!

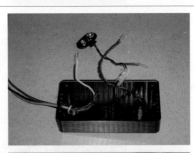

4. TEST YOUR UBER TOOL

Connect the 9V battery to the connector and touch the black and green alligator clips together — you should hear a tone and the light should illuminate. Connect the red wire to a positive 12V (or lower) source and the black to ground, and you should hear a tone and see the light illuminate. If so, your wiring is properly connected.

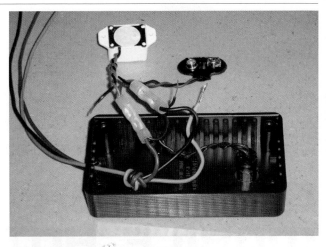

5. REINFORCE THE ALLIGATOR CLIPS

Now would be a good time to put some extra solder on the jumper wires inside the alligator clips. Usually they are merely crimped on, which will not provide a strong enough connection to allow you to let your tester dangle. Just pull back the protective covers on the jumpers, clip the jumpers to a piece of cardboard to keep them from moving around, and load up the solder under where the wires are crimped.

6. FINISHING UP

Now it's time to stuff the buzzer, battery, and wires into the gadget box. Put the cotton balls in the leftover space to keep the tester from sounding like a baby's rattle.

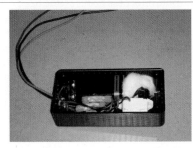

To prevent your battery from dying, keep the red wire from shorting on the black or green wires by pulling the insulation hood over the metal. You should see years of service from your battery since it is rarely used.

12V UBER TESTER WIRE GUIDE
Red/Black
9V output and speaker polarity pop test
Black/Green
Continuity test
Green/Red
Voltage presence — constant or switched

Memorize this guide, or copy it and paste it onto your gadget box.

Dave Mathews has been installing car stereos since his first go-kart in the 1980s and has never stopped tinkering with aftermarket audio. While this creation came to fruition in 1993, more stories and video clips on technology he is currently peering into can be found at davemathews.com.

3

STOMPBOX MOBILE HOTSPOT

TURN YOUR CAR INTO A WI-FI HOTSPOT, THEN USE GPS AND WEBCAM INPUT TO MAP YOUR CURRENT LOCATION ONLINE AND AUTO-GENERATE A PHOTO TRAVELOG. BY TOR AMUNDSON

When Verizon released its new BroadbandAccess service, I had to give it a try. $80 a month for DSL-like speeds in many cities? Yes, please!

I subscribed, bought the 5220 PCMCIA card, and it worked. But the card's antenna was so weak that I often had to rotate my laptop. Also, it only served one computer. What if I wanted to run multiple machines or share connectivity with friends? Then it hit me: I could build the card into an access point and install it in my car with some beefy antennas. I called the project StompBox because this little box would bring my own network "stomping grounds" with me wherever I went.

OVERVIEW

In technical terms, the StompBox is a cellular router. Like all routers, its job is to push data back and forth between multiple interfaces. Ours has two interfaces: Verizon's BroadbandAccess and Wi-Fi.

I decided to base StompBox on embedded hardware rather than on a laptop or full-blown PC. This makes it cheaper, smaller, more reliable, and better for "just plug and unplug" vehicular use. For the platform, I chose Pebble Linux, a Debian distribution from NYC Wireless. For hardware, I used the Soekris 4521, a compact embedded computer that runs on 12VDC and has a great user community. All the software needed — Pebble Linux plus drivers plus code — fits onto a 128MB Compact Flash card

that functions like a computer hard drive but is tougher and more vibration-resistant.

The Wi-Fi interface is simple. The Linux driver HostAP can run an access point from any 802.11 card with a Prism2 chipset. You simply configure your choice of SSID and WEP key and turn it on.

As for the Verizon interface, wireless networking pioneer Phil Karn wrote a how-to on getting the 5220 card to work under Linux. His solution is to make the system treat the 5220 card as a USB modem, and then send it the proper Hayes AT commands to dialup Verizon's network.

With this foundation in place, you can expand your access point's capabilities by adding a USB card, a GPS unit, a webcam, and other devices. I hooked up a GPS and did some Google Maps hacking to create an auto-updated web page (hosted on my home network to avoid wireless traffic jams) that tracked my car's route and showed its current location against a satellite photo. Unfortunately, this and many other bleeding-edge Google Maps hacks were rendered inoperative last May when Google changed the unsupported beta API that hackers' code had been relying on. Then in late June, a Google Maps API beta was officially released, so now old map hacks can be repaired and new hacks written in a more stable environment. Meanwhile, the StompBox's web page also shows photos captured by a webcam pointing out of the car window.

SET UP.

HARDWARE

Soekris net4521 (with case)
From soekris.com

» **Verizon 5220 EVDO/1xRTT card**
» **Pigtail adapter (Orinoco to N-Female, panel mount)**
» **N-Male to FME-Female antenna adapter cable**
» **CDMA antenna (800/1900MHz dual-band)**
For cellular network

» **Senao 2511 Mini-PCI 802.11 card (or any Prism2-chip-set equivalent)**
» **Pigtail adapter (Hirose U.FL to N-Female, panel mount)**
Wi-Fi antenna (2400MHz band)
For Wi-Fi

128MB Compact Flash card

An old mousepad, or other piece of foam less than ¼-inch thick
For vibration-proofing the cards

Serial cable (null modem, DB9-F to DB9-F)

12VDC car power to M-type (5.5mm) plug cable

» **Any GPS unit with standard NMEA output via USB or serial port. If serial, add a USB-to-serial adapter, prefer-ably one that uses the Prolific PL-2303 chip.**
» **2-port USB PCMCIA card, low profile so that it fits inside the router**
For GPS enhancement

D-Link DCS-900W wireless internet camera
12VDC to 6VDC power transformer
For webcam enhancement

12V battery power supply (optional)
(I used a Xantrex P400)

12V car power two-outlet splitter (optional)
To offload car battery while engine isn't running (power enhancement)

SOFTWARE

StompBox .IMG file
Ready-to-go "disk image" for the CF card; downloadable at makezine.com/03/stompbox

Debian 2.4.26 distribution (optional)
Necessary if you want to compile extra code

TOOLS

Phillips screwdriver

Small and large pliers

Drill press (preferred) or drill

Stepper bit
For drilling antenna holes in the case

Dremel tool
For cutting an opening to the USB card

A host computer with a terminal emulator installed and a 9-pin serial port (or a USB-to-serial adapter, listed in previous column for GPS)
For setup. I cover Linux and Macintosh in this article. You could probably also use a Windows machine, but I haven't tested this yet.

Compact Flash reader/writer
For setup

12VDC power supply (deep-cycle battery, lab-bench supply, wall wart, or even your car battery) 1.5 amps or higher, with a 5.5mm "M" plug
For testing

STEPPER BIT

The stepper bit, with its Christmas-tree shape, lets you bore wide, round holes in sheet metal without having to swap in successively larger bits. Hole diameters are marked on the inside of the bit's hollow, making it easy to set a drill press to stop at the right point. They're great for drilling out metal cases and panels, for installing new ports. They're also safer than regular drill bits and not very expensive.

MAKE. IT.

START >> Time: **Two Weekends** Complexity: **Linux-Geek Easy**

1. CONFIGURE THE CF CARD

StompBox's software resides on the Compact Flash card, and I've wrapped it all up in a ready-to-go image downloadable at makezine.com/03/stompbox. While all the software components on this card are open source and easily available, the specifics of configuring it are beyond the scope of this article. If you want to make your own setup from scratch so you can control exactly what's in your system, see moro.fbrtech. com/~tora/EVDO/cfimg.html.

1a. Download the StompBox .IMG file to your host computer, then insert the CF card into the reader. It doesn't matter what's on the card for now; we're going to overwrite it with the image.
LINUX: If the card was not blank, the system may try to auto-mount it. If so, please unmount the drive before continuing, using the umount command.
Mac OS X: If the card was blank, you'll get a "not readable" pop up; click "Ignore." Otherwise, it will auto-mount on the desktop. Run the Disk Utility and repartition the drive into "free space," and then reinsert. Once OS X "ignores" the drive, you can continue.

1b. Find out your system's physical address for the CF card. On my PowerBook, it's usually /dev/disk2, but it differs for every system. Warning: Be sure you have the correct address before proceeding, or you can damage your other drives!

1c. Open a terminal window (Applications/Utilities/Terminal on a Mac) and copy the image to the card using the standard Linux command:
dd if=/files/stompbox.img of=/dev/disk2 bs=8192
Change the input file (IF) and output file (OF) parameters above to match your system. Copying can take up to six minutes, during which you should see the light on the CF reader blink. Once it's done, remove the drive.

---

2. ASSEMBLE THE HARDWARE

2a. Remove the four screws on the bottom of the Soekris net4521 case, and then slide the top off.

2b. Remove all six screws in the motherboard and the two serial-port socket screws to detach the motherboard from the bottom case. Lift it out, and put it someplace safe.

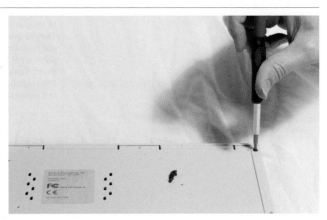

Photography by Tor Amundson

2c. Hold the pigtail adapters' N-Female plugs against the existing antenna holes in the Soekris case. Trace around them with a pencil, and then use the drill and stepper bit to make holes just large enough to accommodate the plugs. Brush away any burrs.

Antenna holes drilled to fit pigtail adapters.

2d. Mount the Wi-Fi pigtail on the right-hand side of the case (the side closest to the power jack). This is the one with the tiny U.FL connector on the end of the cable. Insert the N-Female connector through the case hole and put the lockwasher on the outside. Then screw on the nut and tighten down, using two sets of pliers: one to hold the N-Female jack in place, and the other to turn the nut. Don't over-tighten!

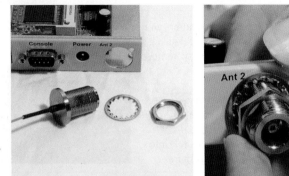

2e. Repeat step 2d above, this time on the left-hand side for the 5220 pigtail. When you are done, both N-Female jacks should be mounted to the case, the 802.11b on the right and the 5220 on the left.

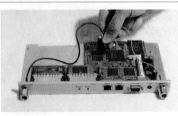

2f. Insert the Mini-PCI card (Senao 2511) into the Soekris motherboard. It angles into the connector and then snaps downwards into place.

2g. Connect the Wi-Fi antenna pigtail's U.FL connector to the antenna port closest to the mini-PCI socket. These connectors are tricky; you may have to use a tiny screwdriver or metal stick to gently snap it down into the socket. Be careful; these connectors are fragile.

2h. Insert the 5220 card into PCMCIA Slot 0 on the Soekris (the one on the right, closest to the mini-PCI socket). It should slide in smoothly, parallel to the motherboard.

2i. Look at the 5220 card end-on; you should see the connector socket. If it's covered by a small plastic cap, remove the cap.

2j. Connect the CDMA antenna pigtail's Orinoco connector to the tip of the card, just below the fold-out antenna. It should snap securely into place with just fingertip pressure.

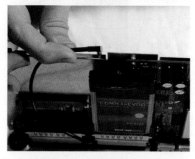

2k. For GPS, insert the USB PCMCIA card into PCMCIA slot 1 (the remaining slot, away from the mini-PCI socket).

2l. Insert the Compact Flash card into the Soekris' CF slot and secure with the bumper screw included in the case.

2m. Use the Dremel to cut two holes in the top of the case, just above each of the USB card's ports and big enough to connect USB cables through.

2n. Cut a small strip of foam and put it inside the upper case where the PCMCIA cards will touch. Once the case is closed, this will snug the cards into place and keep them from vibrating loose.

2o. Reassemble the case, sliding it together carefully. Secure it with the four screws on the bottom.

3. CONFIGURE THE PLATFORM

On the Soekris, the default speed of the BIOS (which does things like count memory, self-test, and handles machine-level communication) is 19,200 bps, while the higher-level Linux OS normally talks at 9,600 bps. For the very first boot up, we'll need to change the speed of the Soekris BIOS to 9,600 bps. That way, we can run our host computer's terminal emulator at one speed and be able to configure both the BIOS and OS.

3a. Plug the Soekris into your 12VDC power source, using an M-type (5.5mm) plug.

3b. Hook your host computer up to the Soekris using the serial cable and an USB-to-serial adapter, if necessary.

3c. Run the terminal program on your host, and set the baud rate to 19,200 8-N-1 (19,200bps, 8 data bits, no parity, 1 stop bit). I used ZTerm.

3d. Plug in the Soekris and watch for the BIOS screen. As it counts through memory, quickly press CTRL-P to access the BIOS set-up. You should see: comBIOS Monitor. Press ? for help. Now, reset the speed by typing: set conspeed = 9600

3e. Enter Reboot to restart the system. You should see the BIOS boot up, followed by the operating system. This will take one to two minutes. If your system does not boot, the CF card might not have been written correctly

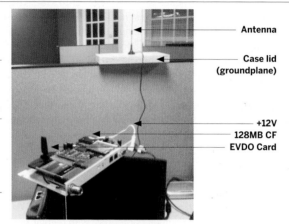

Antenna

Case lid (groundplane)

+12V
128MB CF
EVDO Card

or isn't compatible with the Soekris (rare, but it happens). Other possibilities are that you have weak power supply or faulty hardware.

3f. You should now see the prompt: Debian GNU/Linux 3.0 pebble ttyS0 pebble login: Log in as root, with a null password (just hit Return). Your access point is now ready to configure; you can start by changing to a better password.

4. CONFIGURE THE ROUTER

4a. If you don't have one already, create a dynamic DNS account. Browse to dyndns.org from your host (or any other) computer and follow the instructions for setting up an account for the DynDNS service.

4b. From your terminal, run the router configuration script on the Soekris by entering: /usr/local/bin/initial-configuration
Answer all the questions, and the script will set up the system. You'll need to know your DynDNS account info and the Verizon phone number for your 5220 card. Also, be ready to set some names, passwords, and private network IP addresses for wireless and wired clients (example: 192.168.1.0 and 192.168.2.0). Its last step is to generate new SSH keys, which takes several minutes. When the script completes, it prompts you to reboot; go ahead, and then unplug the Soekris. The router is now ready to go online!

5. GET ONLINE

Plug the router back in, and it should now attempt to log on to the 3G network automatically, a few moments after it boots. You can see if it worked by logging in via terminal and checking the "pppd" messages (from Linux's point-to-point protocol daemon) in the "messages" logfile at /var/log/messages. Use the commands tail or grep to see the most recent messages or to search the entire file.

If it worked, you'll see pppd messages that list your network's IP addresses. Otherwise, check for an incorrect phone number, an inactive account with the provider, or poor reception. When you're too far from coverage, or have bad antenna connections, you'll see a "pppd: LCP timeout" message. You can check your coverage by swapping the 5220 card back into your host computer.

Any Wi-Fi client devices should now be able to join your new network, or you can physically plug Ethernet-enabled devices into the router's ETH1 port.

6. INSTALL IT IN YOUR VEHICLE

Now that your device is online and powered, it's time to put it in the car!

6a. Decide where you want to put your antennas. With regular, ground-plane antennas, reception will be best when they are centered on the largest available expanse of metal, such as your car's roof. But they should not be positioned close to each other; keep the antennas at least 20cm apart, preferably farther. If you get terrible reception (or don't have a metal roof), you may want to select a ground-plane-less antenna, like the kind that mount on truck mirrors. Visit a good ham radio store or local amateur radio club for advice.

6b. Using the N-Male to FME-Female adapter, connect the 3G antenna to the N-Female jack on the left of the router, away from the power jack. Hook the Wi-Fi antenna to the other jack, on the right. Then position and mount both antennas.

6c. Secure the router inside the vehicle where it won't get knocked around, then hook it up to the cigarette lighter via the car power cable.

You now have the heart of the StompBox — a Wi-Fi access point that you can drive anywhere! The next page describes some fun ways to expand on it.

3G COVERAGE

3G is a general term for several types of wireless broadband service, including EVDO, 1xRTT, EDGE, UTMS, and others. Verizon's BroadbandAccess is EVDO/1xRTT. In a 1xRTT area, you'll get ISDN-like speeds, and in an EVDO-capable area, you'll get DSL-like speeds. In both cases, your upload will be slower than your download.

If you drive out of range, your link slows as more packets are lost, until you finally drop offline. Your range depends on the type of antenna and how well it's positioned on your car.

You can greatly improve that range with an auto-gain two-way amplifier. I use Wilson Antennas model #811201. Make sure you can scale down the auto gain as you get closer, or your signal can get too noisy in a good coverage area!

For more advice, ask an amateur radio buff about working at 800/1900 dual band (for 3G) and 2400 MHz (for Wi-Fi).

END

GOING FURTHER.

EXPANSION: ADDING GPS

Since this is a standard x86 platform running Linux, it's open for expansion. Add a GPS unit, and you get vehicular tracking. The router runs Linux's gpsd in the background to poll data from an attached GPS unit and publish it as a network service. I connected my GPS serial port to the router using a USB-to-serial adapter with a Prolific PL-2303 chip. The software on the CF card finds the PL-2303 and attaches the gpsd program, but if you use a USB-connected GPS or a different adapter, you'll need to manually softlink the device to /dev/gps.

Since I wanted something flashy and fun for the vehicle-tracking page, I used Google Maps as the interface. When I first did this last spring, the Google Maps API hadn't been officially published yet, so this was still pretty new and shaky territory. And my scripts broke later, along with many other Google Maps hacks, when Google changed their API out from under us map hackers. But now that the Google Maps API beta has been released officially, it is a much more stable way of using Google's mapping systems, and it gives beautiful results.

I didn't want internet visitors curious about StompBox to flood and bog down my mobile network, so I used gpsd in combination with the

dynamic DNS system to offload traffic onto my home network. I set up a PHP script on one of my servers at home to periodically telnet into Stomp-Box's GPSd port (2947), then query it for position and speed. It then plugged those numbers into an XML document and fed them into the Google Maps Hacking tool set, generating a map. You can see the scripts that did the work on the StompBox website, makezine.com/03/stompbox.

Note that this setup is designed for internet users to track the StompBox remotely, not for in-car, real-time GPS navigation! For that, you're far better off using a dedicated navigation device or software.

GPS how-to: moro.fbrtech.com/~tora/howto/gps.html

EXPANSION PART II: ADDING A WEBCAM SCRIPT

Another fun trick is to add an onboard camera. I chose the easy route: a wireless web camera, the D-Link DCS-900W, which has its own onboard server and a wireless link. Plug one in, configure it, and you can pull still or moving images from it using any browser. It also points out of any window, sticking in place with a suction cup.

Our CF card software is configured to use a DCS-900W camera that's set up on the IP address ending in ".8" on your network. So, for example, if you set 192.168.1.0 as your network, the camera should get the address 192.168.1.8. Set the camera's SSID and WEP keys as you defined for your network.

You can view the camera input from within your StompBox network by going to that IP address. With the firewall rules I added, you can also view it from the public internet by visiting the dynamic DNS name of your system, at port 81

(i.e., http://192.168.1.8:81). Assign a password to limit demand on the camera, if word gets out!

To further avoid network traffic jams, use the getcam.sh script, also on the CF card. This offloads image serving onto a home server. Edit the script to point to your server, and an automatic process (known as a cron job) will upload the latest

images every five minutes. Change the interval by editing the file /etc/cron.d/getcam and add cameras by duplicating the script.

Tor Amundson is a nomadic mad scientist based in Silicon Valley who's studying our side of the temporary dimensional rift he discovered.

MAKING BIODIESEL

THE BEST WAY TO LEARN HOW TO MAKE YOUR OWN BACKYARD BIODIESEL IS TO START WITH A ONE-LITER BATCH. BY ROB ELAM

It's easy to make a small batch of biodiesel that will work in any diesel engine. You don't need any special equipment — an old juice bottle will serve as the "reactor" vessel — and on such a small scale you can quickly refine your technique and perform further experiments. After a few liters' worth of experience, you'll know if you've been bitten by the biodiesel bug.

The principle behind biodieseling is to take vegetable oil (either new or used), and process it into a fuel that's thin enough to spray from a regular diesel engine's fuel-injection system. This is done chemically, by converting the oil into two types of compounds: biodiesel, which shares the original oil's combustibility, and glycerin, which retains the oil's thick, viscous properties. Drain away the glycerin, and you're left with a fuel that you can pour into any diesel vehicle with no further modification.

Once you get to the far side of the learning curve, making biodiesel is very much like cooking. In fact, a commercial biodiesel production plant shares more in common with a large-scale bakery than a petroleum refinery. There's organic chemistry involved in baking a cake, but most bakers wouldn't consider themselves organic chemists.

BIODIESEL CHEMISTRY

Vegetable oil is a triglyceride, which means that its molecule consists of a glycerin "backbone" with three fatty acids attached, forming a shape like a capital letter E. To make biodiesel, we add lye and methanol. The highly caustic lye breaks the three fatty acid branches off of the glycerin backbone. These free fatty acids then bond with the methanol, which turns them into fatty acid methyl esters — otherwise known as biodiesel. The freed glycerin, which is heavier, sinks to the bottom, leaving the fuel (and lye) on top. Wash the lye out of the upper layer, and you have pure biodiesel.

But it's not that simple. With some triglyceride molecules, only one or two fatty-acid branches break off, which leaves mono- or di-glyceride molecules (shaped like capital Ts or Fs), rather than free glycerin. At the same time, mixing methanol and lye produces some water — and oil, water, and lye mixed together make soap.

With all of these incomplete and competing chemical reactions, your batch will inevitably contain soap, water, leftover lye, methanol, and mono- and di-glycerides, along with the nice biodiesel and glycerin. Mono- and di-glycerides are emulsifiers, so they prevent mixed liquids from separating, making it harder to extract biodiesel. The picture gets even muddier when you use waste vegetable oil rather than pure oil, since it contains free fatty acids, water, and countless random contaminants from all those French fries.

These by-products are bad for an engine, potentially causing micro-abrasions that damage fuel injectors or clog fuel filters. But you can remove them by washing or cooking the biodiesel in various ways, or by processing the incompletely converted biodiesel again, as if it were vegetable oil. In extreme cases, you'll end up with a thick, soapy mass that never separates. All biodieselers wind up with a batch of this glop sooner or later. Fortunately, you can use it to make a good, grease-cutting soap — which is something that all biodiesel homebrewers need to have on hand.

Photography by Sara Huston

MATERIALS

[A] At least one liter of vegetable oil (you can double this recipe if desired, using a larger bottle).
Either new or waste vegetable oil is fine. If you are using waste oil, try making batches with samples from different restaurants' grease barrels.

[B] One bottle methanol gasoline treatment, such as Heet (in the yellow bottle) or Pyroil brand.
Sold at auto parts stores. You can buy larger quantities in bulk from local auto-racing suppliers, petroleum distributors, and chemical suppliers.

[C] One bottle isopropyl alcohol gasoline treatment, such as Iso-HEET (in the red bottle) or Pyroil brand.
Carried by the same retailers as methanol. You won't need much of this, even if you start making larger batches.

[D] Between 5 and 10 grams of lye, quantity explained below. You can use regular lye (sodium hydroxide, NaOH) or potash lye (potassium hydroxide, KOH). NaOH is easier to find, but KOH is easier to work with.
NaOH is widely available as Red Devil Lye drain cleaner. You can buy KOH from local soapmaking and tanning craft suppliers, or from braintan.com.

[E] Phenolphthalein solution.
Available from beer and winemaking and lab chemical suppliers. Also contained in many educational chemistry sets. Should be fresh.

[F] One or two gallons of distilled water.
Used for washing the fuel.

[G] Vinegar.
To neutralize discarded lye.

TOOLS

[H] Metric gram scale, sensitive to at least ½ gram.
Available at some tobacco and "head" shops, or look for a triple-beam scale at pawn shops and flea markets.

[I] Two syringes, eyedroppers, or pipettes, calibrated in milliliters. They should each hold up to 10ml, and be marked in increments no larger than .2ml.
Graduated eyedroppers and oral syringes are available in drugstores, sometimes with the baby supplies. You'll use these for different chemicals that you shouldn't mix up, so it helps to get one syringe and one eyedropper.

[J] Candy thermometer.

[K] Measuring cup, beaker, or other way of measuring 220ml and one liter of liquid.

[L] Two 1-pint glass Mason jars with tight lids, and three or more small Mason or babyfood-size glass jars.

[M] Two 2-liter bottles or larger, glass or PET/PETE (#1) plastic. Favor juice, water, or milk bottles over soda, since their wider mouths are easier to funnel into.

Two-liter pot, and a hot plate or electric burner (not gas).

Cheesecloth, clean rags, and a bucket.

Funnel and plastic spoon.

Masking tape or labels, and a marker for labeling.

[N] Safety goggles and gloves.

Litmus strips or electronic pH meter (optional).

START ›› Time: **Six Hours over Three Days** Complexity: **Low**

1. FILTER AND DE-WATER YOUR OIL

If you're using new oil, you can skip to Step #2. But if you're starting with waste oil from a restaurant fryer, it will contain food particles, water, and free fatty acids (FFAs) — contaminants that you need to remove or adjust for. The FFAs make the oil more acidic, (a.k.a. rancid), which counters the effect of the lye. You can compensate for this by adding more lye into the main reaction later, but you need to perform a titration test beforehand in order to determine how much extra lye you'll need.

BIODIESEL HOMEBREWING SAFETY

While biodiesel is safe to handle and store, the homebrewing process involves flammable, poisonous, and caustic chemicals, alcohols, and lye. Keep all chemicals clearly labeled, sealed, and out of reach of children and pets. When handling methanol and lye, wear long sleeves, safety glasses, and gloves made out of nitrile — or, even better, PVC. Wash the gloves after each use, and be careful not to touch your skin or eyes. Keep a water hose nearby in case of skin contact. Methanol can be absorbed through the skin, so wash immediately with water if contact occurs. Immediately flush lye off skin with water or vinegar. Methanol fumes are poisonous, so wear a mask, or hold your breath while pouring, and work outside or with good ventilation.

1a. Start with more than one liter of oil, since the following steps will slightly reduce your oil's volume. Warm the oil to about 95°F in a pot on an electric hot plate (don't use a gas burner, here or anywhere else in this project), then filter it through a few layers of cheesecloth in a funnel (or use a coffee filter).

1b. Heat the oil to 140°F and maintain the temperature for 15 minutes. The water will fall to the bottom, so you'll risk steam explosions if the temperature gets too high. Pour the oil into a bottle or other vessel and let it settle for at least 24 hours. This removes water, which would produce soap in your batch. If you see water at the bottom (it will be dirty, not clear), don't pour it back out with the oil.

2. TEST YOUR OIL
Determining the acidity of the vegetable oil.

2a. Dissolve one gram of lye in one liter of distilled water (0.1% lye solution), or use an equivalent ratio to make a smaller amount. This is your reference test solution, which you can store sealed and re-use for later batches.

2b. In a small jar, dissolve 1ml of slightly warm oil in 10ml of isopropyl alcohol. Stir until clear, then add two drops of phenolphthalein solution.

2c. Using a graduated syringe or dropper, add your reference test solution drop-by-drop into the oil-alcohol solution, keeping track of how much you're using. The more acidic the oil, the more you'll need to add. Stir constantly, and continue adding solution until the mixture stays pink for ten seconds. Note the number of milliliters of lye solution you used; this is the number of extra grams of lye you'll need to add per liter of oil.

This process is called "titration," and it's a standard method of determining a solution's acidity.

MOD YOUR ROD

3. PROCESS THE OIL

This is the main chemical reaction that produces the biodiesel.

3a. Determine how much lye you need. If you're using new oil, use 5 grams of NaOH or 7 grams of KOH per liter. With used oil, use these amounts plus one gram for every milliliter of solution you used in the titration step 2c. For example, if it took 1.5ml of lye solution to turn the mixture pink, use 6.5g of NaOH or 8.5g of KOH.

3b. Measure your lye into a clean Mason jar. Add 220ml of methanol, cover securely, and tip the jar to make sure the lid doesn't leak. Then swirl or shake the jar gently until the lye dissolves fully. This will take a few minutes, and the jar will become slightly warm in the process. This mixture is the methoxide solution, and it's dangerous stuff; you'll need to wash the Mason jar lid after you're done with your batch, or its seal will dissolve. (Some regular homebrewers prepare methoxide ahead of time and store it in #2 HDPE plastic.)

3c. Warm a liter of your oil up to 130°F. Let it cool down if the temperature gets too high.

3d. Pour the oil into a large bottle, add the methoxide solution, cap tightly, and shake like crazy for about five minutes. The contents might change color a couple of times.

3e. Set this mixture aside, and admire. In half an hour or so, you should see a darker, dirty, glycerin layer start to sink toward the bottom, and a larger, lighter, biodiesel layer rise to the top. This is a good time to clean up. If you're sure your bottle won't leak, you may want to let it settle upside-down, so you can drain the glycerin out by cracking the bottlecap. Or you can lay it sideways to make it easier to pour off the biodiesel.

3f. Let the liquids continue to settle overnight.

4. SEPARATE, WASH, AND DRY THE BIODIESEL

Your bottle now contains biodiesel, glycerin, mono- and di-glycerides, soap, methanol, lye, and possibly a little leftover oil (triglycerides). The glycerides are all oil-soluble, so they'll reside predominantly in the upper, biodiesel layer. The thin layer of glycerin, which is water-soluble, will sink. Depending on the oil and catalyst you used, it might be either liquid or solid. Soap, methanol, and lye, which are also water-soluble, will be mixed throughout both layers — although some of the soap can sometimes form its own thin layer between the biodiesel and glycerin.

If you see more than two layers, or only one, then something's wrong — possibly excessive soap or mono-glyceride formation. These are both emulsifiers, and in sufficient quantities they will prevent separation. In this case, check your scales, measurements, and temperatures. You can reprocess the biodiesel with more methoxide, or try again with fresher oil (or new oil). If you can, shake the bottle even harder next time. In an engine, glycerin droplets in biodiesel will clog fuel filters, soap can form ash that will damage injectors, and lye can also abrade fuel injectors. Meanwhile, methanol has toxic and combustible fumes that make biodiesel dangerous to store. You don't want any of these contaminants in your biodiesel. If you left your biodiesel to settle undisturbed for several weeks, these water-soluble impurities would slowly fall out of the biodiesel (except for the methanol). Washing your biodiesel with water removes the harmful impurities, including the methanol, much faster.

Unfortunately, washing will not remove the invisible, oil-soluble mono- and di-glycerides. These are a problem in rare instances when large amounts of certain types of monoglycerides crystallize. This can clog fuel filters and injectors, and cause hard starts, especially in cold weather. High-quality, commercial biodiesel has very low levels of mono- and di-glycerides, which is the ideal for biodiesel homebrewing. You can roughly test for the presence of mono- and di-glycerides in your own batch by processing it a second time, as if it were vegetable oil. If more glycerin drops out, then your first reaction left some unfinished business behind.

4a. Pour the biodiesel layer off the top, into another bottle. Don't pour off any of the glycerin, as it makes washing difficult; better to leave a little biodiesel behind. If you let the bottle settle upside-down, drain the glycerin from the bottom.

4b. Gently add some warm distilled water to the biodiesel.

4c. Rotate the bottle end over end, until the water starts to take on a little bit of white soapiness, which may take a few minutes. Do not shake the bottle. You want to bring water and biodiesel into contact, without mixing it too vigorously. The biodiesel contains soap, and if you overdo the agitation, the soap, biodiesel, and water will make a stable emulsion that won't separate.

4d. Turn the bottle upside-down, crack the cap, and drain away the soapy water. If you're using a soft drink bottle with a narrow neck, you can plug the opening with your thumb instead of using the cap.

4e. Add more warm water and keep repeating the sloshing and draining process. Each time there will be less soap, and you can mix a little more vigorously. If you go too far and get a pale-colored emulsion layer between the biodiesel and white, soapy water, don't drain it away; it's mostly biodiesel. Just keep washing and diluting until the water becomes clear and separates out quickly. It takes a lot of water. But if the emulsification layer persists, try applying heat, adding salt, and adding vinegar, in that order.

4f. After draining the last wash water away, let the biodiesel sit to dry in open air until it's perfectly clear, which may take up to a couple of days. In general, the better your washing, the faster the fuel will clear. If you're in a hurry, you can dry the fuel faster by heating it at a low temperature. As with the evaporation method, the fuel is done when it clears. If you can read a newspaper through the biodiesel, it's dry and ready to pour into a vehicle. Congratulations — you're done!

END

USE IT.

You can put your liter of biodiesel into a jar to pass around to your admiring friends, or use it in a diesel engine. We generally filter fuel before adding it to the vehicle to remove micro-abrasives, but a liter or two probably won't do any damage. To be safe, you can filter your liter through several coffee filters. Since it's a small quantity, don't worry about whether it's OK to add your homebrew to your tank, so long as it's clear. Even if your liter contains a lot of mono- and di-glycerides, they'll be safely diluted by the rest of the fuel in the car.

Litmus strips or a pH meter will test the fuel's acidity — one indication of how clean it is. Biodiesel should measure a neutral 7, with a higher number indicating soap or leftover lye. To test for the presence of glycerin, you can use the Gly-Tek test kit (gly-tek.com), which detects leaked antifreeze in motor oil.

If you start using biodiesel regularly, however, you should have the fuel tested, and change your fuel filter often. Biodiesel can free petroleum residue stuck in the fuel system, which can cause clogs. Also, watch for old fuel lines, which may get sticky and need replacement. Biodiesel can degrade the natural rubber used in older cars' hoses, but it's fine with now-standard synthetic rubber.

Biodiesel will normally store for months, but, like petrodiesel, it can be attacked by certain bacteria. To prevent this, you can mix in the same biocide chemicals that are widely used for petrodiesel storage. Also note that biodiesel made from animal fats can gel in cold weather; if this might be a problem, test a sample in the freezer before filling your tank. Biodiesel made from canola oil has the lowest solidification point.

GOING FURTHER

There are numerous recipes for biodiesel, generally characterized by number and type of main reactions. The method described here is a single-stage process, which uses an alkali to catalyze the main reaction, and relies on titration to determine the proper amount. Two-stage methods dispense with titration and run reactions twice to achieve complete conversion. The two-stage, base-base method repeats the lye reaction, while the two-stage, acid-base method uses sulfuric acid for the first stage. These methods, while more complicated, are also more foolproof and better suited for large batches.

Homebrewers make and wash larger batches of biodiesel in a variety of tanks; one clever design uses a 55-gallon drum that's tipped on its edge to make a drainage point at the bottom, which is fitted with a valve. This makes it easy to drain glycerin or water. Various recipes can turn the dirty leftover glycerin into soap, 40-weight oil, paintbrush cleaner, and other useful products. High-volume operations can buy a $31,000 glycerin-purification distiller from Recycling Sciences (rescience.com), which will convert dirty biodiesel glycerin into nice, clear, commercially valuable glycerin.

Popular, high-volume washing methods include mist washing, in which water droplets sprayed on the surface of biodiesel settle down through the liquid; bubble washing, which uses an aquarium air pump and air stone to gently agitate water and biodiesel; and bulk washing, in which water and biodiesel are agitated manually. For final filtration, down to 5 microns, biodiesel can be pumped through standard, under-cabinet, household water filters.

RESOURCES

This article is partly based on a series of articles by Maria "Mark" Alovert that first appeared in the Energy Self-Sufficiency Newsletter: rebelwolf.com

Mark Alovert's site: localb100.com

Homebrewing forums: biodiesel.infopop.cc, veggieavenger.com/media

Biodiesel policy and activism forum: biodieselnow.com

Biodiesel processor designs and other info: journeytoforever.org/biodiesel.html

Biodiesel stations and industry info: nbb.org

VW Diesel forum: tdiclub.com

Setting up a commercial biodiesel pump in your town: propelfuels.com

Rob Elam is a founder of Propel Fuels, a Seattle-based biodiesel services and fuel distribution company.

BIOFUELS TODAY

THE NATION'S #1 MOCK-NEWSPAPER FOR MAKE READERS INTERESTED IN ALTERNATIVE FUELS.

Grassroots Network Offers Alternative to Big Oil

By Dan Gonsiorowski

SEATTLE — Spotting an old, diesel Volvo with a bumper sticker promoting a cause is not unusual on the streets of Seattle. Spotting two or three of them parked on the same block is a little strange, however — especially when all of them advertise the same cause. But that's the scene on the street outside of Dr. Dan's Alternative Fuel Werks in Seattle, and the stickers all broadcast some variation of "Biodiesel-Powered." The Fuel Werks promotes and sells biodiesel to a segment of the population that Dan Freeman (a.k.a. Dr. Dan) affectionately calls "affluent crackpots." They are people willing to pay $3.67 a gallon to power their cars with a renewable energy source and decrease the amount of pollution released into the atmosphere from the burning of fossil fuels.

Oil and automobile companies have been slow to offer alternatives to the petroleum-fueled car, but conscientious drivers left out in the cold by those industries are addressing the issue themselves. Grassroots distributors like the Fuel Werks are popping into existence nationwide, to serve both private consumers and neighborhood biodiesel cooperatives who pitch in on large, shared storage tanks. Co-ops form via word-of-mouth, flyers posted at local garages, or online queries. One biodieseler, Eric Forrer, has been trolling craigslist for interested individuals in Seattle's North End area. After he finds enough members, Dr. Dan will install a 270-gallon tank in a highly visible location near Forrer's home, and keep it filled.

The biodiesel distribution chain favors local, independent businesses. Consumers buy fuel from co-ops or small distributors like Dr. Dan's, while the distributors buy it from anyone who can make high-quality stuff, from backyard homebrewers to local startups like Seattle Biodiesel, which, in turn, gets the raw vegetable oil they use to produce biodiesel (mostly soybean) from, well, Iowa, for now.

We can seemingly leave lefty political stereotypes aside while viewing the National Biodiesel Board's map of biodiesel distributors, at nbb.org/buyingbiodiesel/distributors. The vast majority are in the Midwest, where vegetable oil is grown and processed, rather than in coastal Sierra Club strongholds such as Seattle, San Francisco, and Berkeley. But it isn't simply a matter of economics, as evidenced by the involvement of musician and farm advocate Willie Nelson. His company, Willie Nelson's Biodiesel, markets biofuel as a way to empower independent farmers to plant crops that will let them participate in the energy market.

Back in Seattle, Dan Freeman hopes that the Fuel Werks will, one day, sell biodiesel that is not only processed, but also grown in-state — and all without the backing, consent or assistance of automobile manufacturers or oil companies.

Deep-Fried Ride: Veggie Oil Inside

By Xeni Jardin

FLORENCE, MA — Do-it-yourselfers are turning to an offbeat solution to rising fuel prices: they're modding cars and trucks to run on discarded cooking oil. With a Greasecar system installed, any diesel vehicle can run on vegetable oil.

Instead of fueling at gas stations, Greasecar devotees tank up at restaurants, where deep-fryer oil is tossed out daily. Since many establishments pay disposal fees, they're happy to give the stuff away.

Greasecar Vegetable Fuel Systems (greasecar.com) sells conversion kits for about $800, around the same price as its Missouri-based competitor Greasel Conversions (greasel.com).

Converted vehicles are actually dual-fuel hybrids, capable of running on both diesel and vegetable oil. The engine can draw fuel from either the regular tank, or a heated oil tank in the trunk. The vehicle starts up in diesel, then switches over to pure veggie power once the grease is warm enough to flow through the fuel injection. This approach allows the grease-burners to work just fine in cold climates.

Bonus benefits: The air coming out of the tailpipe will smell faintly of French fries, and you can slap on a "Drive Vegan" bumper sticker. Greasecar representatives say converted vehicles experience no change in fuel economy, either — though they admit you may need to change your unit of measurement to "fries per gallon."

Slippery Characters Steal Fryer Oil, Leave Mess
By Paul Spinrad

SPRINGFIELD, MO — The multibillion-dollar, used cooking oil recycling business is taking a beating from a new kind of criminal: the grease bandit, as reported in the *Springfield News-Leader*. In some regions, the growth of the biodiesel industry has given new value to old grease. As a result, bandits now cruise the backs of restaurants, sucking the substance out of discard bins with homemade pumper trucks. Many restaurant workers don't realize that they're the wrong trucks, but Keith Wendorf, spokesman for oil recycler Griffin Industries, says that his company's shiny 18-wheelers steam-clean the pickup areas, while the thieves leave pools of grease.

How Eco Is Bio?

Percentage of total environmental impact relative to gas-powered car.

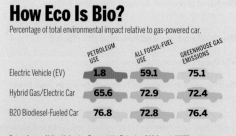

	PETROLEUM USE	ALL FOSSIL-FUEL USE	GREENHOUSE GAS EMISSIONS
Electric Vehicle (EV)	1.8	59.1	75.1
Hybrid Gas/Electric Car	65.6	72.9	72.4
B20 Biodiesel-Fueled Car	76.8	72.8	76.4

Source: Argonne National Laboratory Transportation Technology R&D Center's GREET model, www.transportation.anl.gov/software/GREET

By Polly Powledge

CHICAGO — Hybrids, EVs, or biodiesel — which is best for the environment? Figuring out the answer can be tricky. Diesel gets better mileage than gas, but it takes more petroleum to produce. Electric vehicles (EVs) don't use gasoline at all, but if you plug them into the grid, you're using energy from coal-burning power plants, which pollute. B20 biodiesel (a common blend containing 20% pure biodiesel and 80% petrodiesel) reduces fossil-fuel use, but a rigorous analysis must account for the diesel that's used to truck the vegetable oil to the biodiesel processing plant.

Fortunately, scientists at Argonne National Laboratory have constructed a computer model that lets you do apples-to-apples comparisons of a variety of fuel technologies, and see how they rate ecologically. The results of these "well-to-wheel" calculations can be surprising. For example, while an EV consumes very little petroleum over its lifetime, hybrids are actually slightly better at reducing greenhouse gases.

Photograph by Phillip Torrone

HARDWIRED IPOD

WANT A BETTER WAY TO PLAY MP3s THROUGH YOUR CAR
STEREO THAN USING AN FM TRANSMITTER OR A
CASSETTE ADAPTER? CUT THE STATIC BY CONNECTING
YOUR iPOD TO YOUR STEREO'S AUX JACK.
BY DAMIEN STOLARZ

If you have an iPod (or any portable music device, for that matter), and you drive, it's a no-brainer that you want to bring that audio experience into your car.

Here are my four goals for making a car-based portable music system:
1. **I want to play the music through my car's speakers.**
2. **I want to control the functions safely while I'm driving.**
3. **I want to mount it securely so it doesn't fly around.**
4. **I want to power and charge it so it doesn't run out of juice while I'm driving.**

Each of these problems has a number of solutions depending on how much money and effort you are willing to spend. But clearly, the main problem to solve is: how do you get that audio signal into your car speakers?

To help you decide the best way to get your iPod into your vehicle (and there are probably a million possible permutations of device + car), we're going to look at the different ways you can (or should be able to) integrate a portable MP3 player into a car.

The focus of all these approaches is to put the audio into the head unit. The head unit is the more technical term for the "radio" or "tuner" or "stereo" or "cassette deck" in your car; that is, the head unit is the thing in the dashboard with the dials and buttons that acts as the "analog hub" for your car's audio media experience.

GOAL: PLAY YOUR IPOD THROUGH YOUR CAR SPEAKERS

You can get audio into your car's receiver in one of four ways: FM transmitters, FM modulators, tape adapters, or directly into the head unit via an AUX in, CD, or satellite radio interface. This decision tree will help you find which method is right for you.

MONEY VS. EFFORT:
iPOD-TO-CAR OPTIONS

$$$	iPod bus input iPod aux adapter
	FM transmitter
	FM modulator
	Cassette adapter
$	Patch cable

Easy   Difficult

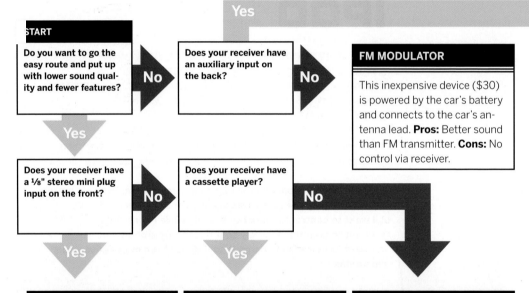

START

Do you want to go the easy route and put up with lower sound quality and fewer features? → **No** → **Does your receiver have an auxiliary input on the back?** → **No** →

↓ **Yes**

Does your receiver have a ⅛" stereo mini plug input on the front? → **No** → **Does your receiver have a cassette player?** → **No** →

↓ **Yes**   ↓ **Yes**

Yes →

FM MODULATOR

This inexpensive device ($30) is powered by the car's battery and connects to the car's antenna lead. **Pros:** Better sound than FM transmitter. **Cons:** No control via receiver.

PATCH CABLE

A male-male mini plug cable costs $2 and connects your iPod's headphone output to the stereo's input jack. **Pros:** Dirt cheap, sounds good. **Cons:** No charging, no control via receiver.

CASSETTE ADAPTER

Translates audio signals from your iPod to the magnetic pulses that your tape deck expects to read off a cassette tape. **Best choice:** Griffin's SmartDeck ($29), which plugs into the top of the iPod and can translate the fast forward/rewind buttons on your tape deck into next track/last track controls on your head unit. **Pros:** Cheap. **Cons:** SmartDeck doesn't charge your iPod.

FM TRANSMITTTER

A $20 to $70 mini-radio station that beams your iPod output to your car's antenna. **Best choice:** Monster iCarPlay Wireless Plus FM transmitter. **Pros:** Charges batteries, tunable across FM spectrum, on-cable controls. **Cons:** Sound quality is not as good as any other solution offered here.

IPOD AUX ADAPTER

Several companies make adapters that interface between your iPod and the CD changer jack on your car's receiver. These companies either reverse-engineer stereo systems or go under non-disclosure agreements with receiver manufacturers to create sub-$100 black boxes with the proprietary CD changer interface on one side and RCA jacks on the other. These boxes trick the receiver into thinking that a CD changer is plugged in, allowing the iPod's signal to play through the speakers.

Some companies make a single, omnibus adapter and let you choose your car via dual-inline package (DIP) switches on the bottom of the unit. You also have to buy a $5 adapter cable that fits your car's jack. Other vendors sell a separate adapter box for each brand of OEM vehicle.

Most adapters give you limited control of your iPod via the receiver. Some brand name car stereo companies have iPod adapters that will show artist/song information on the receiver's display. Monster Cable's iCruze system has an optional dash mount LCD that displays artist/song information. They range in price from $100–$500. Sources: mp3yourcar.com, theistore.com, crutchfield.com, installer.com/aux, logjamelectronics. com/auxinpconv.html, monstercable.com/icruze.

Pros: Charges batteries, best sound quality. **Cons:** Expensive, installation can be difficult for certain cars, receivers, and adapters.

Getting Audio Out of the iPod

It's worth noting that there are two ways to get audio out of the iPod: top or bottom. The top connector on the iPod is a variable output, amplified signal. It's designed to power your headphones. The amplification is controlled with the volume function of the scroll wheel. When using the variable output with any of the adapters listed here, you wind up with two ways to increase volume: the iPod or the head unit. This may seem like a good idea, but it can cause distortion.

The best way to get audio out of the iPod is from the bottom, where there is a line level output. This provides a nice, fixed level designed for amplification by another device (i.e. your head unit). Keep this in mind when buying any adapter equipment.

The $50,000 Solution

If you're about to buy a new BMW, you can skip this decision tree entirely and simply plug your iPod into the manufacturer-supplied cable. In 2004, iPod made a deal with BMW and created the first iPod integration adapters. These first adapters were really a hack that made the iPod emulate a CD changer, but they've since added steering wheel and head unit control of the iPod.

Since that time, Apple has started making more deals with auto manufacturers, which are implementing different levels of iPod display and integration depending on the model of the car. In the coming years, the iPod's proprietary bottom-connector may become a de facto standard for in-car integration. So when you buy a new car, you may be able to simply ask for the "iPod integration" option. **Pros:** Excellent sound and control. **Cons:** You have to buy a BMW for it to work.

MOUNTING, CONTROL, AND CHARGING SOLUTIONS

MOUNTING IT

After deciding how to connect your iPod to your receiver, you need to figure out how to mount it. If you are using a system that lets you control your iPod through your stereo's buttons and dials, then you should stow it in your glove compartment or other out-of-the-way place. If you want to use the iPod's controls, then you should mount it on the dash or other accessible location.

The cheapest and most space-age mounting solution is good old Velcro. If it's good enough for securing objects in the zero-G environment of the space shuttle, it's good enough to affix your iPod securely to your dashboard. You'll want some sort of iPod case or plastic belt clip, just to make sure you don't gum up the back of your iPod. You can get sticky Velcro at craft and hardware stores.

The Apple iPod dock that comes with some models of iPod provides audio output and recharging. Since you already have it, you can also Velcro or tape this to the center console of your vehicle, if you can find a level surface.

Because iPods are about the size of some mobile phones, you can also use a mobile phone holder. Mobile phone holders come in a variety of forms — some clip to AC vents, while others stick to the dash. Check your local car wash for a deal.

MacMice PodBuddy offers FM transmitting, power, and mounting (dvforge. com/podbuddy.shtml).

There are dozens of mounting systems designed specifically for the iPod. Initially, these were all cast in white plastic to match the iPod, but recent versions from Belkin are being made in black to better match the interior of most cars. The iPod holders made by ProClip blend in almost seamlessly with the finish of most cars.

CONTROL IT

While the iPod has a great interface, it's certainly not designed with automobile drivers — who must keep their eyes on the road — in mind.

The wired remote control that comes with the various generations of iPod can serve as an effective hands-free remote. You could even affix it on the steering wheel (given enough wire length) so that you have a "hands-free" remote.

Learn about the Apple iPod remote control protocol at maushammer.com/ systems/ipod-remote/ ipod-remote.html.

If you're interested in wiring up your own remote wired controller for the iPod, perhaps to connect it to your steering wheel controls or your own dash-mounted interface, the protocol and pinouts are available at Maushammer (see above for URL).

POWER AND CHARGE IT

You can buy a FireWire-to-cigarette lighter adapter for less than $10. But if you must, you can build one. First, put 12 volts through a 1-amp fuse to the power pins on a six-pin FireWire cable. You simply put the +12V on pin 1 and the ground on pin 2. A FireWire has a flexible voltage range from 8V to 30V, so you can put the 12V from the car straight into it.

Since the FireWire end connector is fairly hard to solder, it's probably easier to just cut a FireWire cable in half and put the voltage on the two relevant pins.

A great place to learn more about mounting and charging your iPod in the car is iPodlounge (ipodlounge.com).

The $30 SiK imp iPod charger (sik.com) has line level audio output.

URBAN CAMOUFLAGE

With the right accessories, your vehicle can always be on "official business."
By Todd Lappin

Photograph by Todd Lappin

It started as a scam to find on-street parking. One day, I had a social-engineering epiphany — if I decorated my car to make it look like a commercial vehicle, I'd be able to park in yellow-curb loading zones without getting tickets from the overzealous parking control officers who regularly swarmed my neighborhood.

After a trip to an art-supply store to buy some vinyl lettering, a fake company name was created, decals were applied, and indeed, no parking tickets were received.

That was 18 years ago. Since then, all my vehicles have been decked out in corporate camouflage, even as my needs have evolved. In addition to basic parking acquisition, my current car, a 1999 Jeep Cherokee I ordered new from the factory in fleet-service white, has been configured for urban vandalism deterrence,

high-alpine winter driving, assisting stranded motorists, and infiltrating the abandoned military bases where I like to take photographs.

In these pages, I'll reveal the sources and methods I employed to create an effective vehicular disguise. With this covert knowledge, however, comes great responsibility. I don't condone using any of this for nefarious purposes. But like the old Hollywood cop-show cliché about the diaper-service delivery van that functions as a stake-out surveillance post, just remember — the company vehicle you see parked on your street or bearing down in your rearview mirror may not be quite what it seems.

> With a "company" vehicle, you'll enjoy special benefits and respectful admirers.

Push Bumper:

This is the same push bumper found on many police cars and highway service vehicles. Manufactured by Setina (*www.setina.com*), it's rubberized, so it's ideal for gently pushing stalled vehicles and protecting the front end during parallel parking maneuvers.

Police-Style Spotlight:

Manufactured by Unity (*www.unityusa.com*), my police-style spotlight looks official, rotates 360 degrees, and emits a retina-burning beam. It's legal in most states, and when used with discretion, it's ideal for rendering roadside assistance, finding street addresses, and scaring the bejesus out of teenagers parked on Lover's Lane. Mounting the spotlight requires significant drilling.

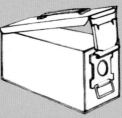

Funeral Sign

If used sparingly, a funeral visor sign can be very effective in eliciting sympathy from motorists and parking authorities. I had my sign laminated and Velcro-backed to preserve its fresh appearance.

Emergency Supplies

I carry these tools to get myself — and other hapless souls — unstuck: a 10,000-pound nylon tow strap, jumper cables, wrenches and screwdrivers, two Maglite flashlights, a hand-held CB radio, duct tape, cable ties, a whistle, leather gloves, and a first-aid kit.

Fire Extinguisher:

My first car — a battered Dodge Monaco — burned itself to a hollow crisp alongside a Rhode Island interstate. Here's what I learned: car fires usually start small, but without a fire extinguisher, the flames quickly grow big. A 2.5LB extinguisher will do the trick ... don't leave home without it.

The Urban Camouflage
Vehicle Makeover

Amber Hazard Flashers:
Mounted inside the vehicle for stealth purposes, these amber flashers light up like a Christmas tree at the flick of a switch. They're useful for traffic diversion and communication with freeway tailgaters. I bought mine from AW Direct (www. awdirect.com), my favorite catalog for tow-truck equipment.

Hazard Striping:
Nothing says "Keep Back" more subliminally than yellow-and-black caution stripes. I hired a local body shop to apply mine using automotive-grade paint, but vinyl tape is a cheaper alternative. Seton (www. seton.com) sells durable OSHA Safety Tape in a variety of widths and color combinations.

E54130

E54130

Illustrations by Damien Scogin

Fleet Vehicle Markings:
A cumbersome fleet number applied to each corner of the vehicle conveys the impersonal sprawl of a big corporation. Most office supply stores sell vinyl letters and numbers, in a variety of sizes, fonts, and colors.

D-Ring Shackle:
This handy little shackle from Warn Industries (www. gowarn.com) mounts in a standard trailer hitch. I use it as an anchor point when extracting vehicles stuck in heavy snow, but it also provides a rugged impact point for parallel parking.

CAUTION: DO NOT OVERWIND

Toys by LeDuc

CALIFORNIA 2007
2XXZ745
LAUGHTER'S HEALTHY!

WINDUP CAR

Mod a cute convertible with a big rotating key. By Sunny Armas (with Paul LeDuc)

When my pickup truck broke down, a friend lent me her Geo Metro convertible for 2 weeks. Well, I'm a good-sized guy, and at first I felt a bit silly driving this tiny car, but I quickly fell in love with it. It's easy to park, great on gas, and a blast to drive. I decided to buy my own, and I thought that if I'm willing to be seen driving this toy-sized car, why not go all the way and put a big windup key on the back?

I called my metal sculptor friend Paul LeDuc for his help and ideas, and we talked about the project. I knew I wanted a big key that would turn slowly, to make the car look as toy-like as possible. I also wanted the key to be easily removable and fit in the trunk to prevent people from trying to rip it off, and to keep the option of driving a little more low-profile. Also, I had just had the car painted, so I wanted to be careful that any mods would not

damage the nice new finish.

We decided that the key should be about 16" across. For the flat part, we drew a template and used a plasma cutter to cut the shape out of ¼"-thick steel plate. The harder part was the center tube, which needed to mount neatly through the trunk of the car. For this, Paul strolled through the hardware store and found a nice chrome mounting ring for finishing bathroom sink drains. The threaded, 1½" inside-diameter ring had some height that would help guide the key, and included a rubber gasket, washer, and nut to secure it around both sides of an aperture.

Paul used a grinder to remove the strainer from the bottom of the sink mounting ring, then cut a hole in the center of the trunk deck (lid) with a drill and a hole saw. He slid in the drain ring, and fitted the

Photograph by Sam Murphy

A

B

C

Fig. A: To turn the key, a windshield wiper motor is mounted under the trunk deck. Power comes from the dashboard cigarette lighter, via wires that run under the carpet and out the same hole as the brake light wires.

Fig. B: The key to making a cute little car even cuter.
Fig. C: Friction mount made from a foam bicycle handle grip makes the key easily removable, and lets you twist it and stop it without harming the motor.

washer and nut on the back. Then he found some steel tubing in his shop that fit perfectly into the sink ring, cut a length of that to be the key's center tube, and welded it to the flat key shape.

To turn the key, we used a spare windshield wiper motor that I had. These run off a 12VDC car battery, of course, and they turn slowly with a lot of torque. To connect the key to the motor and make it removable, we made a friction mount out of a foam bicycle handle grip. We welded a 2½"×⅛" disk to the shaft of the motor, and welded that to a 5" steel rod that's the same diameter as bike handle-bars (⅞"). The shaft protrudes from the middle of the hole in the trunk, and with the foam grip over it, the key slides over it snugly.

The great thing about the foam connection is that it lets the motor run freely if the key's movement is impaired, like if someone grabs it, and it also lets you turn the key when the motor is off. This means that little kids can turn the key before the car starts, and imagine that they're helping the car go!

To attach the motor, Paul removed the trunk deck, laid it upside down, and welded together a frame-work that would hold it securely and keep it aligned straight up. He made the framework out of 1"- and 2"-square tubing, welded to the deck's ribbing with

minimal damage to the paint, and tapped screw holes in to serve as motor mounts.

The motor is powered from the car's cigarette lighter. To do this, I ran wires under the carpet, starting in front near the dash, between the seats, and threading them back into the trunk through the same hole that the brake light wires run out of.

To connect the wires to power, I cut the fan off an old cigarette-lighter-powered fan, and ran the circuit through an old VW bus light switch, which I mounted next to the car's gearshift. I positioned the switch so that when I shift between neutral and first, my thumb can easily turn the key motor off and on. That way, when I stop the car, the key stops, and when the car starts moving, the key starts up again — bringing smiles to all, or at least to most.

Sunny Armas lives in San Jose and enjoys bringing smiles to others. He has helped Paul LeDuc with many other art projects, including parade floats and metal waterfalls.

Photography by Paul Spinrad

Curie Engine
By John Iovine

Changing the temperature of nickel wire turns heat into motion.

You will need: A bit of low-Curie nickel alloy wire, 1" or larger neodymlum "super magnet," copper wire, birthday candle, brass screw and nut, small steel plate, glue or wood screws, 4"×¾" wooden slat, 2"×1¼"×½" wood block. Nickel wire and magnet, or a full kit, are available for purchase at makezine.com/go/heatengine.

1. Put it together.

Glue or screw the steel plate and 4" slat to the wood block base as shown below. Drill a pilot hole, and screw the brass screw at the top of the slat. Drill a ¼" hole in the base about 7⁄16" from the steel plate end. For complete measurements, go to the website listed above.

Wrap about 1" of nickel alloy wire around a pencil to make a coil. Take 4" of copper wire, and twist the coil onto one end. Make a 90° bend in the copper wire 1½" down from the end with the coil, and cut the other end so there's 1" after the bend. Hook the wire bend onto the brass screw, with the coil on the side nearer the steel plate.

2. Test the wire and magnet.

Place the neodymium magnet onto the metal plate. The wire should swing up to meet the magnet,

pulled by the nickel alloy coil. Push down and release the wire, and it should swing back up. Insert a birthday candle in the base, and adjust the magnet position and wire so the coil is suspended above the wick.

3. Light the candle.

Light the candle. The wire should swing down and up repeatedly, out of and back into the candle flame. Like ordinary iron or steel, the nickel-iron alloy is magnetic, but it has a much lower Curie point — the temperature above which it loses its magnetic properties. (The copper wire and brass screw are not magnetic.) When the coil heats up to this point, it is no longer attracted to the magnet, so it falls away and out of the flame. As it cools, it becomes magnetic again and swings back up, over and over again, until the flame burns too low to touch the coil.

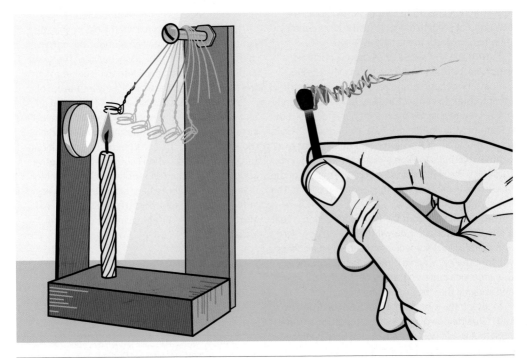

Illustration by Dustin Hostetler

John Iovine wrote the Kirlan Photography article on page 282.

For more info, corrections, and discussion on this piece, please visit makezine.com/09/123_engine

Index

Photography & Illustration Contributors

Douglas Adesko, 138-139, 142-149

David Albertson, 105, 260-263

Dustin Amery, 176

Tor Amundson, 344-349

Wendell Anderson, 204

Gerry Arrington, cover, 244, 266-267

Bill Barminski, 295-297

Melinda Beck, 115

Dennison Bertram, 298-300

Blind Lightnin' Pete, 193

Mark R. Brown, 127, 129

Bill Bumgarner, cover, 119, 121, 123, 125

Bill Coderre, 292

Abe and Josie Connally, 318, 323-326, 328

Thayer Cowdy, 40-41

David Cuartielles, 78

Matthew Dalton, 248-252

Zach DeBord, cover, 164, 166

Dale Dougherty, 92

Mark Frauenfelder, 67, 154-156

Limor Fried, 286-287

William Gurstelle, 194-196, 212, 215-218

Dustin Hostetler, 135, 368

Sara Huston, 351-356

Roger Ibars, cover, 50, 53-58, 59

Tom Igoe, 93-94

Andy Ihnatko, 290

John Iovine, 283-285

Peter Kirn, 206-207

Alek Komarnitsky, 301-303

Timmy Kucynda, 43, 141, 152, 188, 234, 246

Todd Lappin, 16-17, 363

James Larsson, 44-48

Tim Lillis, 167, 214, 257, 320, 322-323, 327

Greg Lipscomb, 61, 62

Steve Lodefink, 225-229, 231

Topher Lucas, cover, 220-221, 227, 230-231

Dave Mathews, 338-341

John Maushammer, 232, 235-242

Michael McDonald, 197

Jake McKenzie, 312-317

Evan McNary, 31, 34-39

Sam Murphy, cover, 186, 189, 270, 366

Jasper Nance, 283

Amy Nightingale, 23

Johnathan Nightingale, 24

Ty Nowotny, 248-252, 312-317

Ross Orr, 273-281

John Edgar Park, 18-22

John Perez, 265

Charles Platt, 10, 63-64

Dave Prochnow, cover, 168-169

Nik Schulz, 12-13, 33, 52, 223, 272, 308, 330-331, 334-335

Damien Scogin, 9, 60, 131-132, 283, 364-365

Bob Scott, 25

Bit Shifter, 100-101

Casimir A. Sienkiewicz, cover, 254-255

Carla Sinclair, cover, 171

Sparkle Labs, 70-73, 75, 77

Paul Spinrad, 367

Liam Staskawicz, 84

Damien Stolarz, 133

Derrick Story, 289

Phillip Torrone, 358

Matt Turner, 336

Robyn Twomey, 87-88, 91-92

Ed Vogel, 190-192

Kirk von Rohr, 66, 96, 150, 153-155, 157-162, 224, 258-259, 306, 310-311

Ren Wang, 79

Mark Watkinson, 103

Sarah Whiting, 174, 177-184

David Williams, 80-81

Susan Williams, 83

Cristiana Yambo, 106-113

Warren Young, 198, 200-203

Nicholas Zambetti, 79